Version 10

Basic Analysis and Graphing

Second Edition

"The real voyage of discovery consists not in seeking new landscapes, but in having new eyes."

Marcel Proust

JMP, A Business Unit of SAS
SAS Campus Drive
Cary, NC 27513

10.0.2

The correct bibliographic citation for this manual is as follows: SAS Institute Inc. 2012. *JMP® 10 Basic Analysis and Graphing, Second Edition.* Cary, NC: SAS Institute Inc.

JMP® 10 Basic Analysis and Graphing, Second Edition

ISBN 978-1-61290-310-1

SAS Institute Inc., SAS Campus Drive, Cary, North Carolina 27513.

1st printing, July 2012

2nd printing, November 2012

Technology License Notices

Contents

JMP Basic Analysis and Graphing

12 Creating Overlay Plots

13 Creating Three-Dimensional Scatterplots

14 Creating Contour Plots

15 Creating Bubble Plots

Chapter 1

Learn about JMP

Documentation and Additional Resources

This chapter covers the following information:

- book conventions
- JMP documentation
- JMP Help
- additional resources, such as the following:
  - other JMP documentation
  - tutorials
  - indexes
  - Web resources

Figure 1.1 The JMP Help Home Window

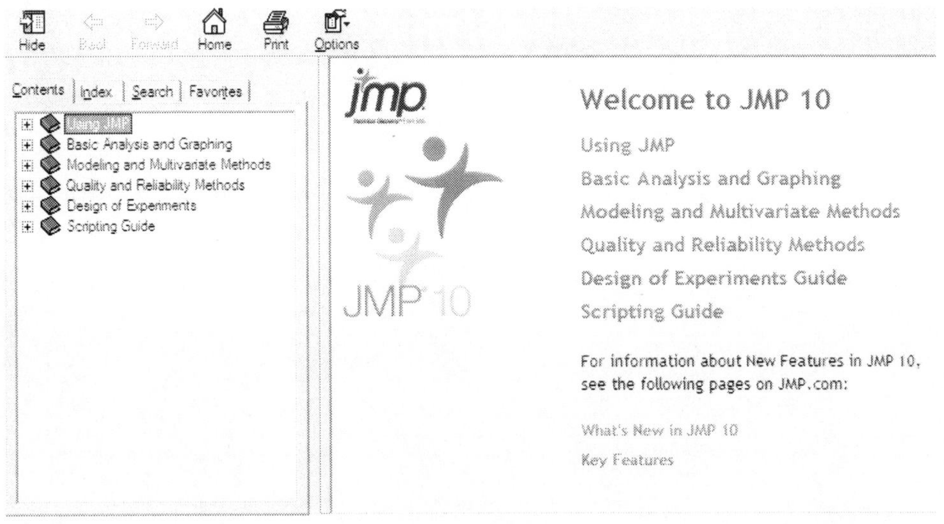

Contents

Book Conventions

The following conventions help you relate written material to information that you see on your screen.

- Sample data table names, column names, path names, file names, file extensions, and folders appear in Helvetica font.
- Code appears in `Lucida Sans Typewriter` font.
- Code output appears in *`Lucida Sans Typewriter`* italic font and is indented further than the preceding code.
- The following items appear in **Helvetica bold**:
  - buttons
  - check boxes
  - commands
  - list names that are selectable
  - menus
  - options
  - tab names
  - text boxes
- The following items appear in italics:
  - words or phrases that are important or have definitions specific to JMP
  - book titles
  - variables
- Features that are for JMP Pro only are noted with the JMP Pro icon JMP PRO.

Note: Special information and limitations appear within a Note.

Tip: Helpful information appears within a Tip.

JMP Documentation

The JMP documentation suite is available by selecting **Help > Books**. You can also request printed documentation through the JMP Web site:

http://support.sas.com/documentation/onlinedoc/jmp/index.html

JMP Help is context-sensitive and searchable.

JMP Documentation Suite

The following table describes the documents in the JMP documentation suite and the purpose of each document.

Document Title	Document Purpose	Document Content
Discovering JMP	If you are not familiar with JMP, start here.	Introduces you to JMP and gets you started using JMP
Using JMP	Learn about JMP data tables and how to perform basic operations.	• general JMP concepts and features that span across all of JMP • material in these JMP Starter categories: File, Tables, and SAS
Basic Analysis and Graphing	Perform basic analysis and graphing functions using this document.	• these Analyze platforms: – Distribution – Fit Y by X – Matched Pairs • these Graph platforms: – Graph Builder – Chart – Overlay Plot – Scatterplot 3D – Contour Plot – Bubble Plot – Parallel Plot – Cell Plot – Tree Map – Scatterplot Matrix – Ternary Plot • material in these JMP Starter categories: Basic and Graph

Document Title	Document Purpose	Document Content
Modeling and Multivariate Methods	Perform advanced modeling or multivariate methods using this document.	• these Analyze platforms: – Fit Model – Screening – Nonlinear – Neural – Gaussian Process – Partition – Time Series – Categorical – Choice – Model Comparison – Multivariate – Cluster – Principal Components – Discriminant – Partial Least Squares – Item Analysis • these Graph platforms: – Profilers – Surface Plot • material in these JMP Starter categories: Model, Multivariate, and Surface

Document Title	Document Purpose	Document Content
Quality and Reliability Methods	Perform quality control or reliability engineering using this document.	• these Analyze platforms: – Control Chart Builder – Measurement Systems Analysis – Variability / Attribute Gauge Chart – Capability – Control Charts – Pareto Plot – Diagram (Ishikawa) – Life Distribution – Fit Life by X – Recurrence Analysis – Degradation – Reliability Forecast – Reliability Growth – Survival – Fit Parametric Survival – Fit Proportional Hazards • material in these JMP Starter window categories: Reliability, Measure, and Control
Design of Experiments	Design experiments using this document.	• everything related to the **DOE** menu • material in this JMP Starter window category: DOE
Scripting Guide	Learn about the JMP Scripting Language (JSL) using this document.	reference guide for using JSL commands

In addition, the *New Features* document is available at http://www.jmp.com/support/downloads/documentation.shtml.

Note: The **Books** menu also contains two reference cards that can be printed: The *Menu Card* describes JMP menus, and the *Quick Reference* describes JMP keyboard shortcuts.

JMP Help

JMP Help is an abbreviated version of the documentation suite providing targeted information. You can access the full-length PDF files from within the Help.

You can access JMP Help in several ways:

- Press the F1 key.
- Get help on a specific part of a data table or report window. Select the Help tool from the **Tools** menu and then click anywhere in a data table or report window to see the Help for that area.
- Within a window, click a **Help** button.
- Search and view JMP Help on Windows using the **Help > Contents**, **Search**, and **Index** options. On Mac, select **Help > JMP Help**.

JMP Books by Users

Additional books about using JMP that are written by JMP users are available on the JMP Web site:

http://www.jmp.com/support/books.shtml

JMPer Cable

The JMPer Cable is a yearly technical publication targeted to users of JMP. The JMPer Cable is available on the JMP Web site:

http://www.jmp.com/about/newsletters/jmpercable/

Additional Resources for Learning JMP

In addition to JMP documentation and JMP Help, you can also learn about JMP using the following resources:

- Tutorials (see "Tutorials on page 25)
- JMP Starter (see "The JMP Starter Window on page 26)
- Sample data tables (see "Sample Data Tables on page 26)
- Indexes (see "Learn about Statistical and JSL Terms on page 26)
- Tip of the Day (see "Learn JMP Tips and Tricks on page 26)
- Web resources (see "Access Resources on the Web on page 27)

Tutorials

You can access JMP tutorials by selecting **Help > Tutorials**. The first item on the **Tutorials** menu is **Tutorials Directory**. This opens a new window with all the tutorials grouped by category.

If you are not familiar with JMP, then start with the **Beginners Tutorial**. It steps you through the JMP interface and explains the basics of using JMP.

The rest of the tutorials help you with specific aspects of JMP, such as creating a pie chart, using Graph Builder, and so on.

The JMP Starter Window

The JMP Starter window is a good place to begin if you are not familiar with JMP or data analysis. Options are categorized and described, and you launch them by clicking a button. The JMP Starter window covers many of the options found in the **Analyze**, **Graph**, **Tables**, and **File** menus.

- To open the JMP Starter window, select **View** (**Window** on the Macintosh) **> JMP Starter**.
- To display the JMP Starter automatically when you open JMP, select **File > Preferences > General**, and then select **JMP Starter** from the Initial JMP Window list.

Sample Data Tables

All of the examples in the JMP documentation suite use sample data. Select **Help > Sample Data** to do the following actions:

- Open the sample data directory.
- Open an alphabetized list of all sample data tables.
- Open sample scripts.
- Find a sample data table within a category.

Sample data tables are installed in the following directory:

On Windows: C:\Program Files\SAS\JMP\<version_number>\Samples\Data

On Macintosh: \Library\Application Support\JMP\<version_number>\Samples\Data

Learn about Statistical and JSL Terms

The **Help** menu contains the following indexes:

Statistics Index Provides definitions of statistical terms.

Scripting Index Lets you search for information about JSL functions, objects, and display boxes. You can also run sample scripts from the Scripting Index.

Learn JMP Tips and Tricks

When you first start JMP, you see the Tip of the Day window. This window provides tips for using JMP.

To turn off the Tip of the Day, clear the **Show tips at startup** check box. To view it again, select **Help > Tip of the Day**. Or, you can turn it off using the Preferences window. See the *Using JMP* book for details.

Tooltips

JMP provides descriptive tooltips when you place your cursor over items, such as the following:

- Menu or toolbar options
- Labels in graphs
- Text results in the report window (move your cursor in a circle to reveal)
- Files or windows in the Home Window
- Code in the Script Editor

Tip: You can hide tooltips in the JMP Preferences. Select **File > Preferences > General** (or **JMP > Preferences > General** on Macintosh) and then deselect **Show menu tips**.

Access Resources on the Web

To access JMP resources on the Web, select **Help > JMP.com** or **Help > JMP User Community**.

The **JMP.com** option takes you to the JMP Web site, and the **JMP User Community** option takes you to JMP online user forums, file exchange, learning library, and more.

Chapter 2

Performing Univariate Analysis

Using the Distribution Platform

The Distribution platform describes the distribution of variables using histograms, additional graphs, and reports. You can examine the distribution of several variables at once. The report content for each variable varies, depending on whether the variable is categorical (nominal or ordinal) or continuous.

The Distribution report window is interactive. Clicking on a histogram bar highlights the corresponding data in any other histograms and in the data table. See Figure 2.1.

Figure 2.1 Example of the Distribution Platform

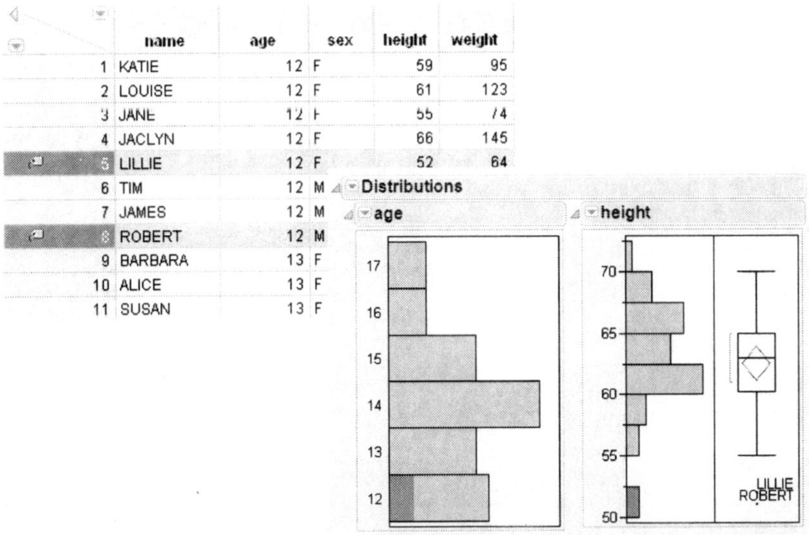

Contents

Overview of the Distribution Platform

The treatment of variables in the Distribution platform is different, depending on the modeling type of variable, which can be categorical (nominal or ordinal) or continuous.

Categorical Variables

For categorical variables, the initial graph that appears is a histogram. The histogram shows a bar for each level of the ordinal or nominal variable. You can also add a divided (mosaic) bar chart.

The reports show counts and proportions. You can add confidence intervals and test the probabilities.

Continuous Variables

For numeric continuous variables, the initial graphs show a histogram and an outlier box plot. The histogram shows a bar for grouped values of the continuous variable. The following options are also available:

- quantile box plot
- normal quantile plot
- stem and leaf plot
- CDF plot

The reports show selected quantiles and summary statistics. Report options are available for the following:

- saving ranks, probability scores, normal quantile values, and so on, as new columns in the data table
- testing the mean and standard deviation of the column against a constant you specify
- fitting various distributions and nonparametric smoothing curves
- performing a capability analysis for a quality control application
- confidence intervals, prediction intervals, and tolerance intervals

Example of the Distribution Platform

Suppose that you have data on 40 students, and you want to see the distribution of age and height among the students.

1. Open the Big Class.jmp sample data table.
2. Select **Analyze > Distribution**.
3. Select age and height and click **Y, Columns**.
4. Click **OK**.

Figure 2.2 Example of the Distribution Platform

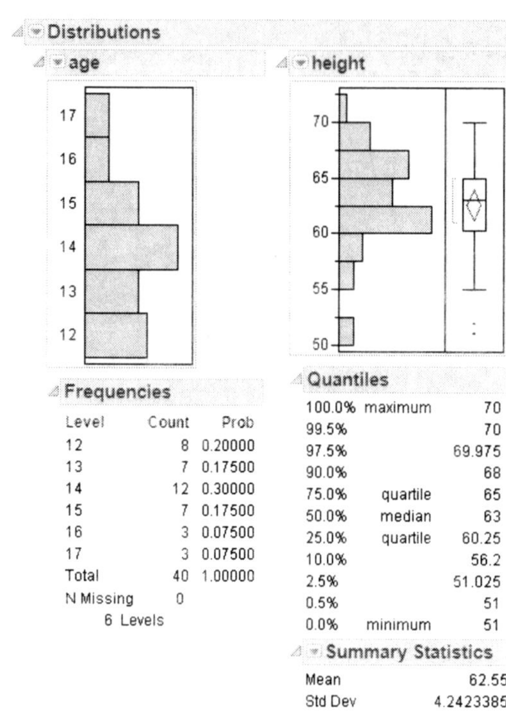

From the histograms, you notice the following:

- The ages are not uniformly distributed.
- For height, there are two points with extreme values (that might be *outliers*).

Click on the bar for 50 in the height histogram to take a closer look at the potential outliers.

- The corresponding ages are highlighted in the age histogram. The potential outliers are age 12.
- The corresponding rows are highlighted in the data table. The names of the potential outliers are Lillie and Robert.

Add labels to the potential outliers in the height histogram.

1. Select both outliers.
2. Right-click on one of the outliers and select **Row Label**.

   Label icons are added to these rows in the data table.
3. (Optional) Resize the box plot wider to see the full labels.

Figure 2.3 Potential Outliers Labeled

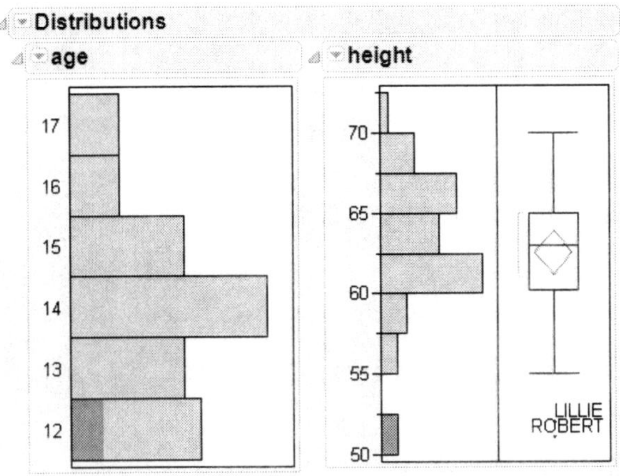

Launch the Distribution Platform

Launch the Distribution platform by selecting **Analyze > Distribution**.

Figure 2.4 The Distribution Launch Window

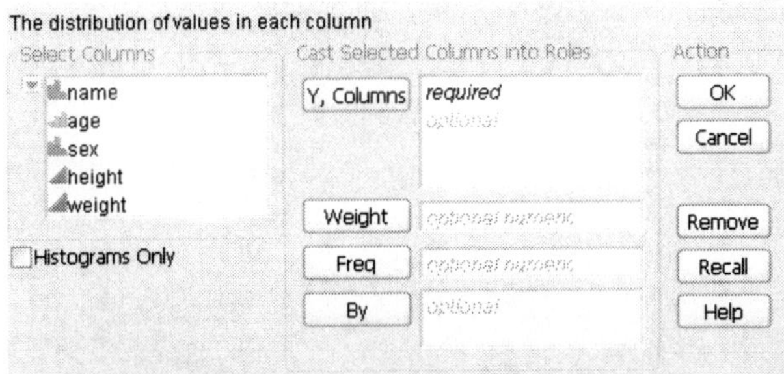

Table 2.1 Description of the Distribution Launch Window

Y, Columns	Assigns the variables that you want to analyze. A histogram and associated reports appear for each variable.
Weight	Assigns a variable to give the observations different weights. Any moment that is based on the Sum Wgts is affected by weights.

Table 2.1 Description of the Distribution Launch Window *(Continued)*

Freq	Assigns a frequency variable to this role. This is useful if you have summarized data. In this instance, you have one column for the Y values and another column for the frequency of occurrence of the Y values. The sum of this variable is included in the overall count appearing in the Summary Statistics report (represented by N). All other moment statistics (mean, standard deviation, and so on) are also affected by the **Freq** variable.
By	Produces a separate report for each level of the **By** variable. If more than one **By** variable is assigned, a separate report is produced for each possible combination of the levels of the **By** variables.
Histograms Only	Removes everything except the histograms from the report window.

For general information about launch windows, see *Using JMP*.

The Distribution Report

Follow the instructions in "Example of the Distribution Platform" on page 31 to produce the report shown in Figure 2.5.

Figure 2.5 The Initial Distribution Report Window

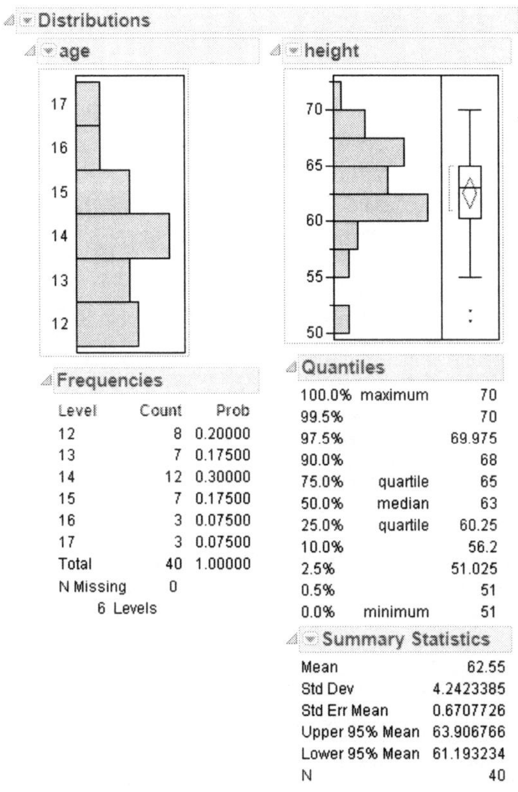

The initial Distribution report contains a histogram and reports for each variable. Note the following:

- To replace a variable in a report, from the Columns panel of the associated data table, drag and drop the variable into the axis of the histogram.
- To insert a new variable into a report, creating a new histogram, drag and drop the variable outside of an existing histogram. The new variable can be placed before, between, or after the existing histograms.

Note: To remove a variable, select **Remove** from the red triangle menu.

- The red triangle menu next to Distributions contains options that affect all of the variables. See “Distribution Platform Options” on page 41.
- The red triangle menu next to each variable contains options that affect only that variable. See “Options for Categorical Variables” on page 42 or “Options for Continuous Variables” on page 44. If you hold down the CTRL key and select a variable option, the option applies to all of the variables that have the same modeling type.
- Histograms visually display your data. See “Histograms” on page 36.

- The initial report for a categorical variable contains a Frequencies report. See "The Frequencies Report" on page 38.
- The initial report for a continuous variable contains a Quantiles and a Summary Statistics report. See "The Quantiles Report" on page 39 and "The Summary Statistics Report" on page 39.

Histograms

Histograms visually display your data. For categorical (nominal or ordinal) variables, the histogram shows a bar for each level of the ordinal or nominal variable. For continuous variables, the histogram shows a bar for grouped values of the continuous variable.

Table 2.2 Histogram Actions

Highlighting data	Click on a histogram bar or an outlying point in the graph. The corresponding rows are highlighted in the data table, and corresponding sections of other histograms are also highlighted, if applicable. See "Highlight Bars and Select Rows" on page 37.
Creating a subset	Double-click on a histogram bar, or right-click on a histogram bar and select **Subset**. A new data table is created that contains only the selected data.
Resizing the entire histogram	Hover over the histogram borders until you see a double-sided arrow. Then click and drag the borders. For more details, see the *Using JMP* book.
Rescaling the axis	(Continuous variables only) Click and drag on an axis to rescale it. Alternatively, hover over the axis until you see a hand. Then, double-click on the axis and set the parameters in the Axis Specification window.
Resizing histogram bars	(Continuous variables only) There are multiple options to resize histogram bars. See "Resize Histogram Bars for Continuous Variables" on page 37.
Specifying your selection	Specify the data that you select in multiple histograms. See "Specify Your Selection in Multiple Histograms" on page 38.

To see additional options for the histogram or the associated data table:

- Right-click on a histogram. See the *Using JMP* book.
- Click on the red triangle next to the variable, and select **Histogram Options**. Options are slightly different depending on the variable modeling type. See "Options for Categorical Variables" on page 42 or "Options for Continuous Variables" on page 44.

Resize Histogram Bars for Continuous Variables

Resize histogram bars for continuous variables by using the following:

- the Grabber (hand) tool
- the **Set Bin Width** option
- the **Increment** option

Use the Grabber Tool

The Grabber tool is a quick way to explore your data.

1. Select **Tools > Grabber**.

Note: (Windows only) To see the menu bar, you might need to hover over the bar below the window title. You can also change this setting in **File > Preferences > Windows Specific**.

2. Place the grabber tool anywhere in the histogram.
3. Click and drag the histogram bars.

Think of each bar as a bin that holds a number of observations:

- Moving the hand to the left increases the bin width and combines intervals. The number of bars decreases as the bar size increases.
- Moving the hand to the right decreases the bin width, producing more bars.
- Moving the hand up or down shifts the bin locations on the axis, which changes the contents and size of each bin.

Use the Set Bin Width Option

The **Set Bin Width** option is a more precise way to set the width for all bars in a histogram. To use the Set Bin Width option, from the red triangle menu for the variable, select **Histogram Options > Set Bin Width**. Change the bin width value.

Use the Increment Option

The **Increment** option is another precise way to set the bar width. To use the **Increment** option, double-click on the axis, and change the Increment value.

Highlight Bars and Select Rows

Clicking on a histogram bar highlights the bar and selects the corresponding rows in the data table. The appropriate portions of all other graphical displays also highlight the selection. Figure 2.6 shows the results of highlighting a bar in the height histogram. The corresponding rows are selected in the data table.

Note: To deselect histogram bars, press the CTRL key and click on the highlighted bars.

Figure 2.6 Highlighting Bars and Rows

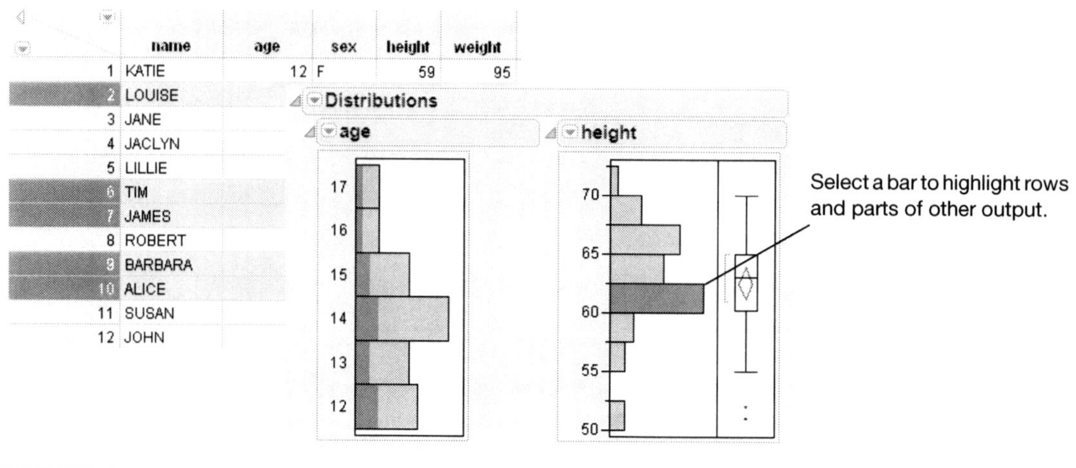

Specify Your Selection in Multiple Histograms

Extend or narrow your selection in histograms as follows:

- To extend your selection, hold down the SHIFT key and select another bar. This is the equivalent of using an *or* operator.
- To narrow your selection, hold down the ALT key and select another bar. This is the equivalent of using an *and* operator.

Related Information

- "Example of Selecting Data in Multiple Histograms" on page 61

The Frequencies Report

For nominal and ordinal variables, the Frequencies report lists the levels of the variables, along with the associated frequency of occurrence and probabilities.

For each level of a categorical (nominal or ordinal) variable, the Frequencies report contains the information described in Table 2.3. Missing values are omitted from the analysis.

Table 2.3 Description of the Frequencies Report

Level	Lists each value found for a response variable.
Count	Lists the number of rows found for each level of a response variable. If you use a Freq variable, the Count is the sum of the Freq variables for each level of the response variable.

Table 2.3 Description of the Frequencies Report *(Continued)*

Prob	Lists the probability (or proportion) of occurrence for each level of a response variable. The probability is computed as the count divided by the total frequency of the variable, shown at the bottom of the table.
StdErr Prob	Lists the standard error of the probabilities. This column might be hidden. To show the column, right-click in the table and select **Columns > StdErr Prob**.
Cum Prob	Contains the cumulative sum of the column of probabilities. This column might be hidden. To show the column, right-click in the table and select **Columns > Cum Prob**.

The Quantiles Report

For continuous variables, the Quantiles report lists the values of selected quantiles (sometimes called percentiles).

Related Information

- "Statistical Details for Quantiles" on page 69

The Summary Statistics Report

For continuous variables, the Summary Statistics report displays the mean, standard deviation, and other summary statistics. You can control which statistics appear in this report by selecting **Customize Summary Statistics** from the red triangle menu next to Summary Statistics.

- Table 2.4 describes the statistics that appear by default.
- Table 2.5 describes additional statistics that you can add to the report using the **Customize Summary Statistics** window.

Table 2.4 Description of the Summary Statistics Report

Mean	Estimates the expected value of the underlying distribution for the response variable, which is the arithmetic average of the column's values. It is the sum of the non-missing values divided by the number of non-missing values.
Std Dev	The normal distribution is mainly defined by the mean and standard deviation. These parameters provide an easy way to summarize data as the sample becomes large: • 68% of the values are within one standard deviation of the mean • 95% of the values are within two standard deviations of the mean • 99.7% of the values are within three standard deviations of the mean

Table 2.4 Description of the Summary Statistics Report *(Continued)*

Std Err Mean	The standard error of the mean, which estimates the standard deviation of the distribution of the mean.
Upper 95% Mean and Lower 95% Mean	Are 95% confidence limits about the mean. They define an interval that is very likely to contain the true population mean.
N	Is the total number of nonmissing values.

Table 2.5 Additional Summary Statistics

Sum Weight	The sum of a column assigned to the role of Weight (in the launch window). Sum Wgt is used in the denominator for computations of the mean instead of *N*.
Sum	The sum of the response values.
Variance	The sample variance, and the square of the sample standard deviation.
Skewness	Measures sidedness or symmetry.
Kurtosis	Measures peakedness or heaviness of tails.
CV	The percent coefficient of variation. It is computed as the standard deviation divided by the mean and multiplied by 100. The coefficient of variation can be used to assess relative variation, for example when comparing the variation in data measured in different units or with different magnitudes.
N Missing	The number of missing observations.
N Zero	The number of zero values.
N Unique	The number of unique values.
Uncorrected SS	The uncorrected sum of squares or sum of values squared.
Corrected SS	The corrected sum of squares or sum of squares of deviations from the mean.
Autocorrelation	(Appears only if you have not specified a Frequency variable.) First autocorrelation that tests if the residuals are correlated across the rows. This test helps detect non-randomness in the data.
Median	Represents the 50th percentile of the data.
Mode	The value that occurs most often in the data. If there are multiple modes, the smallest mode appears.

Table 2.5 Additional Summary Statistics *(Continued)*

Trimmed Mean	(Does not appear if you have specified a Weight variable.)The mean calculated after removing the smallest p% and the largest p% of the data.
Geometric Mean	The *n*th root of the product of the data.
Range	The difference between the maximum and minimum of the data.
Interquartile Range	The difference between the 3rd and 1st quartiles.
Median Absolute Deviation	(Does not appear if you have specified a Weight variable.) The median of the absolute deviations from the median.
Enter (1-alpha) for mean confidence interval	Specify the alpha level for the mean confidence interval.
Enter trimmed mean percent	Specify the trimmed mean percentage. The percentage is trimmed off each side of the data.

Summary Statistics Options

The red triangle menu next to Summary Statistics contains these options:

Customize Summary Statistics Select which statistics you want to appear from the list. You can select or deselect all summary statistics. See Table 2.5.

Show All Modes Shows all of the modes if there are multiple modes.

Related Information

- "Statistical Details for Summary Statistics" on page 69

Distribution Platform Options

The red triangle menu next to Distributions contains options that affect all of the reports and graphs in the Distribution platform.

Table 2.6 Descriptions of Distribution Platform Options

Uniform Scaling	Scales all axes with the same minimum, maximum, and intervals so that the distributions can be easily compared.
Stack	Changes the orientation of the histogram and the reports to horizontal and stacks the individual distribution reports vertically. Deselect this option to return the report window to its original layout.

Table 2.6 Descriptions of Distribution Platform Options *(Continued)*

Save for Adobe Flash platform (.SWF)	Saves the histograms as **.swf** files that are Adobe Flash player compatible. Use these files in presentations and in Web pages. An HTML page is also saved that shows you the correct code for using the resulting **.swf** file. For more information about this option, go to http://www.jmp.com/support/swfhelp/en.
Script	This menu contains options that are available to all platforms. They enable you to redo the analysis or save the JSL commands for the analysis to a window or a file. For more information, see *Using JMP*.

Options for Categorical Variables

The red triangle menus next to each variable in the report window contain additional options that apply to the variable. This section describes the options that are available for categorical (nominal or ordinal) variables.

To see the options that are available for continuous variables, see "Options for Continuous Variables" on page 44.

Table 2.7 Description of Options for Categorical Variables

The **Display Options** sub-menu contains the following options:	
Frequencies	Shows or hides the Frequencies report. See "The Frequencies Report" on page 38.
Horizontal Layout	Changes the orientation of the histogram and the reports to vertical or horizontal.
Axes on Left	Moves the **Count**, **Prob**, and **Density** axes to the left instead of the right. This option is applicable only if **Horizontal Layout** is selected.
The **Histograms** sub-menu contains the following options:	
Histogram	Shows or hides the histogram. See "Histograms" on page 36.
Vertical	Changes the orientation of the histogram from a vertical to a horizontal orientation.
Std Error Bars	Draws the standard error bar on each level of the histogram.
Separate Bars	Separates the histogram bars.
Histogram Color	Changes the color of the histogram bars.

Table 2.7 Description of Options for Categorical Variables *(Continued)*

Count Axis	Adds an axis that shows the frequency of column values represented by the histogram bars.
Prob Axis	Adds an axis that shows the proportion of column values represented by histogram bars.
Density Axis	Adds an axis that shows the length of the bars in the histogram. The count and probability axes are based on the following calculations: prob = (bar width)*density count = (bar width)*density*(total count)
Show Percents	Labels the percent of column values represented by each histogram bar.
Show Counts	Labels the frequency of column values represented by each histogram bar.
Mosaic Plot	Displays a mosaic bar chart for each nominal or ordinal response variable. A mosaic plot is a stacked bar chart where each segment is proportional to its group's frequency count.
Order By	Reorders the histogram, mosaic plot, and Frequencies report in ascending or descending order, by count. To save the new order as a column property, use the **Save > Value Ordering** option.
Test Probabilities	Displays a report that tests hypothesized probabilities. See "Examples of the Test Probabilities Option" on page 62 for more details.
Confidence Interval	This menu contains confidence levels. Select a value that is listed, or select **Other** to enter your own. JMP computes score confidence intervals.
The **Save** sub-menu contains the following options:	
Level Numbers	Creates a new column in the data table called Level <colname>. The level number of each observation corresponds to the histogram bar that contains the observation.
Value Ordering	(Use with the **Order By** option) Creates a new value ordering column property in the data table, reflecting the new order.
Script to log	Displays the script commands to generate the current report in the log window. Select **View > Log** to see the log window.
Remove	Permanently removes the variable and all its reports from the Distribution report.

Options for Continuous Variables

The red triangle menus next to each variable in the report window contain additional options that apply to the variable. This section describes the options that are available for continuous variables.

To see the options that are available for categorical (nominal and ordinal) variables, see "Options for Categorical Variables" on page 42.

Table 2.8 Description of Options for Continuous Variables

The **Display Options** sub-menu contains the following options:	
Quantiles	Shows or hides the Quantiles report. See "The Quantiles Report" on page 39.
Set Quantile Increment	Changes the quantile increment or revert back to the default quantile increment.
Custom Quantiles	Sets custom quantiles by values or by increments. You can also specify the confidence level.
Summary Statistics	Shows or hides the Summary Statistics report. See "The Summary Statistics Report" on page 39.
Customize Summary Statistics	Adds or removes statistics from the Summary Statistics report. See "The Summary Statistics Report" on page 39.
Horizontal Layout	Changes the orientation of the histogram and the reports to vertical or horizontal.
Axes on Left	Moves the **Count**, **Prob**, **Density**, and **Normal Quantile Plot** axes to the left instead of the right. This option is applicable only if **Horizontal Layout** is selected.
The **Histograms** sub-menu contains the following options:	
Histogram	Shows or hides the histogram. See "Histograms" on page 36.
Shadowgram	Replaces the histogram with a shadowgram. To understand a shadowgram, consider that if the bin width of a histogram is changed, the appearance of the histogram changes. A shadowgram overlays histograms with different bin widths. Dominant features of a distribution are less transparent on the shadowgram. Note that the following options are not available for shadowgrams: • **Std Error Bars** • **Show Counts** • **Show Percents**

Table 2.8 Description of Options for Continuous Variables *(Continued)*

Vertical	Changes the orientation of the histogram from a vertical to a horizontal orientation.
Std Error Bars	Draws the standard error bar on each level of the histogram using the standard error. The standard error bar adjusts automatically when you adjust the number of bars with the hand tool. **Related Information** • "Resize Histogram Bars for Continuous Variables" on page 37 • "Statistical Details for Standard Error Bars" on page 68
Set Bin Width	Changes the bin width of the histogram bars. See "Resize Histogram Bars for Continuous Variables" on page 37.
Histogram Color	Changes the color of the histogram bars.
Count Axis	Adds an axis that shows the frequency of column values represented by the histogram bars. **Note:** If you resize the histogram bars, the count axis also resizes.
Prob Axis	Adds an axis that shows the proportion of column values represented by histogram bars. **Note:** If you resize the histogram bars, the probability axis also resizes.
Density Axis	The density is the length of the bars in the histogram. Both the count and probability are based on the following calculations: prob = (bar width)*density count = (bar width)*density*(total count) When looking at density curves that are added by the Fit Distribution option, the density axis shows the point estimates of the curves. **Note:** If you resize the histogram bars, the density axis remains constant.
Show Percents	Labels the proportion of column values represented by each histogram bar.
Show Counts	Labels the frequency of column values represented by each histogram bar.
Normal Quantile Plot	Adds a normal quantile plot that shows the extent to which the variable is normally distributed. See "Normal Quantile Plot" on page 46.
Outlier Box Plot	Adds an outlier box plot that shows the outliers in your data. See "Outlier Box Plot" on page 47.

Table 2.8 Description of Options for Continuous Variables *(Continued)*

Stem and Leaf	Adds a stem and leaf report, which is a variation of the histogram. See "Stem and Leaf" on page 49.
CDF Plot	Adds a plot of the empirical cumulative distribution function. See "CDF Plot" on page 49.
Test Mean	Performs a one-sample test for the mean. See "Test Mean" on page 50.
Test Std Dev	Performs a one-sample test for the standard deviation. See "Test Std Dev" on page 51.
Confidence Interval	Choose confidence intervals for the mean and standard deviation. See "Confidence Intervals for Continuous Variables" on page 52.
Prediction Interval	Choose prediction intervals for a single observation, or for the mean and standard deviation of the next randomly selected sample. See "Prediction Intervals" on page 53.
Tolerance Interval	Computes an interval to contain at least a specified proportion of the population. See "Tolerance Intervals" on page 54.
Capability Analysis	Measures the conformance of a process to given specification limits. See "Capability Analysis" on page 54.
Continuous Fit	Fits distributions to continuous variables. See "Fit Distributions" on page 56.
Discrete Fit	Fits distributions to discrete variables. See "Fit Distributions" on page 56.
Save	Saves information about continuous or categorical variables. See "Save Commands for Continuous Variables" on page 52.
Remove	Permanently removes the variable and all its reports from the Distribution report.

Normal Quantile Plot

Use the **Normal Quantile Plot** option to visualize the extent to which the variable is normally distributed. If a variable is normally distributed, the normal quantile plot approximates a diagonal straight line. This type of plot is also called a quantile-quantile plot, or Q-Q plot.

The normal quantile plot also shows Lilliefors confidence bounds (Conover 1980) and probability and normal quantile scales.

Figure 2.7 Normal Quantile Plot

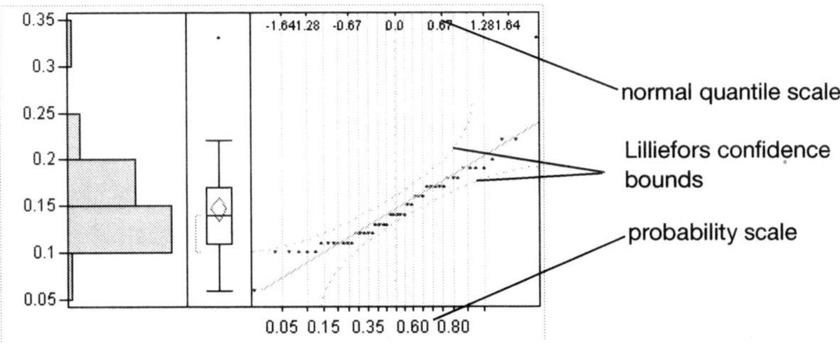

Note the following information:

- The *y*-axis shows the column values.
- The *x*-axis shows the empirical cumulative probability for each value.

Related Information

- "Statistical Details for the Normal Quantile Plot" on page 70

Outlier Box Plot

Use the outlier box plot (also called a Tukey outlier box plot) to see the distribution and identify possible outliers. Generally, box plots show selected quantiles of continuous distributions.

Figure 2.8 Outlier Box Plot

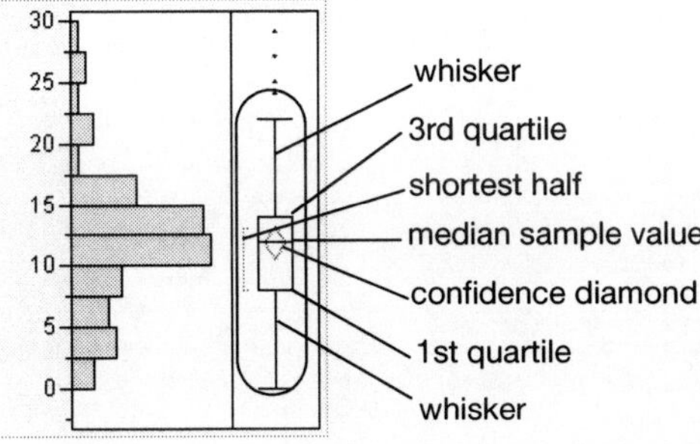

Note the following aspects about outlier box plots:

- The vertical line within the box represents the median sample value.
- The confidence diamond contains the mean and the upper and lower 95% of the mean. If you drew a line through the middle of the diamond, you would have the mean. The top and bottom points of the diamond represent the upper and lower 95% of the mean.

- The ends of the box represent the 25th and 75th quantiles, also expressed as the 1st and 3rd *quartile*, respectively.
- The difference between the 1st and 3rd quartiles is called the *interquartile range.*
- The box has lines that extend from each end, sometimes called *whiskers*. The whiskers extend from the ends of the box to the outermost data point that falls within the distances computed as follows:

  1st quartile - 1.5*(interquartile range)

  3rd quartile + 1.5*(interquartile range)

  If the data points do not reach the computed ranges, then the whiskers are determined by the upper and lower data point values (not including outliers).
- The bracket outside of the box identifies the *shortest half,* which is the most dense 50% of the observations (Rousseuw and Leroy 1987).

Remove Objects from the Outlier Box Plot

To remove the confidence diamond or the shortest half, proceed as follows:

1. Right-click on the outlier box plot and select **Customize.**
2. Click **Box Plot.**
3. Deselect the check box next to **Confidence Diamond** or **Shortest Half.**

For more details about the Customize Graph window, see the *Using JMP* book.

Quantile Box Plot

The Quantile Box Plot displays specific quantiles from the Quantiles report. If the distribution is symmetric, the quantiles in the box plot are approximately equidistant from each other. At a glance, you can see whether the distribution is symmetric. For example, if the quantile marks are grouped closely at one end, but have greater spacing at the other end, the distribution is skewed toward the end with more spacing. See Figure 2.9.

Figure 2.9 Quantile Box Plot

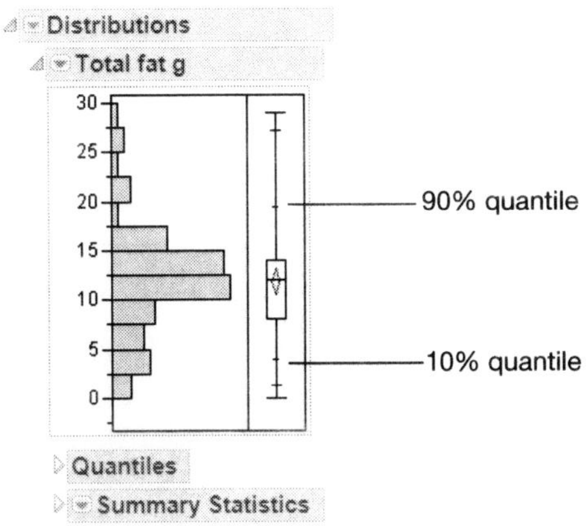

Quantiles are values where the *p*th quantile is larger than *p*% of the values. For example, 10% of the data lies below the 10th quantile, and 90% of the data lies below the 90th quantile.

Stem and Leaf

Each line of the plot has a **Stem** value that is the leading digit of a range of column values. The **Leaf** values are made from the next-in-line digits of the values. You can see the data point by joining the stem and leaf. In some cases, the numbers on the stem and leaf plot are rounded versions of the actual data in the table. The stem-and-leaf plot actively responds to clicking and the brush tool.

Note: The stem-and-leaf plot does not support fractional frequencies.

CDF Plot

The CDF plot creates a plot of the empirical cumulative distribution function. Use the CDF plot to determine the percent of data that is at or below a given value on the *x*-axis.

Figure 2.10 CDF Plot

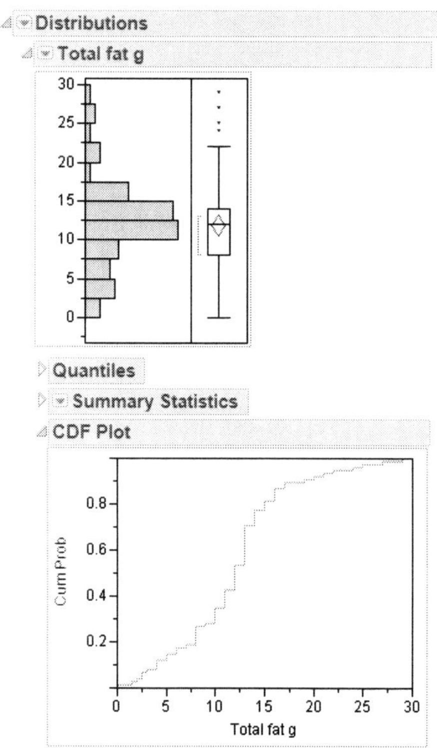

For example, in this CDF plot, approximately 30% of the data is less than a total fat value of 10 grams.

Test Mean

Use the **Test Mean** window to specify options for and perform a one-sample test for the mean. If you specify a value for the standard deviation, a *z*-test is performed. Otherwise, the sample standard deviation is used to perform a *t*-test. You can also request the nonparametric Wilcoxon Signed-Rank test.

Use the **Test Mean** option repeatedly to test different values. Each time you test the mean, a new Test Mean report appears.

Table 2.9 Description of the Test Mean Report

Statistics that are calculated for Test Mean:	
t Test (or z Test)	Lists the value of the test statistic and the *p*-values for the two-sided and one-sided alternatives.

Table 2.9 Description of the Test Mean Report *(Continued)*

Signed-Rank	(Only appears for the Wilcoxon Signed-Rank test) Lists the value of the Wilcoxon signed-rank statistic followed by the *p*-values for the two-sided and one-sided alternatives. The test assumes only that the distribution is symmetric. See "Statistical Details for the Wilcoxon Signed Rank Test" on page 71.
Probability values:	
Prob>|t|	The probability of obtaining an absolute *t*-value by chance alone that is greater than the observed *t*-value when the population mean is equal to the hypothesized value. This is the *p*-value for observed significance of the two-tailed *t*-test.
Prob>t	The probability of obtaining a *t*-value greater than the computed sample *t* ratio by chance alone when the population mean is not different from the hypothesized value. This is the *p*-value for an upper-tailed test.
Prob<t	The probability of obtaining a *t*-value less than the computed sample *t* ratio by chance alone when the population mean is not different from the hypothesized value. This is the *p*-value for a lower-tailed test.

Table 2.10 Descriptions of the Test Mean Options

PValue animation	Starts an interactive visual representation of the p-value. Enables you to change the hypothesized mean value while watching how the change affects the p-value.
Power animation	Starts an interactive visual representation of power and beta. You can change the hypothesized mean and sample mean while watching how the changes affect power and beta.
Remove Test	Removes the mean test.

Test Std Dev

Use the **Test Std Dev** option to perform a one-sample test for the standard deviation (details in Neter, Wasserman, and Kutner 1990). Use the **Test Std Dev** option repeatedly to test different values. Each time you test the standard deviation, a new Test Standard Deviation report appears.

Table 2.11 Description of the Test Std Dev Report

Test Statistic	Provides the value of the Chi-square test statistic. See "Statistical Details for the Standard Deviation Test" on page 71.

Table 2.11 Description of the Test Std Dev Report *(Continued)*

Min PValue	The probability of obtaining a greater Chi-square value by chance alone when the population standard deviation is not different from the hypothesized value. See "Statistical Details for the Standard Deviation Test" on page 71.
Prob>ChiSq	The probability of obtaining a Chi-square value greater than the computed sample Chi-square by chance alone when the population standard deviation is not different from the hypothesized value. This is the *p*-value for observed significance of a one-tailed *t*-test.
Prob<ChiSq	The probability of obtaining a Chi-square value less than the computed sample Chi-square by chance alone when the population standard deviation is not different from the hypothesized value. This is the *p*-value for observed significance of a one-tailed *t*-test.

Confidence Intervals for Continuous Variables

The **Confidence Interval** options display confidence intervals for the mean and standard deviation. The **0.90**, **0.95**, and **0.99** options compute two-sided confidence intervals for the mean and standard deviation. Use the **Confidence Interval > Other** option to select a confidence level, and select one-sided or two-sided confidence intervals. You can also type a known sigma. If you use a known sigma, the confidence interval for the mean is based on *z*-values rather than *t*-values.

The Confidence Intervals report shows the mean and standard deviation parameter estimates with upper and lower confidence limits for $1 - \alpha$.

Save Commands for Continuous Variables

Use the Save menu commands to save information about continuous variables. Each Save command generates a new column in the current data table. The new column is named by appending the variable name (denoted <colname> in the following definitions) to the Save command name. See Table 2.12.

Select the Save commands repeatedly to save the same information multiple times under different circumstances, such as before and after combining histogram bars. If you use a Save command multiple times, the column name is numbered (name1, name2, and so on) to ensure unique column names.

Table 2.12 Descriptions of Save Commands

Command	Column Added to Data Table	Description
Level Numbers	Level <colname>	The level number of each observation corresponds to the histogram bar that contains the observation. The histogram bars are numbered from low to high, beginning with 1.

Table 2.12 Descriptions of Save Commands *(Continued)*

Command	Column Added to Data Table	Description
Level Midpoints	Midpoint <colname>	The midpoint value for each observation is computed by adding half the level width to the lower level bound.
Ranks	Ranked <colname>	Provides a ranking for each of the corresponding column's values starting at 1. Duplicate response values are assigned consecutive ranks in order of their occurrence in the data table.
Ranks Averaged	RankAvgd <colname>	If a value is unique, then the averaged rank is the same as the rank. If a value occurs k times, the average rank is computed as the sum of the value's ranks divided by k.
Prob Scores	Prob <colname>	For N nonmissing scores, the probability score of a value is computed as the averaged rank of that value divided by $N + 1$. This column is similar to the empirical cumulative distribution function.
Normal Quantiles	N-Quantile <colname>	Saves the Normal quantiles to the data table. See "Statistical Details for the Normal Quantile Plot" on page 70.
Standardized	Std <colname>	Saves standardized values to the data table. See "Statistical Details for Saving Standardized Data" on page 71.
Centered	Centered <colname>	Saves values for centering on zero.
Spec Limits	(none)	Stores the specification limits applied in a capability analysis as a column property of the corresponding column in the current data table. Automatically retrieves and displays the specification limits when you repeat the capability analysis.
Script to Log	(none)	Prints the script to the log window. Run the script to recreate the analysis.

Prediction Intervals

Prediction intervals concern a single observation, or the mean and standard deviation of the next randomly selected sample. The calculations assume that the given sample is selected randomly from a normal distribution. Select one-sided or two-sided prediction intervals.

When you select the **Prediction Interval** option for a variable, the Prediction Intervals window appears. Use the window to specify the confidence level, the number of future samples, and either a one-sided or two-sided limit.

Related Information

- "Statistical Details for Prediction Intervals" on page 72
- "Example of Prediction Intervals" on page 64

Tolerance Intervals

A tolerance interval contains at least a specified proportion of the population. It is a confidence interval for a specified proportion of the population, not the mean, or standard deviation. Complete discussions of tolerance intervals are found in Hahn and Meeker (1991) and in Tamhane and Dunlop (2000).

When you select the **Tolerance Interval** option for a variable, the Tolerance Intervals window appears. Use the window to specify the confidence level, the proportion to cover, and either a one-sided or two-sided limit. The calculations are based on the assumption that the given sample is selected randomly from a normal distribution.

Related Information

- "Statistical Details for Tolerance Intervals" on page 72
- "Example of Tolerance Intervals" on page 66

Capability Analysis

The **Capability Analysis** option measures the conformance of a process to given specification limits. When you select the **Capability Analysis** option for a variable, the Capability Analysis window appears. Use the window to enter specification limits, distribution type, and information about sigma.

Note: To save the specification limits to the data table as a column property, select **Save > Spec Limits**. When you repeat the capability analysis, the saved specification limits are automatically retrieved.

The Capability Analysis report is organized into two sections: Capability Analysis and the distribution type (Long Term Sigma, Specified Sigma, and so on).

Capability Analysis Descriptions

The Capability Analysis window, report, and options are described in the following tables.

Table 2.13 Description of the Capability Analysis Window

<Distribution type>	By default, the normal distribution is assumed when calculating the capability statistics and the percent out of the specification limits. To perform a capability analysis on non-normal distributions, see the description of Spec Limits under "Fit Distribution Options" on page 58.

Table 2.13 Description of the Capability Analysis Window *(Continued)*

<Sigma type>	Estimates sigma (σ) using the selected methods. See "Statistical Details for Capability Analysis" on page 73.

Table 2.14 Description of the Capability Analysis Report

Specification	Lists the specification limits.
Value	Lists the values that you specified for each specification limit and the target.
Portion and % Actual	Portion labels describe the numbers in the % Actual column, as follows: • Below LSL gives the percentage of the data that is below the lower specification limit. • Above USL gives the percentage of the data that is above the upper specification limit. • Total Outside gives the total percentage of the data that is either below LSL or above USL.
Capability	Type of process capability indices. See Table 2.19. **Note:** There is a preference for Capability called **Ppk Capability Labeling** that labels the long-term capability output with P_{pk} labels. Open the Preference window (**File > Preferences**), then select **Platforms > Distribution** to see this preference.
Index	Process capability index values.
Upper CI	Upper confidence interval.
Lower CI	Lower confidence interval.
Portion and Percent	Portion labels describe the numbers in the Percent column, as follows: • Below LSL gives the percentage of the fitted distribution that is below the lower specification limit. • Above USL gives the percentage of the fitted distribution that is above the upper specification limit. • Total Outside gives the total percentage of the fitted distribution that is either below LSL or above USL.
PPM (parts per million)	The PPM value is the **Percent** column multiplied by 10,000.
Sigma Quality	Sigma Quality is frequently used in Six Sigma methods, and is also referred to as the *process sigma*. See "Statistical Details for Capability Analysis" on page 73.

Table 2.15 Description of the Capability Analysis Options

Z Bench	Shows the values (represented by **Index**) of the Benchmark Z statistics. According to the AIAG Statistical Process Control manual, *Z* represents the number of standard deviation units from the process average to a value of interest such as an engineering specification. When used in capability assessment, *Z* USL is the distance to the upper specification limit and *Z* LSL is the distance to the lower specification limit. See "Statistical Details for Capability Analysis" on page 73.
Capability Animation	Interactively change the specification limits and the process mean to see the effects on the capability statistics. This option is available only for capability analyses based on the Normal distribution.

Related Information

- "Statistical Details for Capability Analysis" on page 73
- "Example of Capability Analysis" on page 67

Fit Distributions

Use the **Continuous** or **Discrete Fit** options to fit a distribution to a continuous or discrete variable.

A curve is overlaid on the histogram, and a Parameter Estimates report is added to the report window. A red triangle menu contains additional options. See "Fit Distribution Options" on page 58.

Note: The Life Distribution platform also contains options for distribution fitting that might use different parameterizations and allow for censoring. See the *Quality and Reliability Methods* book.

Continuous Fit

Use the **Continuous Fit** options to fit the following distributions to a continuous variable.

- The Normal distribution is often used to model measures that are symmetric with most of the values falling in the middle of the curve.
- The LogNormal distribution is often used to model values that are constrained by zero but have a few very large values. The LogNormal distribution can be obtained by exponentiating the Normal distribution.
- The Weibull, Weibull with threshold, and Extreme Value distributions often provide a good model for estimating the length of life, especially for mechanical devices and in biology.
- The Exponential distribution is especially useful for describing events that randomly occur over time, such as survival data. The exponential distribution might also be useful for modeling elapsed time between the occurrence of non-overlapping events, such as the time between a user's computer query and response of the server, the arrival of customers at a service desk, or calls coming in at a switchboard.

- The Gamma distribution is bound by zero and has a flexible shape.
- The Beta distribution is useful for modeling the behavior of random variables that are constrained to fall in the interval 0,1. For example, proportions always fall between 0 and 1.
- The Normal Mixtures distribution fits a mixture of normal distributions. This flexible distribution is capable of fitting multi-modal data. You can also fit two or more distributions by selecting the **Normal 2 Mixture**, **Normal 3 Mixture**, or **Other** options.
- The Smooth Curve distribution... A smooth curve is fit using nonparametric density estimation (kernel density estimation). The smooth curve is overlaid on the histogram and a slider appears beneath the plot. Control the amount of smoothing by changing the kernel standard deviation with the slider. The initial Kernel Std estimate is formed by summing the normal densities of the kernel standard deviation located at each data point.
- The Johnson Su, Johnson Sb, and Johnson Sl Distributions are useful for its data-fitting capabilities because it supports every possible combination of skewness and kurtosis.
- The Generalized Log (Glog) distribution is useful for fitting data that are rarely normally distributed and often have non-constant variance, like biological assay data.

Comparing All Distributions

The **All** option fits all applicable continuous distributions to a variable. The Compare Distributions report contains statistics about each fitted distribution. Use the check boxes to show or hide a fit report and overlay curve for the selected distribution. By default, the best fit distribution is selected.

The Show Distribution list is sorted by AICc in ascending order.

If your data has negative values, the Show Distribution list does not include those distributions that require data with positive values. If your data has non-integer values, the list of distributions does not include discrete distributions. Distributions with threshold parameters, like Beta and Johnson Sb, are not included in the list of possible distributions.

Related Information

- "Statistical Details for Continuous Fit Distributions" on page 76
- "Statistical Details for Fitted Quantiles" on page 84
- "Statistical Details for Fit Distribution Options" on page 84

Discrete Fit

Use the Discrete Fit options to fit a distribution (such as Poisson or Binomial) to a discrete variable. The available distributions are as follows:

- Poisson
- Binomial
- Gamma Poisson
- Beta Binomial

Related Information

- "Statistical Details for Discrete Fit Distributions" on page 81
- "Statistical Details for Fitted Quantiles" on page 84
- "Statistical Details for Fit Distribution Options" on page 84

Fit Distribution Options

Each fitted distribution report has a red triangle menu that contains additional options.

Table 2.16 Description of Fit Distribution Options

Diagnostic Plot	Creates a quantile or a probability plot. See "Diagnostic Plot" on page 59.
Density Curve	Uses the estimated parameters of the distribution to overlay a density curve on the histogram.
Goodness of Fit	Computes the goodness of fit test for the fitted distribution. See "Goodness of Fit" on page 60.
Fix Parameters	Enables you to fix parameters and re-estimate the non-fixed parameters. An Adequacy LR (likelihood ratio) Test report also appears, which tests your new parameters to determine whether they fit the data.
Quantiles	Returns the un scaled and un centered quantiles for the specific lower probability values that you specify.
Set Spec Limits for K Sigma	Use this option when you do not know the specification limits for a process and you want to use its distribution as a guideline for setting specification limits. Usually specification limits are derived using engineering considerations. If there are no engineering considerations, and if the data represents a trusted benchmark (well behaved process), then quantiles from a fitted distribution are often used to help set specification limits. See "Statistical Details for Fit Distribution Options" on page 84.
Spec Limits	Computes generalizations of the standard capability indices, based on the specification limits and target you specify. See "Spec Limits" on page 61.
Save Fitted Quantiles	Saves the fitted quantile values as a new column in the current data table. See "Statistical Details for Fitted Quantiles" on page 84.

Table 2.16 Description of Fit Distribution Options *(Continued)*

Save Density Formula	Creates a new column in the current data table that contains fitted values that have been computed by the density formula. The density formula uses the estimated parameter values.
Save Spec Limits	Saves the specification limits as a column property. See "Statistical Details for Fit Distribution Options" on page 84.
Save Transformed	Creates a new column and saves a formula. The formula can transform the column to normality using the fitted distribution. This option is available only when one of the Johnson distributions or the Glog distribution is fit.
Remove Fit	Removes the distribution fit from the report window.

Diagnostic Plot

The **Diagnostic Plot** option creates a quantile or a probability plot. Depending on the fitted distribution, the plot is one of four formats.

Table 2.17 Descriptions of Plot Formats

Plot Format	Applicable Distributions
The fitted quantiles versus the data	• Weibull with threshold • Gamma • Beta • Poisson • GammaPoisson • Binomial • BetaBinomial
The fitted probability versus the data	• Normal • Normal Mixtures • Exponential
The fitted probability versus the data on log scale	• Weibull • LogNormal • Extreme Value

Table 2.17 Descriptions of Plot Formats *(Continued)*

Plot Format	Applicable Distributions
The fitted probability versus the standard normal quantile	• Johnson Sl • Johnson Sb • Johnson Su • Glog

Table 2.18 describes the options in the red triangle menu next to Diagnostic Plot.

Table 2.18 Descriptions of the Diagnostic Plot Options

Rotate	Reverses the *x*- and *y*-axes.
Confidence Limits	Draws Lilliefors 95% confidence limits for the Normal Quantile plot, and 95% equal precision bands with a = 0.001 and b = 0.99 for all other quantile plots (Meeker and Escobar (1998)).
Line of Fit	Draws the straight diagonal reference line. If a variable fits the selected distribution, the values fall approximately on the reference line.
Median Reference Line	Draws a horizontal line at the median of the response.

Goodness of Fit

The **Goodness of Fit** option computes the goodness of fit test for the fitted distribution. The goodness of fit tests are not Chi-square tests, but are EDF (Empirical Distribution Function) tests. EDF tests offer advantages over the Chi-square tests, including improved power and invariance with respect to histogram midpoints.

- For Normal distributions, the Shapiro-Wilk test for normality is reported when the sample size is less than or equal to 2000, and the KSL test is computed for samples that are greater than 2000.
- For discrete distributions (such as Poisson distributions) that have sample sizes less than or equal to 30, the Goodness of Fit test is formed using two one-sided exact Kolmogorov tests combined to form a near exact test. For details, see Conover 1972. For sample sizes greater than 30, a Pearson Chi-squared goodness of fit test is performed.

Related Information

- "Statistical Details for Fit Distribution Options" on page 84

Spec Limits

The **Spec Limits** option launches a window requesting specification limits and target, and then computes generalizations of the standard capability indices. This is done using the fact that for the normal distribution, 3σ is both the distance from the lower 0.135 percentile to median (or mean) and the distance from the median (or mean) to the upper 99.865 percentile. These percentiles are estimated from the fitted distribution, and the appropriate percentile-to-median distances are substituted for 3σ in the standard formulas.

Related Information

- "Statistical Details for Fit Distribution Options" on page 84

Additional Examples of the Distribution Platform

This section contains additional examples using the Distribution platform.

Example of Selecting Data in Multiple Histograms

1. Open the Companies.jmp sample data table.
2. Select **Analyze > Distribution**.
3. Select Type and Size Co and click **Y, Columns.**
4. Click **OK**.

   You want to see the type distribution of companies that are small.
5. Click on the bar next to small.

   You can see that there are more small computer companies than there are pharmaceutical companies. To broaden your selection, add medium companies.
6. Hold down the SHIFT key. In the Size Co histogram, click on the bar next to medium.

   You can see the type distribution of small and medium sized companies. See Figure 2.11 at left. To narrow down your selection, you want to see the small and medium pharmaceutical companies only.
7. Hold down the ALT key. In the Type histogram, click in the Pharmaceutical bar.

   You can see how many of the small and medium companies are pharmaceutical companies. See Figure 2.11 at right.

Figure 2.11 Selecting Data in Multiple Histograms

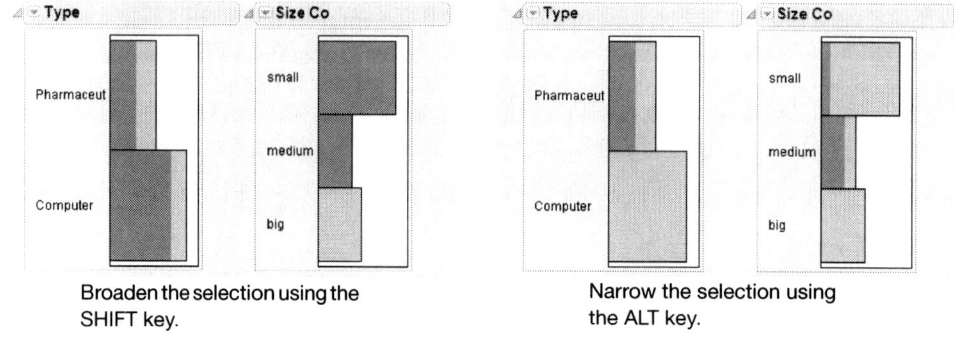

Broaden the selection using the SHIFT key.

Narrow the selection using the ALT key.

Examples of the Test Probabilities Option

Initiate a test probability report for a variable with more than two levels:

1. Open the VA Lung Cancer.jmp sample data table.
2. Select **Analyze > Distribution.**
3. Select Cell Type and click **Y, Columns.**
4. Click **OK.**
5. From the red triangle menu next to Cell Type, select **Test Probabilities.**

   See Figure 2.12 at left.

Initiate a test probability report for a variable with exactly two levels:

1. Open the Penicillin.jmp sample data table.
2. Select **Analyze > Distribution.**
3. Select Response and click **Y, Columns.**
4. Click **OK.**
5. From the red triangle menu next to Response, select **Test Probabilities.**

   See Figure 2.12 at right.

Figure 2.12 Examples of Test Probabilities Options

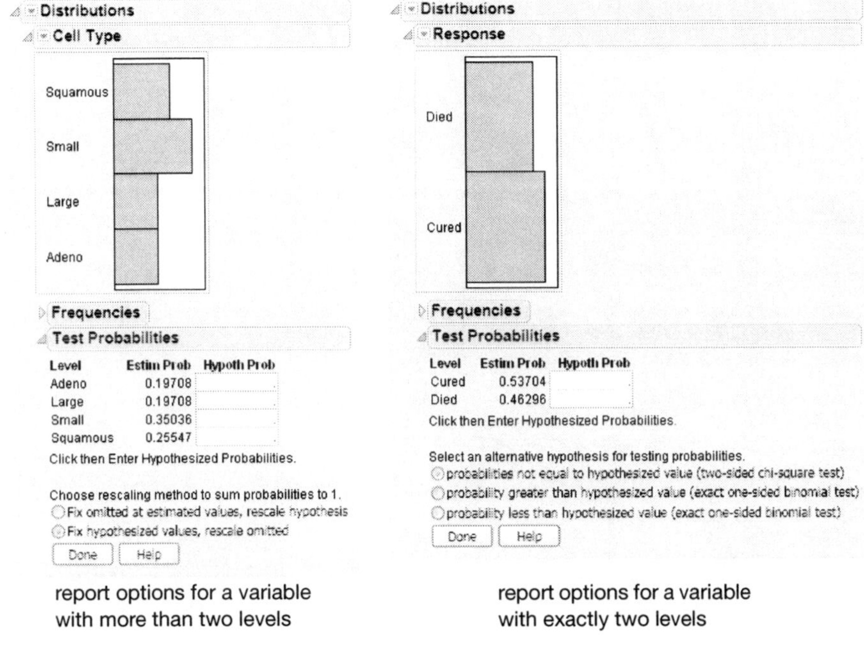

report options for a variable with more than two levels

report options for a variable with exactly two levels

Example of Generating the Test Probabilities Report

To generate a test probabilities report for a variable with more than two levels:

1. Refer to Figure 2.12 at left. Type 0.25 in all four Hypoth Prob fields.
2. Click the **Fix hypothesized values, rescale omitted** button.
3. Click **Done.**

Likelihood Ratio and Pearson Chi-square tests are calculated. See Figure 2.13 at left.

To generate a test probabilities report for a variable with exactly two levels:

1. Refer to Figure 2.12 at right. Type 0.5 in both Hypoth Prob fields.
2. Click the **probability less than hypothesized value** button.
3. Click **Done.**

Exact probabilities are calculated for the binomial test. See Figure 2.13 at right.

Figure 2.13 Examples of Test Probabilities Reports

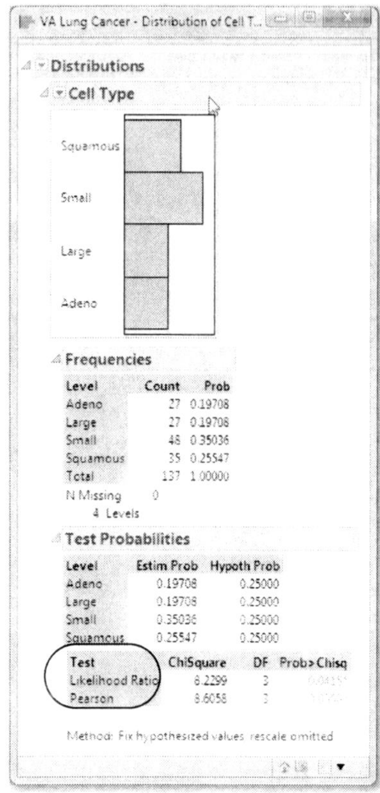

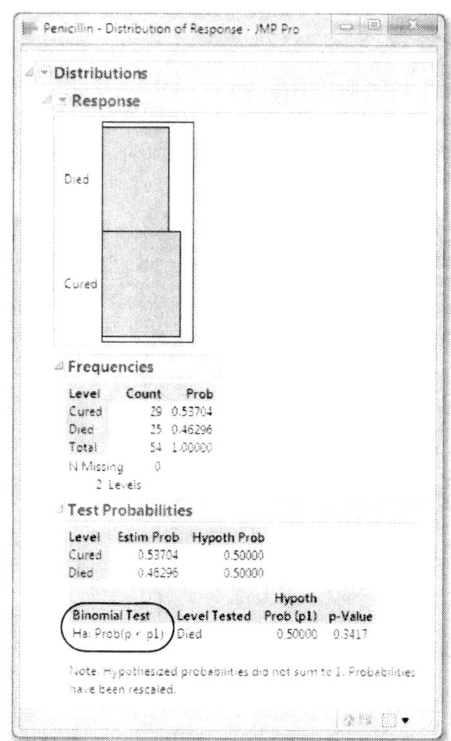

Example of Prediction Intervals

Suppose you are interested in computing prediction intervals for the next 10 observations of ozone level.

1. Open the **Cities.jmp** sample data table.
2. Select **Analyze > Distribution**.
3. Select **OZONE** and click **Y, Columns**.
4. Click **OK**.
5. From the red triangle next to OZONE, select **Prediction Interval**.

Figure 2.14 The Prediction Intervals Window

Enter (1-alpha) for prediction interval 0.95
Enter number of future samples 1

Two-sided
One-sided lower limit
One-sided upper limit

OK Cancel Help

6. In the Prediction Intervals window, type 10 next to **Enter number of future samples.**
7. Click **OK.**

Figure 2.15 Example of a Prediction Interval Report

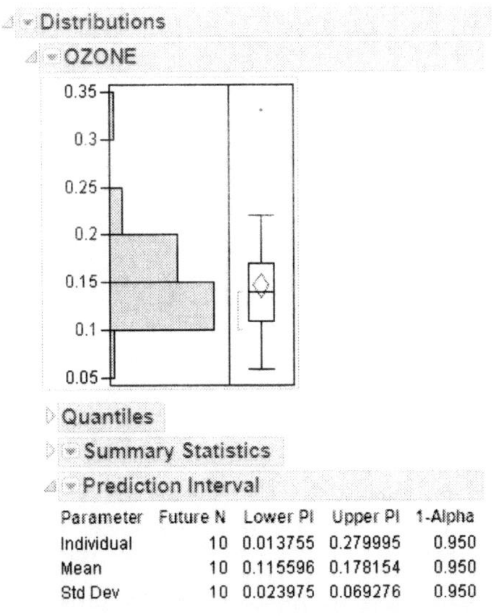

Parameter	Future N	Lower PI	Upper PI	1-Alpha
Individual	10	0.013755	0.279995	0.950
Mean	10	0.115596	0.178154	0.950
Std Dev	10	0.023975	0.069276	0.950

In this example, you can be 95% confident about the following:

- Each of the next 10 observations will be between .013755 and .279995.
- The mean of the next 10 observations will be between .115596 and .178154.
- The standard deviation of the next 10 observations will be between .023975 and .069276.

Example of Tolerance Intervals

Suppose you want to estimate an interval that contains 90% of ozone level measurements.

1. Open the Cities.jmp sample data table.
2. Select **Analyze > Distribution**.
3. Select OZONE and click **Y, Columns**.
4. Click **OK**.
5. From the red triangle menu next to OZONE, select **Tolerance Interval**.

Figure 2.16 The Tolerance Intervals Window

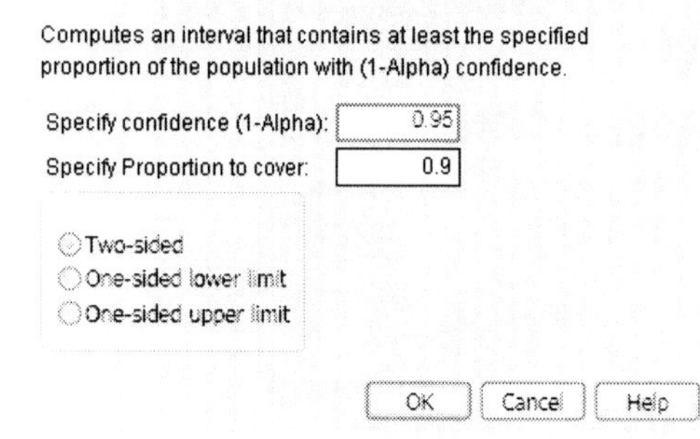

6. Keep the default selections, and click **OK**.

Figure 2.17 Example of a Tolerance Interval Report

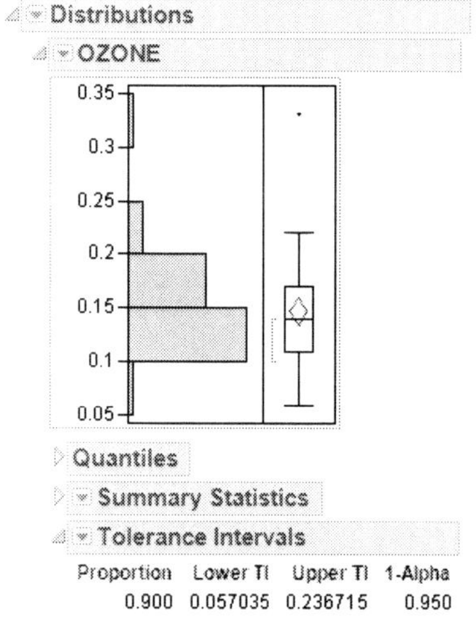

In this example, you can be 95% confident that at least 90% of the population lie between .057035 and .236715, based on the Lower TI (tolerance interval) and Upper TI values.

Example of Capability Analysis

Suppose you want to characterize the abrasion levels of the tires your company manufactures. The lower and upper specification limits are 100 and 200, respectively.

1. Open the Tiretread.jmp sample data table.
2. Select **Analyze > Distribution.**
3. Select ABRASION and click **Y, Columns**.
4. Click **OK**.
5. From the red triangle menu next to ABRASION, select **Capability Analysis.**
6. Type 100 for the **Lower Spec Limit**.
7. Type 200 for the **Upper Spec Limit**.
8. Keep the rest of the default selections, and click **OK**.
9. From the red triangle menu next to ABRASION, select **Histogram Options > Vertical**.

Figure 2.18 Example of the Capability Analysis Report

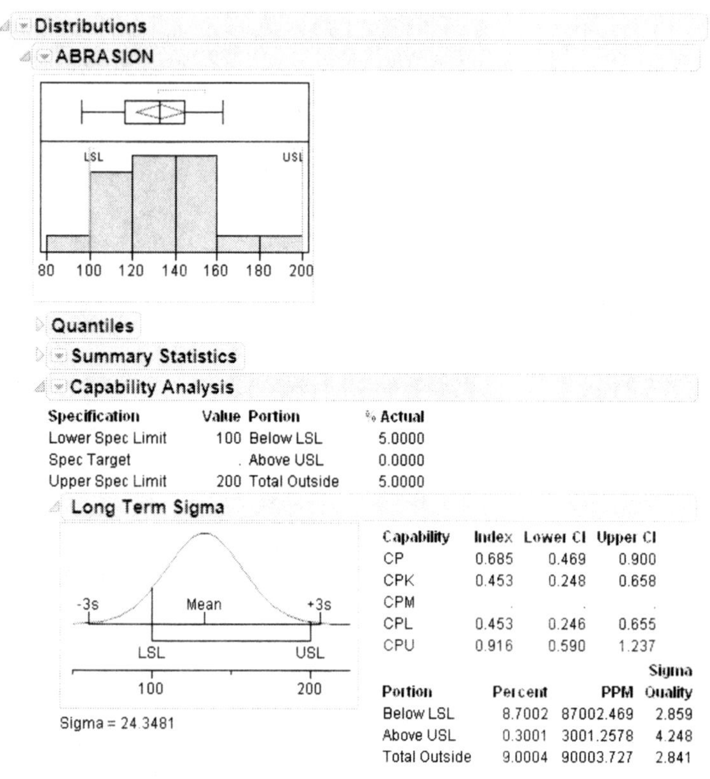

Specification	Value	Portion	% Actual
Lower Spec Limit	100	Below LSL	5.0000
Spec Target	.	Above USL	0.0000
Upper Spec Limit	200	Total Outside	5.0000

Capability	Index	Lower CI	Upper CI
CP	0.685	0.469	0.900
CPK	0.453	0.248	0.658
CPM	.	.	.
CPL	0.453	0.246	0.655
CPU	0.916	0.590	1.237

Portion	Percent	PPM	Sigma Quality
Below LSL	8.7002	87002.469	2.859
Above USL	0.3001	3001.2578	4.248
Total Outside	9.0004	90003.727	2.841

The spec limits are added to the histogram so that the data can be visually compared to the limits. As you can see, some of the abrasion levels are below the lower spec limit, and some are very close to the upper spec limit. The Capability Analysis results are added to the report. The Cpk value is 0.453, indicating a process that is not capable, relative to the given specification limits.

Statistical Details for the Distribution Platform

This section contains statistical details for Distribution options and reports.

Statistical Details for Standard Error Bars

Standard errors bars are calculated using the standard error $\sqrt{np_i(1-p_i)}$ where $p_i=n_i/n$.

Statistical Details for Quantiles

This section describes how quantiles are computed.

To compute the *p*th quantile of *N* non-missing values in a column, arrange the *N* values in ascending order and call these column values $y_1, y_2, \ldots, y_N$. Compute the rank number for the *p*th quantile as $p / 100(N + 1)$.

- If the result is an integer, the *p*th quantile is that rank's corresponding value.
- If the result is not an integer, the *p*th quantile is found by interpolation. The *p*th quantile, denoted q_p, is computed as follows:

  $$q_p = (1 - f)y_i + (f)y_{i+1}$$

  where:

  – n is the number of non-missing values for a variable

  – $y_1, y_2, \ldots, y_n$ represents the ordered values of the variable

  – y_{n+1} is taken to be y_n

  – i is the integer part and f is the fractional part of $(n+1)p$.

  – $(n + 1)p = i + f$

For example, suppose a data table has 15 rows and you want to find the 75th and 90th quantile values of a continuous column. After the column is arranged in ascending order, the ranks that contain these quantiles are computed as follows:

$$\frac{75}{100}(15 + 1) = 12 \text{ and } \frac{90}{100}(15 + 1) = 14.4$$

The value y_{12} is the 75th quantile. The 90th quantile is interpolated by computing a weighted average of the 14th and 15th ranked values as $y_{90} = 0.6y_{14} + 0.4y_{15}$.

Statistical Details for Summary Statistics

This section contains statistical details for specific statistics in the Summary Statistics report.

Mean

The mean is the sum of the non-missing values divided by the number of non-missing values. If you assigned a **Weight** or **Freq** variable, the mean is computed by JMP as follows:

1. Each column value is multiplied by its corresponding weight or frequency.
2. These values are added and divided by the sum of the weights or frequencies.

Std Dev

The standard deviation measures the spread of a distribution around the mean. It is often denoted as s and is the square root of the sample variance, denoted s^2.

$$s = \sqrt{s^2} \text{ where}$$

$$s^2 = \sum_{i=1}^{N} \frac{w_i(y_i - \bar{y}_w)^2}{N-1}$$

$$\bar{y}_w = \text{weighed mean}$$

Std Err Mean

The standard error means is computed by dividing the sample standard deviation, s, by the square root of N. In the launch window, if you specified a column for Weight or Freq, then the denominator is the square root of the sum of the weights or frequencies.

Skewness

Skewness is based on the third moment about the mean and is computed as follows:

$$\sum w_i^{\frac{3}{2}} z_i^3 \frac{N}{(N-1)(N-2)} \text{ where } z_i = \frac{x_i - \bar{x}}{s}$$

and w_i is a weight term (= 1 for equally weighted items)

Kurtosis

Kurtosis is based on the fourth moment about the mean and is computed as follows:

$$\frac{n(n+1)}{(n-1)(n-2)(n-3)} \sum_{i=1}^{n} w_i^2 \left(\frac{x_i - \bar{x}}{s}\right)^4 - \frac{3(n-1)^2}{(n-2)(n-3)}$$

where w_i is a weight term (= 1 for equally weighted items). Using this formula, the Normal distribution has a kurtosis of 0.

Statistical Details for the Normal Quantile Plot

The empirical cumulative probability for each value is computed as follows:

$$\frac{r_i}{N+1}$$

where r_i is the rank of the ith observation, and N is the number of non-missing (and nonexcluded) observations.

The normal quantile values are computed as follows:

$$\Phi^{-1}\left(\frac{r_i}{N+1}\right)$$

where Φ is the cumulative probability distribution function for the normal distribution.

These normal quantile values are Van Der Waerden approximations to the order statistics that are expected for the normal distribution.

Statistical Details for the Wilcoxon Signed Rank Test

The Wilcoxon signed-rank test uses average ranks for ties. The *p*-values are exact for $n \leq 20$ where n is the number of values not equal to the hypothesized value. For $n > 20$ a Student's *t* approximation given by Iman (1974) is used.

Statistical Details for the Standard Deviation Test

Here is the formula for calculating the Test Statistic:

$$\frac{(n-1)s^2}{\sigma^2}$$

The Test Statistic is distributed as a Chi-square variable with $n - 1$ degrees of freedom when the population is normal.

The Min PValue is the *p*-value of the two-tailed test, and is calculated as follows:

2*min(p1,p2)

where p1 is the lower one-tail *p*-value and p2 is the upper one-tail *p*-value.

Statistical Details for Normal Quantiles

The normal quantile values are computed as follows:

$$\Phi^{-1}\left(\frac{r_i}{N+1}\right)$$ where:

- Φ is the cumulative probability distribution function for the normal distribution
- r_i is the rank of the *i*th observation
- N is the number of non-missing observations

Statistical Details for Saving Standardized Data

The standardized values are computed using the following formula:

$$\frac{X-\bar{X}}{S_X}$$ where:

- X is the original column
- $\bar{X}$ is the mean of column X
- S_X is the standard deviation of column X

Statistical Details for Prediction Intervals

The formulas that JMP uses for computing prediction intervals are as follows:

- For m future observations:

$$[\underset{\sim}{y}_m, \tilde{y}_m] = \bar{X} \pm t_{(1-\alpha/2m;n-1)} \times \sqrt{1+\frac{1}{n}} \times s \text{ for } m \geq 1$$

- For the mean of m future observations:

$$[Y_l, Y_u] = \bar{X} \pm t_{(1-\alpha/2, n-1)} \times \sqrt{\frac{1}{m}+\frac{1}{n}} \times s \text{ for } m \geq 1 .$$

- For the standard deviation of m future observations:

$$[s_l, s_u] = \left[s \times \sqrt{\frac{1}{F_{(1-\alpha/2;(n-1, m-1))}}}, s \times \sqrt{F_{(1-\alpha/2;(m-1, n-1))}} \right] \text{ for } m \geq 2$$

where m = number of future observations, and n = number of points in current analysis sample.

- The one-sided intervals are formed by using 1-α in the quantile functions.

For references, see Hahn and Meeker (1991), pages 61-64.

Statistical Details for Tolerance Intervals

This section contains statistical details for one-sided and two-sided tolerance intervals.

One-Sided Interval

The one-sided interval is computed as follows:

Upper Limit $= \bar{x} + g's$

Lower Limit $= \bar{x} - g's$

where

$$g' = t(1-\alpha, n-1, \Phi^{-1}(p) \cdot \sqrt{n}) / \sqrt{n} \text{ from Table 1 of Odeh and Owen (1980).}$$

t is the quantile from the non-central t-distribution, and Φ^{-1} is the standard normal quantile.

Two-Sided Interval

The two-sided interval is computed as follows:

$$[\tilde{T}_{p_L}, \tilde{T}_{p_U}] = [\bar{x} - g_{(1-\alpha/2;p_L,n)}s, \bar{x} + g_{(1-\alpha/2;p_U,n)}s]$$

where

s = standard deviation and $g_{(1-\alpha/2;p,n)}$ is a constant that can be found in Table 4 of Odeh and Owen 1980).

To determine g, consider the fraction of the population captured by the tolerance interval. Tamhane and Dunlop (2000) give this fraction as follows:

$$\Phi\left(\frac{\bar{x}+gs-\mu}{\sigma}\right)-\Phi\left(\frac{\bar{x}-gs-\mu}{\sigma}\right)$$

where Φ denotes the standard normal c.d.f. (cumulative distribution function). Therefore, g solves the following equation:

$$P\left\{\Phi\left(\frac{\bar{X}+gs-\mu}{\sigma}\right)-\Phi\left(\frac{\bar{X}-gs-\mu}{\sigma}\right)\geq 1-\gamma\right\} = 1-\alpha$$

where $1-\gamma$ is the fraction of all future observations contained in the tolerance interval.

More information is given in Tables A.1a, A.1b, A.11a, and A.11b of Hahn and Meeker (1991).

Statistical Details for Capability Analysis

All capability analyses use the same formulas. Options differ in how sigma (σ) is computed:

- **Long-term** uses the overall sigma. This option is used for P_{pk} statistics, and computes sigma as follows:

$$\sigma = \sqrt{\sum_{i=1}^{n}\frac{(x_i-\bar{x})^2}{n-1}}$$

Note: There is a preference for Distribution called **Ppk Capability Labeling** that labels the long-term capability output with P_{pk} labels. This option is found using **File > Preferences**, then select Platforms > Distribution.

- **Specified Sigma** enables you to type a specific, known sigma used for computing capability analyses. Sigma is user-specified, and is therefore not computed.
- **Moving Range** enables you to enter a range span, which computes sigma as follows:

$$\sigma = \frac{\bar{R}}{d_2(n)} \text{ where}$$

  $\bar{R}$ is the average of the moving ranges

  $d_2(n)$ is the expected value of the range of n independent normally distributed variables with unit standard deviation.

- **Short Term Sigma, Group by Fixed Subgroup Size** if r is the number of subgroups of size n_j and each ith subgroup is defined by the order of the data, sigma is computed as follows:

$$\sigma = \sqrt{\frac{1}{r}\sum_{i=1}^{r}\sum_{j=1}^{n_j}\frac{(X_{ij}-\overline{X}_i)}{n_j-1}} \quad \text{where} \quad \overline{X}_i = \frac{1}{n_j}\sum_{j=1}^{n_j} X_{ij}$$

This formula is commonly referred to as the Root Mean Square Error, or RMSE.

Table 2.19 Descriptions of Capability Indices and Computational Formulas

Index	Index Name	Formula
CP	process capability ratio, C_p	(USL - LSL)/6*s* where: • USL is the upper spec limit • LSL is the lower spec limit
CIs for CP	Lower CI on CP	$CP\sqrt{\frac{\chi^2_{\alpha/2,\, n-1}}{n-1}}$
	Upper CI on CP	$CP\sqrt{\frac{\chi^2_{1-\alpha/2,\, n-1}}{n-1}}$
CPK (PPK for AIAG)	process capability index, C_{pk}	min(CPL, CPU)
CIs for CPK See Bissell (1990)	Lower CI	$\hat{C}_{pk}\left[1-\Phi^{-1}(1-\alpha/2)\sqrt{\frac{1}{9n\hat{C}^2_{pk}}+\frac{1}{2(n-1)}}\right]$
	Upper CI	$\hat{C}_{pk}\left[1+\Phi^{-1}(1-\alpha/2)\sqrt{\frac{1}{9n\hat{C}^2_{pk}}+\frac{1}{2(n-1)}}\right]$
CPM	process capability index, C_{pm}	$\frac{\min(\text{target}-\text{LSL},\ \text{USL}-\text{target})}{3\sqrt{s^2+(\text{mean}-\text{target})^2}}$
CIs for CPM	Lower CI on CPM	$CPM\sqrt{\frac{\chi^2_{\alpha/2,\gamma}}{\gamma}}$, where $\gamma = \frac{n\left(1+\left(\frac{\bar{x}-\text{Target}}{s}\right)^2\right)^2}{1+2\left(\frac{\bar{x}-\text{Target}}{s}\right)^2}$
	Upper CI on CPM	$CPM\sqrt{\frac{\chi^2_{1-\alpha/2,\gamma}}{\gamma}}$ where γ = same as above.

Table 2.19 Descriptions of Capability Indices and Computational Formulas *(Continued)*

Index	Index Name	Formula
CPL	process capability ratio of one-sided lower spec	(mean - LSL)/3*s*
CPU	process capability ratio of one-sided upper spec	(USL - mean)/3*s*

- A capability index of 1.33 is considered to be the minimum acceptable. For a normal distribution, this gives an expected number of nonconforming units of about 6 per 100,000.
- Exact 100(1 - α)% lower and upper confidence limits for CPL are computed using a generalization of the method of Chou et al. (1990), who point out that the 100(1 - α) lower confidence limit for CPL (denoted by CPLLCL) satisfies the following equation:

  $$\Pr\{T_{n-1}(\delta = 3\sqrt{n})\text{CPLLCL} \le 3\text{CPL}\sqrt{n}\} = 1 - \alpha$$

  where $T_{n-1}(\delta)$ has a non-central *t*-distribution with *n* - 1 degrees of freedom and noncentrality parameter δ.
- Exact 100(1 - α)% lower and upper confidence limits for CPU are also computed using a generalization of the method of Chou et al. (1990), who point out that the 100(1 - α) lower confidence limit for CPU (denoted CPULCL) satisfies the following equation:

  $$\Pr\{T_{n-1}(\delta = 3\sqrt{n})\text{CPULCL} \ge 3\text{CPU}\sqrt{n}\} = 1 - \alpha$$

  where $T_{n-1}(\delta)$ has a non-central *t*-distribution with *n* - 1 degrees of freedom and noncentrality parameter δ.

Note: Because of a lack of supporting research at the time of this writing, computing confidence intervals for capability indices is not recommended, except for cases when the capability indices are based on the standard deviation.

- Sigma Quality is defined as the following

$$\text{Sigma Quality} = \text{Normal Quantile}\left(1 - \frac{\%\text{ outside}}{100}\right) + 1.5$$

$$\text{Sigma Quality Above} = \text{Normal Quantile}\left(1 - \frac{\%\text{ above}}{100}\right) + 1.5$$

$$\text{Sigma Quality Below} = \text{Normal Quantile}\left(1 - \frac{\%\text{ below}}{100}\right) + 1.5$$

For example, if there are 3 defects in *n*=1,000,000 observations, the formula yields 6.03, or a 6.03 sigma process. The results of the computations of the Sigma Quality Above USL and Sigma Quality Below LSL column values do not sum to the Sigma Quality Total Outside column value because calculating Sigma Quality involves finding normal distribution quantiles, and is therefore not additive.

- Here are the Benchmark Z formulas:

  Z USL = (USL-Xbar)/sigma = 3 * CPU

  Z LSL = (Xbar-LSL)/sigma = 3 * CPL

  Z Bench = Inverse Cumulative Prob(1 - P(LSL) - P(USL))

  where:

  P(LSL) = Prob(X < LSL) = 1 - Cum Prob(Z LSL)

  P(USL) = Prob(X > USL) = 1 - Cum Prob(Z USL).

Statistical Details for Continuous Fit Distributions

This section contains statistical details for the options in the Continuous Fit menu.

Normal

The **Normal** fitting option estimates the parameters of the normal distribution. The normal distribution is often used to model measures that are symmetric with most of the values falling in the middle of the curve. Select the **Normal** fitting for any set of data and test how well a normal distribution fits your data.

The parameters for the normal distribution are as follows:

- μ (the mean) defines the location of the distribution on the *x*-axis
- σ (standard deviation) defines the dispersion or spread of the distribution

The standard normal distribution occurs when $\mu = 0$ and $\sigma = 1$. The Parameter Estimates table shows estimates of μ and σ, with upper and lower 95% confidence limits.

pdf: $$\frac{1}{\sqrt{2\pi\sigma^2}}\exp\left[-\frac{(x-\mu)^2}{2\sigma^2}\right] \quad \text{for} \quad -\infty < x < \infty;\quad -\infty < \mu < \infty;\quad 0 < \sigma$$

$E(x) = \mu$

$Var(x) = \sigma^2$

LogNormal

The **LogNormal** fitting option estimates the parameters μ (scale) and σ (shape) for the two-parameter lognormal distribution. A variable Y is lognormal if and only if $X = \ln(Y)$ is normal. The data must be greater than zero.

pdf: $$\frac{1}{\sigma\sqrt{2\pi}}\frac{\exp\left[\frac{-(\log(x)-\mu)^2}{2\sigma^2}\right]}{x} \quad \text{for} \quad 0 \le x;\quad -\infty < \mu < \infty;\quad 0 < \sigma$$

$E(x) = \exp(\mu + \sigma^2/2)$

$Var(x) = \exp(2(\mu + \sigma^2)) - \exp(2\mu + \sigma^2)$

Weibull, Weibull with Threshold, and Extreme Value

The Weibull distribution has different shapes depending on the values of α (scale) and β (shape). It often provides a good model for estimating the length of life, especially for mechanical devices and in biology. The **Weibull** option is the same as the Weibull with threshold option, with a threshold (θ) parameter of zero. For the Weibull with threshold option, JMP estimates the threshold as the minimum value. If you know what the threshold should be, set it by using the **Fix Parameters** option. See "Fit Distribution Options" on page 58.

The pdf for the Weibull with threshold is as follows:

$$\text{pdf:} \quad \frac{\beta}{\alpha^{\beta}}(x-\theta)^{\beta-1}\exp\left[-\left(\frac{x-\theta}{\alpha}\right)^{\beta}\right] \quad \text{for } \alpha,\beta > 0;\ \theta < x$$

$$E(x) = \theta + \alpha\Gamma\left(1+\frac{1}{\beta}\right)$$

$$\text{Var}(x) = \alpha^2\left\{\Gamma\left(1+\frac{2}{\beta}\right) - \Gamma^2\left(1+\frac{1}{\beta}\right)\right\}$$

where $\Gamma(\cdot)$ is the Gamma function.

The Extreme Value distribution is a two parameter Weibull (α, β) distribution with the transformed parameters $\delta = 1 / \beta$ and $\lambda = \ln(\alpha)$.

Exponential

The exponential distribution is especially useful for describing events that randomly occur over time, such as survival data. The exponential distribution might also be useful for modeling elapsed time between the occurrence of non-overlapping events, such as the time between a user's computer query and response of the server, the arrival of customers at a service desk, or calls coming in at a switchboard.

The Exponential distribution is a special case of the two-parameter Weibull when $\beta = 1$ and $\alpha = \sigma$, and also a special case of the Gamma distribution when $\alpha = 1$.

$$\text{pdf:} \quad \frac{1}{\sigma}\exp(-x/\sigma) \quad \text{for } 0 < \sigma;\ 0 \le x$$

$$E(x) = \sigma$$

$$\text{Var}(x) = \sigma^2$$

Devore (1995) notes that an exponential distribution is *memoryless*. Memoryless means that if you check a component after t hours and it is still working, the distribution of additional lifetime (the conditional probability of additional life given that the component has lived until t) is the same as the original distribution.

Gamma

The **Gamma** fitting option estimates the gamma distribution parameters, $\alpha > 0$ and $\sigma > 0$. The parameter α, called alpha in the fitted gamma report, describes shape or curvature. The parameter σ, called sigma, is the scale parameter of the distribution. A third parameter, θ, called the **Threshold**, is the lower endpoint parameter. It is set to zero by default, unless there are negative values. You can also set its value by using the **Fix Parameters** option. See "Fit Distribution Options" on page 58.

pdf: $$\frac{1}{\Gamma(\alpha)\sigma^{\alpha}}(x-\theta)^{\alpha-1}\exp(-(x-\theta)/\sigma) \quad \text{for} \quad \theta \le x;\ 0 < \alpha,\sigma$$

$$E(x) = \alpha\sigma + \theta$$

$$Var(x) = \alpha\sigma^2$$

- The *standard* gamma distribution has $\sigma = 1$. Sigma is called the scale parameter because values other than 1 stretch or compress the distribution along the *x*-axis.
- The Chi-square $\chi^2_{(\nu)}$ distribution occurs when $\sigma = 2$, $\alpha = \nu/2$, and $\theta = 0$.
- The exponential distribution is the family of gamma curves that occur when $\alpha = 1$ and $\theta = 0$.

The standard gamma density function is strictly decreasing when $\alpha \le 1$. When $\alpha > 1$, the density function begins at zero, increases to a maximum, and then decreases.

Beta

The standard beta distribution is useful for modeling the behavior of random variables that are constrained to fall in the interval 0,1. For example, proportions always fall between 0 and 1. The **Beta** fitting option estimates two shape parameters, $\alpha > 0$ and $\beta > 0$. There are also θ and σ, which are used to define the lower threshold as θ, and the upper threshold as $\theta + \sigma$. The beta distribution has values only for the interval defined by $\theta \le x \le (\theta + \sigma)$. The θ is estimated as the minimum value, and σ is estimated as the range. The standard beta distribution occurs when $\theta = 0$ and $\sigma = 1$.

Set parameters to fixed values by using the **Fix Parameters** option. The upper threshold must be greater than or equal to the maximum data value, and the lower threshold must be less than or equal to the minimum data value. For details about the Fix Parameters option, see "Fit Distribution Options" on page 58.

pdf: $$\frac{1}{B(\alpha,\beta)\sigma^{\alpha+\beta-1}}(x-\theta)^{\alpha-1}(\theta+\sigma-x)^{\beta-1} \quad \text{for} \quad \theta \le x \le \theta+\sigma;\ 0 < \sigma,\alpha,\beta$$

$$E(x) = \theta + \sigma\frac{\alpha}{\alpha+\beta}$$

$$Var(x) = \frac{\sigma^2\alpha\beta}{(\alpha+\beta)^2(\alpha+\beta+1)}$$

where $B(\cdot)$ is the Beta function.

Normal Mixtures

The **Normal Mixtures** option fits a mixture of normal distributions. This flexible distribution is capable of fitting multi-modal data.

Fit a mixture of two or three normal distributions by selecting the **Normal 2 Mixture** or **Normal 3 Mixture** options. Alternatively, you can fit a mixture of *k* normal distributions by selecting the **Other** option. A separate mean, standard deviation, and proportion of the whole is estimated for each group.

$$\text{pdf:} \quad \sum_{i=1}^{k} \frac{\pi_i}{\sigma_i} \phi\left(\frac{x-\mu_i}{\sigma_i}\right)$$

$$E(x) = \sum_{i=1}^{k} \pi_i \mu_i$$

$$\text{Var}(x) = \sum_{i=1}^{k} \pi_i(\mu_i^2 + \sigma_i^2) - \left(\sum_{i=1}^{k} \pi_i \mu_i\right)^2$$

where μ_i, σ_i, and π_i are the respective mean, standard deviation, and proportion for the ith group, and $\phi(\cdot)$ is the standard normal pdf.

Smooth Curve

The **Smooth Curve** option fits a smooth curve using nonparametric density estimation (kernel density estimation). The smooth curve is overlaid on the histogram and a slider appears beneath the plot. Control the amount of smoothing by changing the kernel standard deviation with the slider. The initial Kernel Std estimate is formed by summing the normal densities of the kernel standard deviation located at each data point.

Johnson Su, Johnson Sb, Johnson Sl

The Johnson system of distributions contains three distributions that are all based on a transformed normal distribution. These three distributions are the Johnson Su, which is unbounded for Y; the Johnson Sb, which is bounded on both tails (0 < Y < 1); and the Johnson Sl, leading to the lognormal family of distributions.

Note: The S refers to system, the subscript of the range. Although we implement a different method, information about selection criteria for a particular Johnson system can be found in Slifker and Shapiro (1980).

Johnson distributions are popular because of their flexibility. In particular, the Johnson distribution system is noted for its data-fitting capabilities because it supports every possible combination of skewness and kurtosis.

If Z is a standard normal variate, then the system is defined as follows:

$$Z = \gamma + \delta f(y)$$

where, for the Johnson Su:

$$f(y) = \ln\left(Y + \sqrt{1 + Y^2}\right) = \sinh^{-1} Y$$

$$Y = \frac{X - \theta}{\sigma} \qquad -\infty < X < \infty$$

where, for the Johnson Sb:

$$f(Y) = \ln\left(\frac{Y}{1 - Y}\right)$$

$$Y = \frac{X - \theta}{\sigma} \qquad \theta < X < \theta + \sigma$$

and for the Johnson Sl, where $\sigma = \pm 1$.

$$f(Y) = \ln(Y)$$

$$Y = \frac{X - \theta}{\sigma} \qquad \begin{array}{ll} \theta < X < \infty & \text{if } \sigma = 1 \\ -\infty < X < \theta & \text{if } \sigma = -1 \end{array}$$

Johnson Su

pdf: $$\frac{\delta}{\sigma}\left[1 + \left(\frac{x - \theta}{\sigma}\right)^2\right]^{-1/2} \phi\left[\gamma + \delta \sinh^{-1}\left(\frac{x - \theta}{\sigma}\right)\right] \quad \text{for } -\infty < x, \theta, \gamma < \infty; \ 0 < \theta, \delta$$

Johnson Sb

pdf: $$\phi\left[\gamma + \delta \ln\left(\frac{x - \theta}{\sigma - (x - \theta)}\right)\right]\left(\frac{\delta\sigma}{(x - \theta)(\sigma - (x - \theta))}\right) \quad \text{for } \theta < x < \theta + \sigma; \ 0 < \sigma$$

Johnson Sl

pdf: $$\frac{\delta}{|x - \theta|}\phi\left[\gamma + \delta \ln\left(\frac{x - \theta}{\sigma}\right)\right] \quad \text{for } \theta < x \text{ if } \sigma = 1; \ \theta > x \text{ if } \sigma = -1$$

where $\phi(\cdot)$ is the standard normal pdf.

Note: The parameter confidence intervals are hidden in the default report. Parameter confidence intervals are not very meaningful for Johnson distributions, because they are transformations to normality. To show parameter confidence intervals, right-click in the report and select **Columns > Lower 95%** and **Upper 95%**.

Generalized Log (Glog)

This distribution is useful for fitting data that are rarely normally distributed and often have non-constant variance, like biological assay data. The Glog distribution is described with the parameters μ (location), σ (scale), and λ (shape).

$$\text{pdf:} \quad \phi\left\{\frac{1}{\sigma}\left[\log\left(\frac{x+\sqrt{x^2+\lambda^2}}{2}\right)-\mu\right]\right\}\frac{x+\sqrt{x^2+\lambda^2}}{\sigma(x^2+\lambda^2+x\sqrt{x^2+\lambda^2})}$$

$$\text{for} \quad 0 \le \lambda; \quad 0 < \sigma; \quad -\infty < \mu < \infty$$

The Glog distribution is a transformation to normality, and comes from the following relationship:

$$\text{If } z = \frac{1}{\sigma}\left[\log\left(\frac{x+\sqrt{x^2+\lambda^2}}{2}\right)-\mu\right] \sim N(0,1), \text{ then } x \sim \text{Glog}(\mu,\sigma,\lambda).$$

When $\lambda = 0$, the Glog reduces to the LogNormal (μ,σ).

Note: The parameter confidence intervals are hidden in the default report. Parameter confidence intervals are not very meaningful for the GLog distribution, because it is a transformation to normality. To show parameter confidence intervals, right-click in the report and select **Columns > Lower 95%** and **Upper 95%**.

All

In the Compare Distributions report, the ShowDistribution list is sorted by AICc in ascending order.

The formula for AICc is as follows:

$$\text{AICc} = -2\text{logL} + 2\nu + \frac{2\nu(\nu+1)}{n-(\nu+1)}$$

where:

- logL is the logLikelihood
- n is the sample size
- ν is the number of parameters

If your data has negative values, the ShowDistribution list does not include those distributions that require data with positive values. If your data has non-integer values, the list of distributions does not include discrete distributions. Distributions with threshold parameters, like Beta and Johnson Sb, are not included in the list of possible distributions.

Statistical Details for Discrete Fit Distributions

This section contains statistical details for the options in the Discrete Fit menu.

Poisson

The Poisson distribution has a single scale parameter $\lambda > 0$.

pmf: $\dfrac{e^{-\lambda}\lambda^{x}}{x!}$ for $0 \le \lambda < \infty$; $x = 0,1,2,\ldots$

$E(x) = \lambda$

$Var(x) = \lambda$

Since the Poisson distribution is a discrete distribution, the overlaid curve is a step function, with jumps occurring at every integer.

Gamma Poisson

This distribution is useful when the data is a combination of several Poisson(μ) distributions, each with a different μ. One example is the overall number of accidents combined from multiple intersections, when the mean number of accidents (μ) varies between the intersections.

The Gamma Poisson distribution results from assuming that $x|\mu$ follows a Poisson distribution and μ follows a Gamma(α,τ). The Gamma Poisson has parameters $\lambda = \alpha\tau$ and $\sigma = \tau+1$. The σ is a dispersion parameter. If $\sigma > 1$, there is over dispersion, meaning there is more variation in x than explained by the Poisson alone. If $\sigma = 1$, x reduces to Poisson(λ).

pmf: $\dfrac{\Gamma\left(x + \frac{\lambda}{\sigma - 1}\right)}{\Gamma(x+1)\Gamma\left(\frac{\lambda}{\sigma - 1}\right)}\left(\dfrac{\sigma - 1}{\sigma}\right)^{x}\sigma^{-\frac{\lambda}{\sigma - 1}}$ for $0 < \lambda$; $1 \le \sigma$; $x = 0,1,2,\ldots$

$E(x) = \lambda$

$Var(x) = \lambda\sigma$

where $\Gamma(\cdot)$ is the Gamma function.

Remember that $x|\mu \sim$ Poisson(μ), while $\mu \sim$ Gamma(α,τ). The platform estimates $\lambda = \alpha\tau$ and $\sigma = \tau+1$. To obtain estimates for α and τ, use the following formulas:

$$\tau = \sigma - 1$$

$$\hat{\alpha} = \frac{\ddot{\lambda}}{\hat{\tau}}$$

If the estimate of σ is 1, the formulas do not work. In that case, the Gamma Poisson has reduced to the Poisson(λ), and $\hat{\lambda}$ is the estimate of λ.

If the estimate for α is an integer, the Gamma Poisson is equivalent to a Negative Binomial with the following pmf:

$$p(y) = \binom{y + r - 1}{y} p^{r}(1 - p)^{y} \quad \text{for } 0 \le y$$

with $r = \alpha$ and $(1-p)/p = \tau$.

Binomial

The **Binomial** option accepts data in two formats: a constant sample size, or a column containing sample sizes.

pmf: $\binom{n}{x} p^x(1-p)^{n-x} \quad \text{for} \quad 0 \le p \le 1\,; \quad x = 0,1,2,\ldots,n$

$E(x) = np$

$\text{Var}(x) = np(1\text{-}p)$

where n is the number of independent trials.

Note: The confidence interval for the binomial parameter is a Score interval. See Agresti (1998).

Beta Binomial

This distribution is useful when the data is a combination of several Binomial(p) distributions, each with a different p. One example is the overall number of defects combined from multiple manufacturing lines, when the mean number of defects (p) varies between the lines.

The Beta Binomial distribution results from assuming that $x|\pi$ follows a Binomial(n,π) distribution and π follows a Beta(α,β). The Beta Binomial has parameters $p = \alpha/(\alpha+\beta)$ and $\delta = 1/(\alpha+\beta+1)$. The δ is a dispersion parameter. When $\delta > 0$, there is over dispersion, meaning there is more variation in x than explained by the Binomial alone. When $\delta < 0$, there is under dispersion. When $\delta = 0$, x is distributed as Binomial(n,p). The Beta Binomial only exists when $n \ge 2$.

$$\text{pmf:} \quad \binom{n}{x} \frac{\Gamma\left(\frac{1}{\delta}-1\right)\Gamma\left[x+p\left(\frac{1}{\delta}-1\right)\right]\Gamma\left[n-x+(1-p)\left(\frac{1}{\delta}-1\right)\right]}{\Gamma\left[p\left(\frac{1}{\delta}-1\right)\right]\Gamma\left[(1-p)\left(\frac{1}{\delta}-1\right)\right]\Gamma\left(n+\frac{1}{\delta}-1\right)}$$

$$\text{for} \quad 0 \le p \le 1\,; \quad \max\left(-\frac{p}{n-p-1}, -\frac{1-p}{n-2+p}\right) \le \delta \le 1\,; \quad x = 0,1,2,\ldots,n$$

$E(x) = np$

$\text{Var}(x) = np(1\text{-}p)[1+(n\text{-}1)\delta]$

where $\Gamma(\cdot)$ is the Gamma function.

Remember that $x|\pi \sim \text{Binomial}(n,\pi)$, while $\pi \sim \text{Beta}(\alpha,\beta)$. The parameters $p = \alpha/(\alpha+\beta)$ and $\delta = 1/(\alpha+\beta+1)$ are estimated by the platform. To obtain estimates of α and β, use the following formulas:

$$\hat{\alpha} = \hat{p}\left(\frac{1-\hat{\delta}}{\hat{\delta}}\right)$$

$$\hat{\beta} = (1-\hat{p})\left(\frac{1-\hat{\delta}}{\hat{\delta}}\right)$$

If the estimate of δ is 0, the formulas do not work. In that case, the Beta Binomial has reduced to the Binomial(n,p), and $\hat{p}$ is the estimate of p.

The confidence intervals for the Beta Binomial parameters are profile likelihood intervals.

Comparing All Distributions

The ShowDistribution list is sorted by AICc in ascending order.

The formula for AICc is as follows:

$$\text{AICc} = -2\log L + 2\nu + \frac{2\nu(\nu+1)}{n-(\nu+1)}$$

where:

- logL is the logLikelihood
- n is the sample size
- ν is the number of parameters

Statistical Details for Fitted Quantiles

The fitted quantiles in the Diagnostic Plot and the fitted quantiles saved with the **Save Fitted Quantiles** command are formed using the following method:

1. The data are sorted and ranked. Ties are assigned different ranks.
2. Compute the $p_{[i]} = \text{rank}_{[i]}/(n+1)$.
3. Compute the $\text{quantile}_{[i]} = \text{Quantile}_d(p_{[i]})$ where Quantile_d is the quantile function for the specific fitted distribution, and i = 1,2,...,n.

Statistical Details for Fit Distribution Options

This section describes Goodness of Fit tests for fitting distributions and statistical details for specification limits pertaining to fitted distributions.

Goodness of Fit

Table 2.20 Descriptions of JMP Goodness of Fit Tests

Distribution	Parameters	Goodness of Fit Test
Normal[a]	μ and σ are unknown	Shapiro-Wilk (for $n \le 2000$) Kolmogorov-Smirnov-Lillefors (for $n > 2000$)
	μ and σ are both known	Kolmogorov-Smirnov-Lillefors
	either μ or σ is known	(none)

Table 2.20 Descriptions of JMP Goodness of Fit Tests *(Continued)*

Distribution	Parameters	Goodness of Fit Test
LogNormal	μ and σ are known or unknown	Kolmogorov's *D*
Weibull	α and β known or unknown	Cramér-von Mises W^2
Weibull with threshold	α, β and θ known or unknown	Cramér-von Mises W^2
Extreme Value	α and β known or unknown	Cramér-von Mises W^2
Exponential	σ is known or unknown	Kolmogorov's *D*
Gamma	α and σ are known	Cramér-von Mises W^2
	either α or σ is unknown	(none)
Beta	α and β are known	Kolmogorov's *D*
	either α or β is unknown	(none)
Binomial	ρ is known or unknown and *n* is known	Kolmogorov's *D* (for n ≤ 30) Pearson χ^2 (for n > 30)
Beta Binomial	ρ and δ known or unknown	Kolmogorov's *D* (for n ≤ 30) Pearson χ^2 (for n > 30)
Poisson	λ known or unknown	Kolmogorov's *D* (for n ≤ 30) Pearson χ^2 (for n > 30)
Gamma Poisson	λ or σ known or unknown	Kolmogorov's *D* (for n ≤ 30) Pearson χ^2 (for n > 30)

a. For the three Johnson distributions and the Glog distribution, the data are transformed to Normal, then the appropriate test of normality is performed.

Spec Limits

Writing *T* for the target, *LSL*, and *USL* for the lower and upper specification limits, and P_α for the $\alpha*100^{th}$ percentile, the generalized capability indices are as follows:

$$C_{pl} = \frac{P_{0.5} - LSL}{P_{0.5} - P_{0.00135}}$$

$$C_{pu} = \frac{USL - P_{0.5}}{P_{0.99865} - P_{0.5}}$$

$$C_p = \frac{USL - LSL}{P_{0.99865} - P_{0.00135}}$$

$$C_{pk} = \min\left(\frac{P_{0.5} - LSL}{P_{0.5} - P_{0.00135}}, \frac{USL - P_{0.5}}{P_{0.99865} - P_{0.5}}\right)$$

$$K = 2 \times \frac{\left|\frac{1}{2}(USL + LSL) - P_{0.5}\right|}{USL - LSL}$$

$$C_{pm} = \frac{\min\left(\frac{T - LSL}{P_{0.5} - P_{0.00135}}, \frac{USL - T}{P_{0.99865} - P_{0.5}}\right)}{\sqrt{1 + \left(\frac{\mu - T}{\sigma}\right)^2}}$$

If the data are normally distributed, these formulas reduce to the formulas for standard capability indices. See Table 2.19.

Set Spec Limits for K Sigma

Type a K value and select one-sided or two-sided for your capability analysis. Tail probabilities corresponding to K standard deviations are computed from the Normal distribution. The probabilities are converted to quantiles for the specific distribution that you have fitted. The resulting quantiles are used for specification limits in the capability analysis. This option is similar to the **Quantiles** option, but you provide K instead of probabilities. K corresponds to the number of standard deviations that the specification limits are away from the mean.

For example, for a Normal distribution, where K=3, the 3 standard deviations below and above the mean correspond to the 0.00135^{th} quantile and 0.99865^{th} quantile, respectively. The lower specification limit is set at the 0.00135^{th} quantile, and the upper specification limit is set at the 0.99865^{th} quantile of the fitted distribution. A capability analysis is returned based on those specification limits.

Chapter 3

Introduction to the Fit Y by X Platform

Performing Four Types of Analyses

The Fit Y by X platform analyzes the pair of X and Y variables that you specify, by context, based on modeling type.

Here are the four types of analyses:

- Bivariate fitting
- One-way analysis of variance
- Logistic regression
- Contingency table analysis

Figure 3.1 Examples of Four Types of Analyses

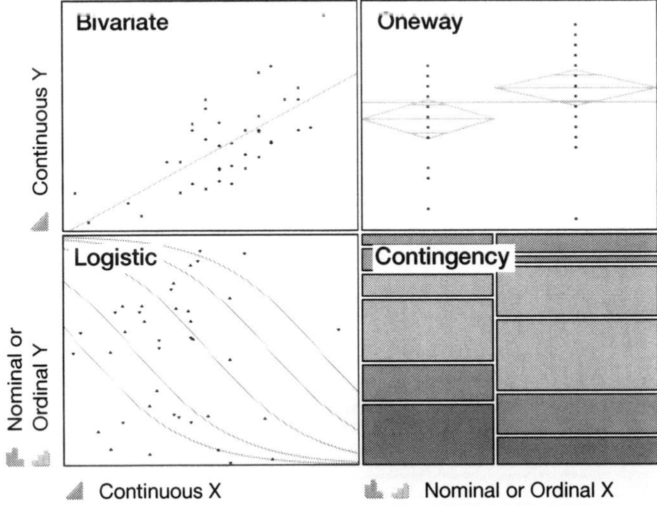

Contents

Overview of the Fit Y by X Platform

The Fit Y by X platform is a collection of four specific platforms (or types of analyses).

Specific Platform	Modeling Types	Description	Reference
Bivariate	Continuous Y by continuous X	Analyzes the relationship between two continuous variables	See *Performing Bivariate Analysis*.
Oneway	Continuous Y by nominal or ordinal X	Analyzes how the distribution of a continuous Y variable differs across groups defined by a categorical X variable	See *Performing Oneway Analysis*.
Logistic	Nominal or ordinal Y by continuous X	Fits the probabilities for response categories to a continuous X predictor	See *Performing Simple Logistic Regression*.
Contingency	Nominal or ordinal Y by nominal or ordinal X	Analyzes the distribution of a categorical response variable Y as conditioned by the values of a categorical X factor	See *Performing Contingency Analysis*.

Launch the Fit Y by X Platform

Launch the **Fit Y by X** platform by selecting **Analyze > Fit Y by X**.

Figure 3.2 The Fit Y by X Launch Window

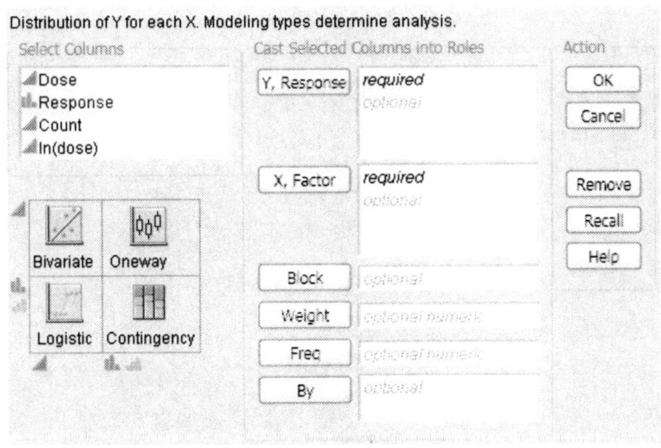

Table 3.1 Description of the Fit Y by X Launch Window

Bivariate, Oneway, Logistic, Contingency	This grid shows which analysis results from the different combinations of data types. Once you have assigned your columns, the applicable platform appears as a label above the grid.
Block	(Optional, for Oneway and Contingency only): • For the Oneway platform, identifies a second factor, which forms a two-way analysis without interaction. The data should be balanced and have equal counts in each block by group cell. If you specify a **Block** variable, the data should be balanced and have equal counts in each block by group cell. • For the Contingency platform, identifies a second factor and performs a Cochran-Mantel-Haenszel test.

For more information about launch windows, see *Using JMP*.

Launch Specific Analyses from the JMP Starter Window

From the JMP Starter window, you can launch a specific analysis (**Bivariate**, **Oneway**, **Logistic**, or **Contingency**). If you select this option, specify the correct modeling types (Y and X variables) for the analysis. See Table 3.2.

To launch a specific analysis from the JMP Starter Window, proceed as follows:

1. Click on the **Basic** category.
2. Click on the specific analysis that you want to launch.

Most of the platform launch options are the same. However, the naming for some of the Y and X platform buttons is tailored for the specific analysis that you are performing.

Table 3.2 Platforms and Buttons

Platform or Analysis	Y Button	X Button
Fit Y by X	Y, Response	X, Factor
Bivariate	Y, Response	X, Regressor
Oneway	Y, Response	X, Grouping
Logistic	Y, Categorical Response	X, Continuous Regressor
Contingency	Y, Response Category	X, Grouping Category

Chapter 4

Performing Bivariate Analysis

Using the Fit Y by X or Bivariate Platform

The Bivariate platform is the *continuous by continuous* personality of the Fit Y by X platform. Bivariate analysis shows the relationship between two continuous variables. The analysis results appear in a scatterplot. You can interactively add other types of fits, such as simple linear regression, polynomial regression, and so on.

Figure 4.1 Example of Bivariate Analysis

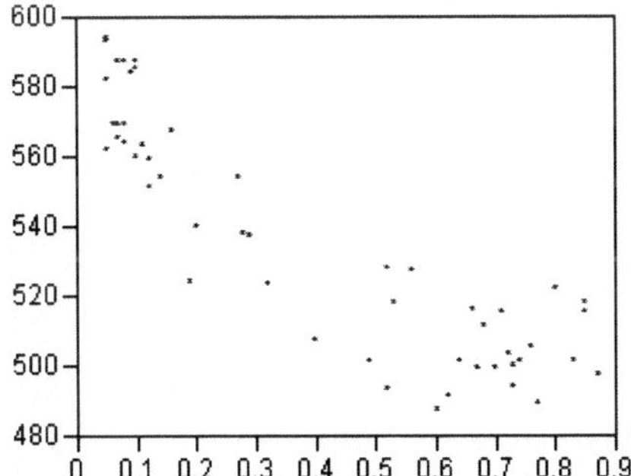

Contents

Example of Bivariate Analysis

This example uses the **SAT.jmp** sample data table. SAT test scores for students in the 50 U.S. states, plus the District of Columbia, are divided into two areas: verbal and math. You want to find out how the percentage of students taking the SAT tests is related to verbal test scores for 2004.

1. Open the SAT.jmp sample data table.
2. Select **Analyze > Fit Y by X**.
3. Select 2004 Verbal and click **Y, Response**.
4. Select % Taking (2004) and click **X, Factor**.
5. Click **OK**.

Figure 4.2 Example of SAT Scores by Percent Taking

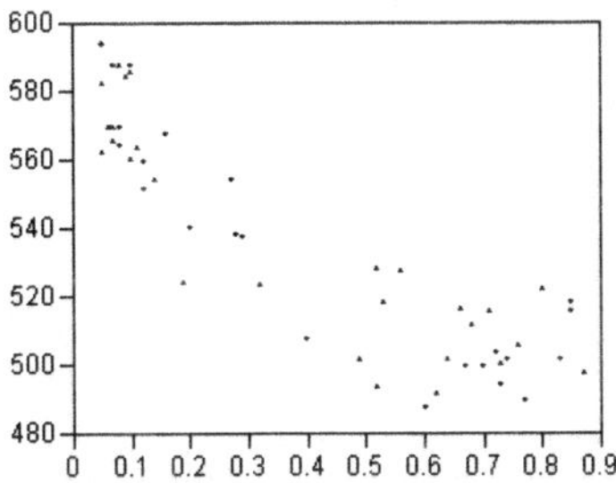

You can see that the verbal scores were higher when a smaller percentage of the population took the test.

Launch the Bivariate Platform

You can perform a bivariate analysis using either the Fit Y by X platform or the Bivariate platform. The two approaches give equivalent results.

- To launch the Fit Y by X platform, select **Analyze > Fit Y by X**.

  or

- To launch the Bivariate platform, from the JMP Starter window, click on the **Basic** category and click **Bivariate**.

Figure 4.3 The Bivariate Launch Window

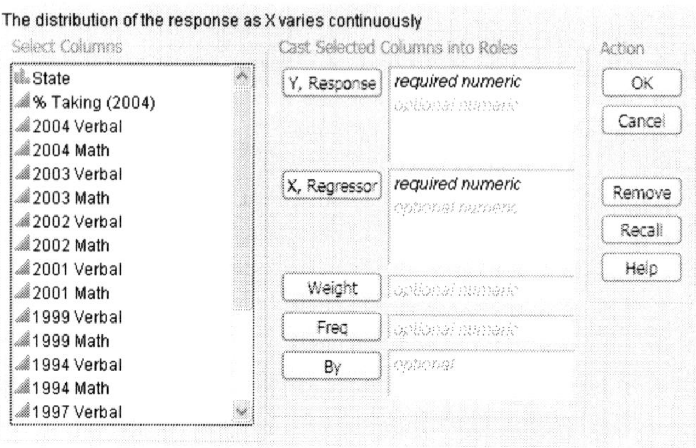

For information about this launch window, see "Introduction to the Fit Y by X Platform" chapter on page 87.

The Bivariate Plot

To produce the plot shown in Figure 4.4, follow the instructions in "Example of Bivariate Analysis" on page 93.

Figure 4.4 The Bivariate Plot

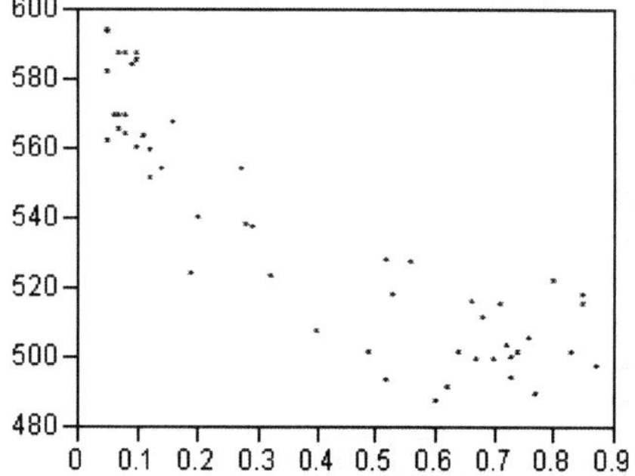

The Bivariate report begins with a plot for each pair of *X* and *Y* variables. Replace variables in the plot by dragging and dropping a variable, in one of two ways: swap existing variables by dragging and dropping a variable from one axis to the other axis; or, click on a variable in the Columns panel of the associated data table and drag it onto an axis.

You can interact with this plot just as you can with other JMP plots (for example, resizing the plot, highlighting points with the arrow or brush tool, and labeling points). For details about these features, see the *Using JMP* book.

You can fit curves on the plot and view statistical reports and additional menus using the fitting commands that are located within the red triangle menu. See "Fitting Commands and General Options" on page 95.

Fitting Commands and General Options

The Bivariate Fit red triangle menu contains the fitting commands and general options.

Figure 4.5 Fitting Commands and General Options

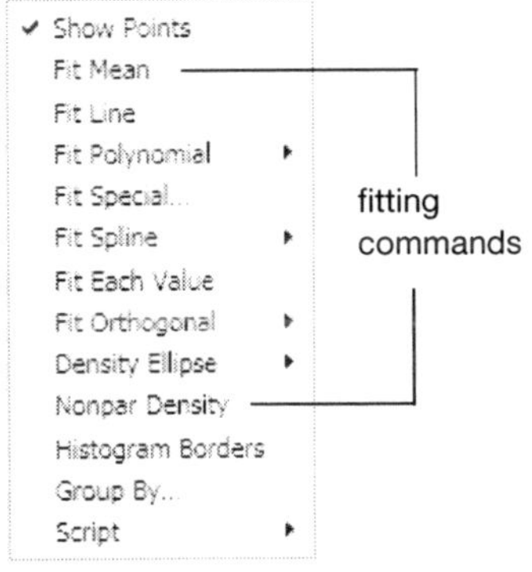

Table 4.1 Descriptions of the General Options

Show Points	Hides or shows the points in the scatterplot. A check mark indicates that points are shown.
Histogram Borders	Attaches histograms to the *x*- and *y*-axes of the scatterplot. A check mark indicates that histogram borders are turned on. See "Histogram Borders" on page 114.
Group By	Lets you select a classification (or grouping) variable. A separate analysis is computed for each level of the grouping variable, and regression curves or ellipses are overlaid on the scatterplot. See "Group By" on page 114.
Script	Contains options that are available to all platforms. These options enable you to redo the analysis or save the JSL commands for the analysis to a window or a file. For more information, see *Using JMP*.

Each fitting command adds the following:

- a line, curve, or distribution to the scatterplot

- a red triangle menu to the report window
- a specific report to the report window

Figure 4.6 Example of the Fit Mean Fitting Command

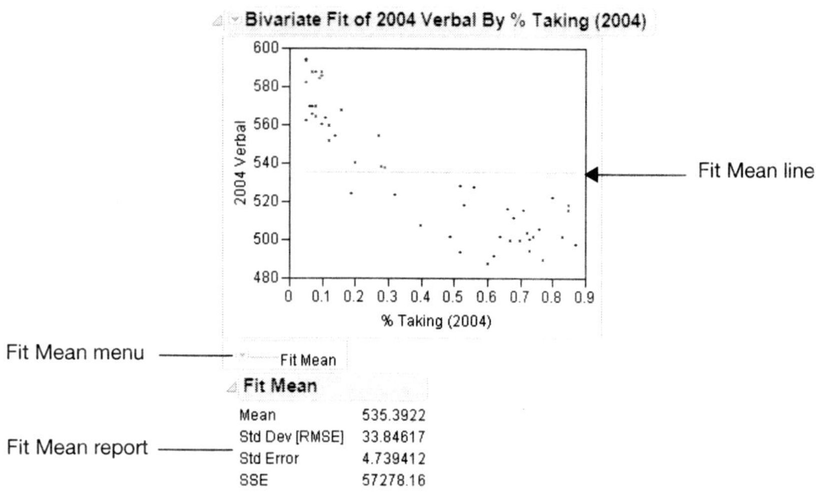

Table 4.2 Descriptions of the Fitting Commands

Fit Mean	Adds a horizontal line to the scatterplot that represents the mean of the Y response variable.	See "Fit Mean" on page 98.
Fit Line	Adds straight line fits to your scatterplot using least squares regression.	See "Fit Line and Fit Polynomial" on page 99.
Fit Polynomial	Fits polynomial curves of a certain degree using least squares regression.	See "Fit Line and Fit Polynomial" on page 99.
Fit Special	Transforms *Y* and *X*. Transformations include: log, square root, square, reciprocal, and exponential. You can also turn off center polynomials, constrain the intercept and the slope, and fit polynomial models.	See "Fit Special" on page 106.
Fit Spline	Fits a smoothing spline that varies in smoothness (or flexibility) according to the lambda (λ) value. The λ value is a tuning parameter in the spline formula.	See "Fit Spline" on page 107.
Fit Each Value	Fits a value to each unique *X* value, which can be compared to other fitted lines, showing the concept of lack of fit.	See "Fit Each Value" on page 109.

Table 4.2 Descriptions of the Fitting Commands *(Continued)*

Fit Orthogonal	Fits lines that adjust for variability in *X* as well as *Y*.	See "Fit Orthogonal" on page 110.
Density Ellipse	Draws an ellipse that contains a specified mass of points.	See "Density Ellipse" on page 111.
Nonpar Density	Shows patterns in the point density, which is useful when the scatterplot is so darkened by points that it is difficult to distinguish patterns.	See "Nonpar Density" on page 113.

Note: You can remove a fit using the **Remove Fit** command. For details, see "Fitting Menu Options" on page 115.

Fitting Command Categories

Fitting command categories include regression fits and density estimation.

Category	Description	Fitting Commands
Regression Fits	Regression methods fit a curve through the points. The curve is an equation (a model) that is estimated using least squares, which minimizes the sum of squared differences from each point to the line (or curve). Regression fits assume that the *Y* variable is distributed as a random scatter above and below a line of fit.	**Fit Mean** **Fit Line** **Fit Polynomial** **Fit Special** **Fit Spline** **Fit Each Value** **Fit Orthogonal**
Density Estimation	Density estimation fits a bivariate distribution to the points. You can either select a bivariate normal density, characterized by elliptical contours, or a general nonparametric density.	**Fit Density Ellipse** **Nonpar Density**

Fit the Same Command Multiple Times

You can select the same fitting command multiple times, and each new fit is overlaid on the scatterplot. You can try fits, exclude points and refit, and you can compare them on the same scatterplot.

To apply a fitting command to multiple analyses in your report window, hold down the CTRL key and select a fitting option.

Fit Mean

Using the **Fit Mean** command, you can add a horizontal line to the scatterplot that represents the mean of the *Y* response variable. You can start by fitting the mean and then use the mean line as a reference for other fits (such as straight lines, confidence curves, polynomial curves, and so on).

Figure 4.7 Example of Fit Mean

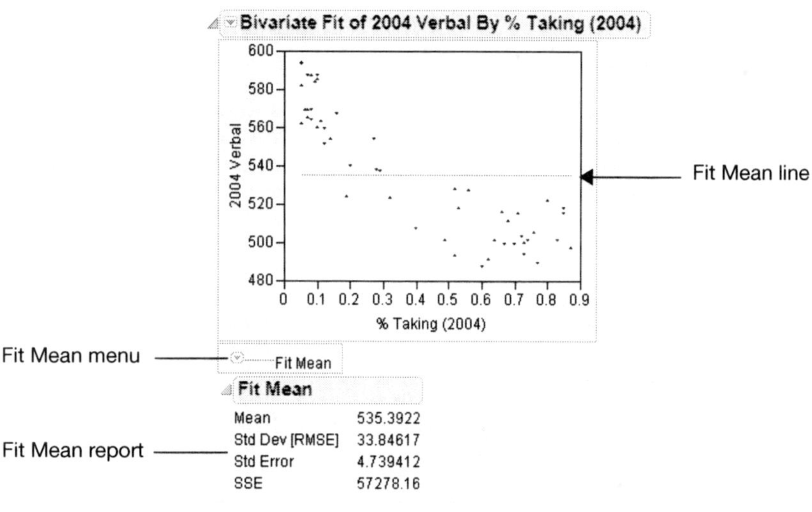

Fit Mean Report

The Fit Mean report shows summary statistics about the fit of the mean.

Table 4.3 Description of the Fit Mean Report

Mean	Mean of the response variable. The predicted response when there are no specified effects in the model.
Std Dev [RMSE]	Standard deviation of the response variable. Square root of the mean square error, also called the root mean square error (or RMSE).
Std Error	Standard deviation of the response mean. Calculated by dividing the RMSE by the square root of the number of values.
SSE	Error sum of squares for the simple mean model. Appears as the sum of squares for **Error** in the analysis of variance tables for each model fit.

Related Information

- "Fitting Menus" on page 115

Fit Line and Fit Polynomial

Using the **Fit Line** command, you can add straight line fits to your scatterplot using least squares regression. Using the **Fit Polynomial** command, you can fit polynomial curves of a certain degree using least squares regression.

Figure 4.8 Example of Fit Line and Fit Polynomial

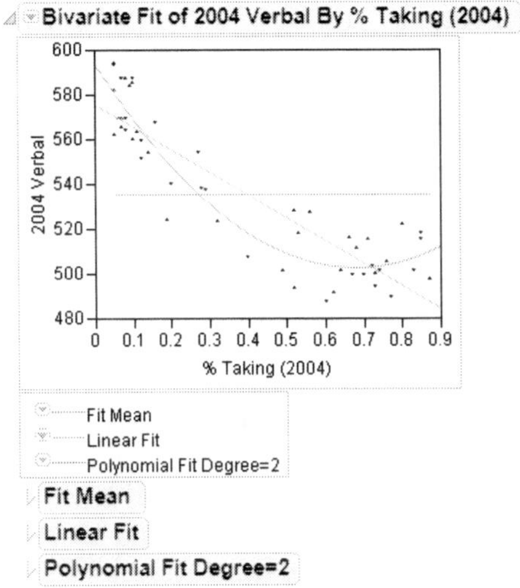

Figure 4.8 shows an example that compares a linear fit to the mean line and to a degree 2 polynomial fit.

Note the following information:

- The **Fit Line** output is equivalent to a polynomial fit of degree 1.
- The **Fit Mean** output is equivalent to a polynomial fit of degree 0.

Linear Fit and Polynomial Fit Reports

The Linear Fit and Polynomial Fit reports begin with the equation of fit.

Figure 4.9 Example of Equations of Fit

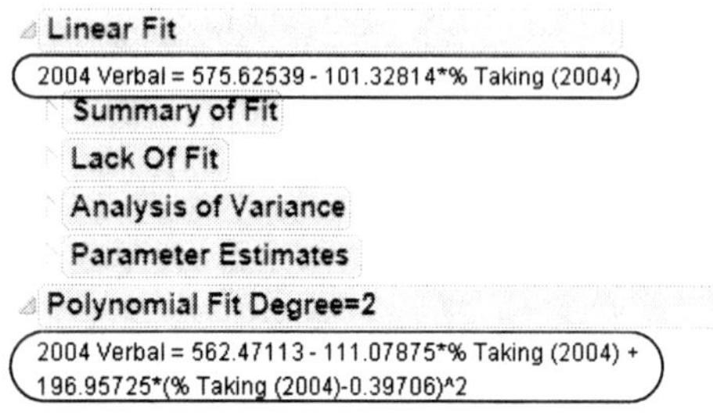

Note: You can edit the equation by clicking on it.

Each Linear and Polynomial Fit Degree report contains at least three reports. A fourth report, Lack of Fit, appears only if there are *X* replicates in your data.

Summary of Fit Report

The Summary of Fit reports show the numeric summaries of the response for the linear fit and polynomial fit of degree 2 for the same data. You can compare multiple Summary of Fit reports to see the improvement of one model over another, indicated by a larger Rsquare value and smaller Root Mean Square Error.

Figure 4.10 Summary of Fit Reports for Linear and Polynomial Fits

Linear Fit

2004 Verbal = 575.62539 - 101.32814*% Taking (2004)

Summary of Fit

RSquare	0.790274
RSquare Adj	0.785994
Root Mean Square Error	15.65752
Mean of Response	535.3922
Observations (or Sum Wgts)	51

Polynomial Fit Degree=2

2004 Verbal = 562.47113 - 111.07875*% Taking (2004) +
196.95725*(% Taking (2004)-0.39706)^2

Summary of Fit

RSquare	0.879413
RSquare Adj	0.874389
Root Mean Square Error	11.99566
Mean of Response	535.3922
Observations (or Sum Wgts)	51

Table 4.4 Description of the Summary of Fit Report

RSquare	Measures the proportion of the variation explained by the model. The remaining variation is not explained by the model and attributed to random error. The Rsquare is 1 if the model fits perfectly. The Rsquare values in Figure 4.10 indicate that the polynomial fit of degree 2 gives a small improvement over the linear fit. See "Statistical Details for the Summary of Fit Report" on page 125.

Table 4.4 Description of the Summary of Fit Report *(Continued)*

RSquare Adj	Adjusts the Rsquare value to make it more comparable over models with different numbers of parameters by using the degrees of freedom in its computation. See "Statistical Details for the Summary of Fit Report" on page 125.
Root Mean Square Error	Estimates the standard deviation of the random error. It is the square root of the mean square for Error in the Analysis of Variance report. See Figure 4.12.
Mean of Response	Provides the sample mean (arithmetic average) of the response variable. This is the predicted response when no model effects are specified.
Observations	Provides the number of observations used to estimate the fit. If there is a weight variable, this is the sum of the weights.

Lack of Fit Report

Note: The Lack of Fit report appears only if there are multiple rows that have the same *x* value.

Using the Lack of Fit report, you can estimate the error, regardless of whether you have the right form of the model. This occurs when multiple observations occur at the same *x* value. The error that you measure for these exact replicates is called *pure error.* This is the portion of the sample error that cannot be explained or predicted no matter what form of model is used. However, a lack of fit test might not be of much use if it has only a few degrees of freedom for it (few replicated *x* values).

Figure 4.11 Examples of Lack of Fit Reports for Linear and Polynomial Fits

Linear Fit

2004 Verbal = 575.62539 - 101.32814*% Taking (2004)

Summary of Fit

Lack Of Fit

Source	DF	Sum of Squares	Mean Square	F Ratio
Lack Of Fit	36	9,662.983	268.416	1.4850
Pure Error	13	2,349.750	180.750	Prob > F
Total Error	49	12,012.733		0.2252
				Max RSq
				0.9590

Polynomial Fit Degree=2

2004 Verbal = 562.47113 - 111.07875*% Taking (2004) + 196.95725*(% Taking (2004)-0.39706)^2

Summary of Fit

Lack Of Fit

Source	DF	Sum of Squares	Mean Square	F Ratio
Lack Of Fit	35	4,557.2473	130.207	0.7204
Pure Error	13	2,349.7500	180.750	Prob > F
Total Error	48	6,906.9973		0.7862
				Max RSq
				0.9590

The difference between the residual error from the model and the pure error is called the *lack of fit error.* The lack of fit error can be significantly greater than the pure error if you have the wrong functional form of the regressor. In that case, you should try a different type of model fit. The Lack of Fit report tests whether the lack of fit error is zero.

Table 4.5 Description of the Lack of Fit Report

Source	The three sources of variation: Lack of Fit, Pure Error, and Total Error.

Table 4.5 Description of the Lack of Fit Report *(Continued)*

DF	The *degrees of freedom* (DF) for each source of error. • The **Total Error** DF is the degrees of freedom found on the **Error** line of the Analysis of Variance table (shown under the "Analysis of Variance Report" on page 103). It is the difference between the **Total** DF and the **Model** DF found in that table. The **Error** DF is partitioned into degrees of freedom for lack of fit and for pure error. • The **Pure Error** DF is pooled from each group where there are multiple rows with the same values for each effect. See "Statistical Details for the Lack of Fit Report" on page 125. • The **Lack of Fit** DF is the difference between the **Total Error** and **Pure Error** DF.
Sum of Squares	The sum of squares (SS for short) for each source of error. • The **Total Error** SS is the sum of squares found on the **Error** line of the corresponding Analysis of Variance table, shown under "Analysis of Variance Report" on page 103. • The **Pure Error** SS is pooled from each group where there are multiple rows with the same value for the *x* variable. This estimates the portion of the true random error that is not explained by model *x* effect. See "Statistical Details for the Lack of Fit Report" on page 125. • The **Lack of Fit** SS is the difference between the **Total Error** and **Pure Error** sum of squares. If the lack of fit SS is large, the model might not be appropriate for the data. The *F*-ratio described below tests whether the variation due to lack of fit is small enough to be accepted as a negligible portion of the pure error.
Mean Square	The sum of squares divided by its associated degrees of freedom. This computation converts the sum of squares to an average (mean square). *F*-ratios for statistical tests are the ratios of mean squares.
F Ratio	The ratio of mean square for lack of fit to mean square for Pure Error. It tests the hypothesis that the lack of fit error is zero.
Prob > F	The probability of obtaining a greater *F*-value by chance alone if the variation due to lack of fit variance and the pure error variance are the same. A high *p* value means that there is not a significant lack of fit.
Max RSq	The maximum R^2 that can be achieved by a model using only the variables in the model. See "Statistical Details for the Lack of Fit Report" on page 125.

Analysis of Variance Report

Analysis of variance (ANOVA) for a regression partitions the total variation of a sample into components. These components are used to compute an *F*-ratio that evaluates the effectiveness of the model. If the probability associated with the *F*-ratio is small, then the model is considered a better statistical fit for the data than the response mean alone.

The Analysis of Variance reports in Figure 4.12 compare a linear fit (**Fit Line**) and a second degree (**Fit Polynomial**). Both fits are statistically better from a horizontal line at the mean.

Figure 4.12 Examples of Analysis of Variance Reports for Linear and Polynomial Fits

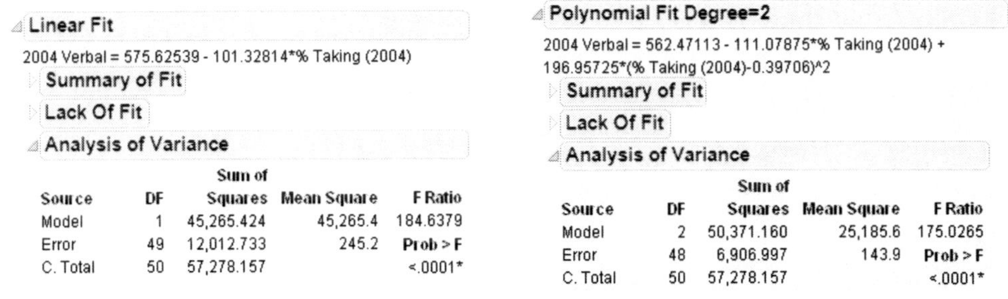

Linear Fit

2004 Verbal = 575.62539 - 101.32814*% Taking (2004)

Summary of Fit

Lack Of Fit

Analysis of Variance

Source	DF	Sum of Squares	Mean Square	F Ratio
Model	1	45,265.424	45,265.4	184.6379
Error	49	12,012.733	245.2	Prob > F
C. Total	50	57,278.157		<.0001*

Polynomial Fit Degree=2

2004 Verbal = 562.47113 - 111.07875*% Taking (2004) + 196.95725*(% Taking (2004)-0.39706)^2

Summary of Fit

Lack Of Fit

Analysis of Variance

Source	DF	Sum of Squares	Mean Square	F Ratio
Model	2	50,371.160	25,185.6	175.0265
Error	48	6,906.997	143.9	Prob > F
C. Total	50	57,278.157		<.0001*

Table 4.6 Description of the Analysis of Variance Report

Source	The three sources of variation: **Model**, **Error**, and **C. Total**.
DF	The degrees of freedom (DF) for each source of variation: • A degree of freedom is subtracted from the total number of non missing values (*N*) for each parameter estimate used in the computation. The computation of the total sample variation uses an estimate of the mean. Therefore, one degree of freedom is subtracted from the total, leaving 50. The total corrected degrees of freedom are partitioned into the Model and Error terms. • One degree of freedom from the total (shown on the **Model** line) is used to estimate a single regression parameter (the slope) for the linear fit. Two degrees of freedom are used to estimate the parameters (β_1 and β_2) for a polynomial fit of degree 2. • The **Error** degrees of freedom is the difference between **C. Total** df and **Model** df.

Table 4.6 Description of the Analysis of Variance Report *(Continued)*

Sum of Squares	The sum of squares (SS for short) for each source of variation: • In this example, the total (**C. Total**) sum of squared distances of each response from the sample mean is 57,258.157, as shown in Figure 4.12. That is the sum of squares for the base model (or simple mean model) used for comparison with all other models. • For the linear regression, the sum of squared distances from each point to the line of fit reduces from 12,012.733. This is the residual or unexplained **(Error)** SS after fitting the model. The residual SS for a second degree polynomial fit is 6,906.997, accounting for slightly more variation than the linear fit. That is, the model accounts for more variation because the model SS are higher for the second degree polynomial than the linear fit. The **C. total** SS less the **Error** SS gives the sum of squares attributed to the model.
Mean Square	The sum of squares divided by its associated degrees of freedom. The *F*-ratio for a statistical test is the ratio of the following mean squares: • The **Model** mean square for the linear fit is 45,265.424. This value estimates the error variance, but only under the hypothesis that the model parameters are zero. • The **Error** mean square is 245.2. This value estimates the error variance.
F Ratio	The model mean square divided by the error mean square. The underlying hypothesis of the fit is that all the regression parameters (except the intercept) are zero. If this hypothesis is true, then both the mean square for error and the mean square for model estimate the error variance, and their ratio has an *F*-distribution. If a parameter is a significant model effect, the *F*-ratio is usually higher than expected by chance alone.
Prob > F	The observed significance probability (*p*-value) of obtaining a greater *F*-value by chance alone if the specified model fits no better than the overall response mean. Observed significance probabilities of 0.05 or less are often considered evidence of a regression effect.

Parameter Estimates Report

The terms in the Parameter Estimates report for a linear fit are the intercept and the single *x* variable.

For a polynomial fit of order *k*, there is an estimate for the model intercept and a parameter estimate for each of the *k* powers of the *X* variable.

Figure 4.13 Examples of Parameter Estimates Reports for Linear and Polynomial Fits

Linear Fit

2004 Verbal = 575.62539 - 101.32814*% Taking (2004)

Summary of Fit

Lack Of Fit

Analysis of Variance

Parameter Estimates

Term	Estimate	Std Error	t Ratio	Prob>|t|
Intercept	575.62539	3.684288	156.24	<.0001*
% Taking (2004)	-101.3281	7.457094	-13.59	<.0001*

Polynomial Fit Degree=2

2004 Verbal = 562.47113 - 111.07875*% Taking (2004) + 196.95725*(% Taking (2004)-0.39706)^2

Summary of Fit

Lack Of Fit

Analysis of Variance

Parameter Estimates

Term	Estimate	Std Error	t Ratio	Prob>|t|
Intercept	562.47113	3.583843	156.95	<.0001*
% Taking (2004)	-111.0788	5.942967	-18.69	<.0001*
(% Taking (2004)-0.39706)^2	196.95725	33.06487	5.96	<.0001*

Table 4.7 Description of the Parameter Estimates Report

Term	Lists the name of each parameter in the requested model. The intercept is a constant term in all models.
Estimate	Lists the parameter estimates of the linear model. The prediction formula is the linear combination of these estimates with the values of their corresponding variables.
Std Error	Lists the estimates of the standard errors of the parameter estimates. They are used in constructing tests and confidence intervals.
t Ratio	Lists the test statistics for the hypothesis that each parameter is zero. It is the ratio of the parameter estimate to its standard error. If the hypothesis is true, then this statistic has a Student's *t*-distribution.
Prob>|t|	Lists the observed significance probability calculated from each *t*-ratio. It is the probability of getting, by chance alone, a *t*-ratio greater (in absolute value) than the computed value, given a true null hypothesis. Often, a value below 0.05 (or sometimes 0.01) is interpreted as evidence that the parameter is significantly different from zero.

To reveal additional statistics, right-click in the report and select the **Columns** menu. Statistics not shown by default are as follows:

Lower 95% The lower endpoint of the 95% confidence interval for the parameter estimate.

Upper 95% The upper endpoint of the 95% confidence interval for the parameter estimate.

Std Beta The standardized parameter estimate. It is useful for comparing the effect of *X* variables that are measured on different scales. See "Statistical Details for the Parameter Estimates Report" on page 126.

VIF The variance inflation factor.

Design Std Error The design standard error for the parameter estimate. See "Statistical Details for the Parameter Estimates Report" on page 126.

Related Information

- "Statistical Details for Fit Line" on page 124
- "Fitting Menus" on page 115

Fit Special

Using the **Fit Special** command, you can transform Y and X. Transformations include the following: log, square root, square, reciprocal, and exponential. You can also constrain the slope and intercept, fit a polynomial of specific degree, and center the polynomial.

Table 4.8 Description of the Specify Transformation or Constraint Window

Y Transformation	Use these options to transform the Y variable.
X Transformation	Use these options to transform the X variable.
Degree	Use this option to fit a polynomial of the specified degree.
Centered Polynomial	To turn off polynomial centering, deselect the **Centered Polynomial** check box. See Figure 4.20. Note that for transformations of the *X* variable, polynomial centering is not performed. Centering polynomials stabilizes the regression coefficients and reduces multicollinearity.
Constrain Intercept to	Select this check box to constrain the model intercept to be the specified value.
Constrain Slope to	Select this check box to constrain the model slope to be the specified value.

Fit Special Reports and Menus

Depending on your selections in the Fit Special window, you see different reports and menus. The flowchart in Figure 4.14 shows you what reports and menus you see depending on your choices.

Figure 4.14 Example of Fit Special Flowchart

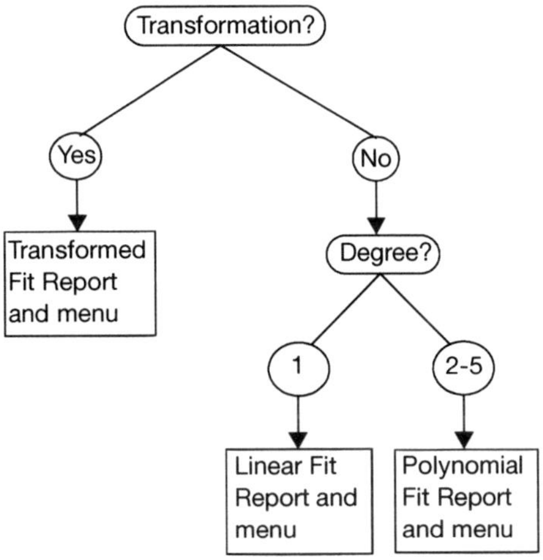

Transformation?
Yes
No
Transformed Fit Report and menu
Degree?
1
2-5
Linear Fit Report and menu
Polynomial Fit Report and menu

Transformed Fit Report

The Transformed Fit report contains the reports described in "Linear Fit and Polynomial Fit Reports" on page 99.

However, if you transformed *Y*, the Fit Measured on Original Scale report appears. This shows the measures of fit based on the original *Y* variables, and the fitted model transformed back to the original scale.

Related Information

- "Example of the Fit Special Command" on page 118
- "Linear Fit and Polynomial Fit Reports" on page 99
- "Fitting Menus" on page 115

Fit Spline

Using the **Fit Spline** command, you can fit a smoothing spline that varies in smoothness (or flexibility) according to the lambda (λ) value. The lambda value is a tuning parameter in the spline formula. As the value of λ decreases, the error term of the spline model has more weight and the fit becomes more flexible and curved. As the value of λ increases, the fit becomes stiff (less curved), approaching a straight line.

Note the following information:

- The smoothing spline can help you see the expected value of the distribution of *Y* across *X*.
- The points closest to each piece of the fitted curve have the most influence on it. The influence increases as you lower the value of λ, producing a highly flexible curve.

- If you want to use a lambda value that is not listed on the menu, select **Fit Spline > Other**. If the scaling of the *X* variable changes, the fitted model also changes. To prevent this from happening, select the **Standardize X** option. This option guarantees that the fitted model remains the same for either the original x variable or the scaled *X* variable.
- You might find it helpful to try several λ values. You can use the **Lambda** slider beneath the Smoothing Spline report to experiment with different λ values. However, λ is not invariant to the scaling of the data. For example, the λ value for an *X* measured in inches, is not the same as the λ value for an *X* measured in centimeters.

Smoothing Spline Fit Report

The Smoothing Spline Fit report contains the R-Square for the spline fit and the Sum of Squares Error. You can use these values to compare the spline fit to other fits, or to compare different spline fits to each other.

Table 4.9 Description of the Smoothing Spline Fit Report

R-Square	Measures the proportion of variation accounted for by the smoothing spline model. For more information, see "Statistical Details for the Smoothing Fit Reports" on page 126.
Sum of Squares Error	Sum of squared distances from each point to the fitted spline. It is the unexplained error (*residual*) after fitting the spline model.
Change Lambda	Enables you to change the λ value, either by entering a number, or by moving the slider.

Related Information

- "Fitting Menus" on page 115
- "Statistical Details for Fit Spline" on page 124

Kernel Smoother

The **Kernel Smoother** command produces a curve formed by repeatedly finding a locally weighted fit of a simple curve (a line or a quadratic) at sampled points in the domain. The many local fits (128 in total) are combined to produce the smooth curve over the entire domain. This method is also called *Loess* or *Lowess*, which was originally an acronym for Locally Weighted Scatterplot Smoother. See Cleveland (1979).

Use this method to quickly see the relationship between variables and to help you determine the type of analysis or fit to perform.

Local Smoother Report

The Local Smoother report contains the **R-Square** for the kernel smoother fit and the **Sum of Squares Error**. You can use these values to compare the kernel smoother fit to other fits, or to compare different kernel smoother fits to each other.

Table 4.10 Description of the Local Smoother Report

R-Square	Measures the proportion of variation accounted for by the kernel smoother model. For more information, see "Statistical Details for the Smoothing Fit Reports" on page 126.
Sum of Squares Error	Sum of squared distances from each point to the fitted kernel smoother. It is the unexplained error (*residual*) after fitting the kernel smoother model.
Local Fit (lambda)	Select the polynomial degree for each local fit. Quadratic polynomials can track local bumpiness more smoothly. Lambda is the degree of certain polynomials that are fitted by the method. Lambda can be 1 or 2.
Weight Function	Specify how to weight the data in the neighborhood of each local fit. Loess uses tri-cube. The weight function determines the influence that each xi and yi has on the fitting of the line. The influence decreases as xi increases in distance from x and finally becomes zero.
Smoothness (alpha)	Controls how many points are part of each local fit. Use the slider or type in a value directly. Alpha is a smoothing parameter. It can be any positive number, but typical values are 1/4 to 1. As alpha increases, the curve becomes smoother.
Robustness	Reweights the points to deemphasize points that are farther from the fitted curve. Specify the number of times to repeat the process (number of passes). The goal is to converge the curve and automatically filter out outliers by giving them small weights.

Related Information

- "Fitting Menus" on page 115

Fit Each Value

The **Fit Each Value** command fits a value to each unique *X* value. The fitted values are the means of the response for each unique *X* value.

Fit Each Value Report

The Fit Each Value report shows summary statistics about the model fit.

Table 4.11 Description of the Fit Each Value Report

Number of Observations	Gives the total number of observations.
Number of Unique Values	Gives the number of unique X values.
Degrees of Freedom	Gives the pure error degrees of freedom.
Sum of Squares	Gives the pure error sum of squares.
Mean Square	Gives the pure error mean square.

Related Information

• "Fitting Menus" on page 115

Fit Orthogonal

The **Fit Orthogonal** command fits lines that adjust for variability in *X* as well as *Y*.

Fit Orthogonal Options

The following table describes the available options to specify a variance ratio.

Univariate Variances, Prin Comp	Uses the univariate variance estimates computed from the samples of *X* and *Y*. This turns out to be the standardized first principal component. This option is not a good choice in a measurement systems application since the error variances are not likely to be proportional to the population variances.
Equal Variances	Uses 1 as the variance ratio, which assumes that the error variances are the same. Using equal variances is equivalent to the non-standardized first principal component line. Suppose that the scatterplot is scaled the same in the *X* and *Y* directions. When you show a normal density ellipse, you see that this line is the longest axis of the ellipse.
Fit X to Y	Uses a variance ratio of zero, which indicates that *Y* effectively has no variance.
Specified Variance Ratio	Lets you enter any ratio that you want, giving you the ability to make use of known information about the measurement error in *X* and response error in *Y*.

Orthogonal Regression Report

The Orthogonal Regression report shows summary statistics about the orthogonal regression model.

The following table describes the Orthogonal Regression report.

Table 4.12 Description of the Orthogonal Regression Report

Variable	Gives the names of the variables used to fit the line.
Mean	Gives the mean of each variable.
Std Dev	Gives the standard deviation of each variable.
Variance Ratio	Gives the variance ratio used to fit the line.
Correlation	Gives the correlation between the two variables.
Intercept	Gives the intercept of the fitted line.
Slope	Gives the slope of the fitted line.
LowerCL	Gives the lower confidence limit for the slope.
UpperCL	Gives the upper confidence limit for the slope.
Alpha	Enter the alpha level used in computing the confidence interval.

Related Information

- "Fitting Menus" on page 115
- "Statistical Details for Fit Orthogonal" on page 124
- "Example Using the Fit Orthogonal Command" on page 120

Density Ellipse

Using the **Density Ellipse** option, you can draw an ellipse (or ellipses) that contains the specified mass of points. The number of points is determined by the probability that you select from the **Density Ellipse** menu).

Figure 4.15 Example of Density Ellipses

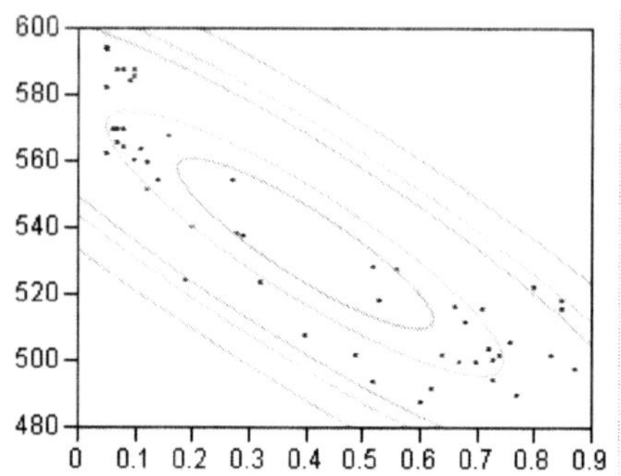

The density ellipsoid is computed from the bivariate normal distribution fit to the *X* and *Y* variables. The bivariate normal density is a function of the means and standard deviations of the *X* and *Y* variables and the correlation between them. The **Other** selection lets you specify any probability greater than zero and less than or equal to one.

These ellipses are both density contours and confidence curves. As confidence curves, they show where a given percentage of the data is expected to lie, assuming the bivariate normal distribution.

The density ellipsoid is a good graphical indicator of the correlation between two variables. The ellipsoid collapses diagonally as the correlation between the two variables approaches either 1 or –1. The ellipsoid is more circular (less diagonally oriented) if the two variables are less correlated.

Correlation Report

The Correlation report that accompanies each **Density Ellipse** fit shows the correlation coefficient for the *X* and *Y* variables.

Note: To see a matrix of ellipses and correlations for many pairs of variables, use the **Multivariate** command in the **Analyze > Multivariate Methods** menu.

Table 4.13 Description of the Correlation Report

Variable	Gives the names of the variables used in creating the ellipse
Mean	Gives the average of both the *X* and *Y* variable.
Std Dev	Gives the standard deviation of both the *X* and *Y* variable. A discussion of the mean and standard deviation are in the section "The Summary Statistics Report" on page 39 in the "Performing Univariate Analysis" chapter.

Table 4.13 Description of the Correlation Report *(Continued)*

Correlation	The Pearson correlation coefficient. If there is an exact linear relationship between two variables, the correlation is 1 or –1 depending on whether the variables are positively or negatively related. If there is no relationship, the correlation tends toward zero. For more information, see "Statistical Details for the Correlation Report" on page 126.
Signif. Prob	Probability of obtaining, by chance alone, a correlation with greater absolute value than the computed value if no linear relationship exists between the *X* and *Y* variables.
Number	Gives the number of observations used in the calculations.

Related Information

- "Fitting Menus" on page 115

Nonpar Density

When a plot shows thousands of points, the mass of points can be too dark to show patterns in density. Using the **Nonpar Density** (nonparametric density) option makes it easier to see the patterns.

Bivariate density estimation models a smooth surface that describes how dense the data points are at each point in that surface. The plot adds a set of contour lines showing the density (Figure 4.16). The contour lines are quantile contours in 5% intervals. This means that about 5% of the points are below the lowest contour, 10% are below the next contour, and so on. The highest contour has about 95% of the points below it.

Figure 4.16 Example of Nonpar Density

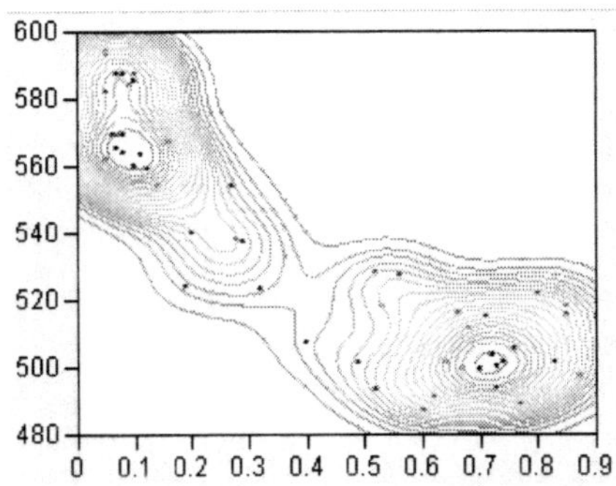

Nonparametric Bivariate Density Report

The nonparametric bivariate density report shows the kernel standard deviations used in creating the nonparametric density.

Related Information

- "Fitting Menus" on page 115

Histogram Borders

The **Histogram Borders** option appends histograms to the *x*- and *y*-axes of the scatterplot. You can use the histograms to visualize the marginal distributions of the X and Y variables.

Figure 4.17 Example of Histogram Borders

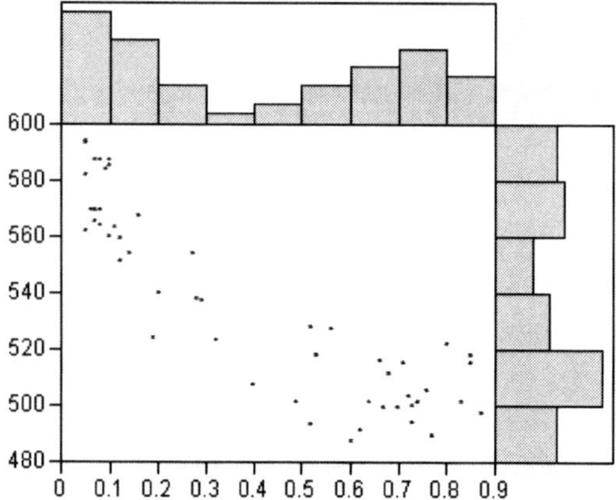

Group By

Using the **Group By** option, you can select a classification (grouping) variable. When a grouping variable is in effect, the Bivariate platform computes a separate analysis for each level of the grouping variable. Regression curves or ellipses then appear on the scatterplot. The fit for each level of the grouping variable is identified beneath the scatterplot, with individual popup menus to save or remove fitting information.

The **Group By** option is checked in the Fitting menu when a grouping variable is in effect. You can change the grouping variable by first selecting the **Group By** option to remove (uncheck) the existing variable. Then, select the **Group By** option again and respond to its window as before.

You might use the Group By option in these different ways:

- An overlay of linear regression lines lets you compare slopes visually.
- An overlay of density ellipses can show clusters of points by levels of a grouping variable.

Related Information

- "Example of Group By Using Density Ellipses" on page 122
- "Example of Group By Using Regression Lines" on page 123

Fitting Menus

In addition to a report, each fitting command adds a fitting menu to the report window. The following table shows the fitting menus that correspond to each fitting command.

Fitting Command	Fitting Menu
Fit Mean	Fit Mean
Fit Line	Linear Fit
Fit Polynomial	Polynomial Fit Degree=X*
Fit Special	Linear Fit Polynomial Fit Degree=X* Transformed Fit X* Constrained Fits
Fit Spline	Smoothing Spline Fit, lambda=X*
Fit Each Value	Fit Each Value
Fit Orthogonal	Orthogonal Fit Ratio=X*
Density Ellipse	Bivariate Normal Ellipse P=X*
Nonpar Density	Quantile Density Colors

*X=variable character or number

Fitting Menu Options

The following table describes the options in the Fitting menus.

Confid Curves Fit	Displays or hides the confidence limits for the expected value (mean). This option is not available for the Fit Spline, Density Ellipse, Fit Each Value, and Fit Orthogonal fits and is dimmed on those menus.
Confid Curves Indiv	Displays or hides the confidence limits for an individual predicted value. The confidence limits reflect variation in the error and variation in the parameter estimates. This option is not available for the Fit Mean, Fit Spline, Density Ellipse, Fit Each Value, and Fit Orthogonal fits and is dimmed on those menus.
Line Color	Lets you select from a palette of colors for assigning a color to each fit.
Line of Fit	Displays or hides the line of fit.

Line Style	Lets you select from the palette of line styles for each fit.
Line Width	Gives three line widths for the line of fit. The default line width is the thinnest line.
Report	Turns the fit's text report on and off.
Save Predicteds	Creates a new column in the current data table called Predicted colname where colname is the name of the *Y* variable. This column includes the prediction formula and the computed sample predicted values. The prediction formula computes values automatically for rows that you add to the table. This option is not available for the Fit Each Value and Density Ellipse fits and is dimmed on those menus. **Note:** You can use the **Save Predicteds** and **Save Residuals** commands for each fit. If you use these commands multiple times or with a grouping variable, it is best to rename the resulting columns in the data table to reflect each fit.
Save Residuals	Creates a new column in the current data table called Residuals colname where colname is the name of the *Y* variable. Each value is the difference between the actual (observed) value and its predicted value. Unlike the **Save Predicteds** command, this command does not create a formula in the new column. This option is not available for the Fit Each Value and Density Ellipse fits and is dimmed on those menus. **Note:** You can use the **Save Predicteds** and **Save Residuals** commands for each fit. If you use these commands multiple times or with a grouping variable, it is best to rename the resulting columns in the data table to reflect each fit.
Remove Fit	Removes the fit from the graph and removes its text report.
Linear Fits, Polynomial Fits, and Fit Special Only:	
Confid Shaded Fit	Draws the same curves as the **Confid Curves Fit** command and shades the area between the curves.
Confid Shaded Indiv	Draws the same curves as the **Confid Curves Indiv** command and shades the area between the curves.
Plot Residuals	Produces four diagnostic plots: residual by predicted, actual by predicted, residual by row, and a normal quantile plot of the residuals. See "Diagnostics Plots" on page 118.
Set Alpha Level	Prompts you to enter the alpha level to compute and display confidence levels for line fits, polynomial fits, and special fits.

Smoothing Spline Fit and Local Smoother Only:	
Save Coefficients	Saves the spline coefficients as a new data table, with columns called X, A, B, C, and D. The X column gives the knot points. A, B, C, and D are the intercept, linear, quadratic, and cubic coefficients of the third-degree polynomial. These coefficients span from the corresponding value in the X column to the next highest value.
Bivariate Normal Ellipse Only:	
Shaded Contour	Shades the area inside the density ellipse.
Select Points Inside	Selects the points inside the ellipse.
Select Points Outside	Selects the points outside the ellipse.
Quantile Density Contours Only:	
Kernel Control	Displays a slider for each variable, where you can change the kernel standard deviation that defines the range of *X* and *Y* values for determining the density of contour lines.
5% Contours	Shows or hides the 5% contour lines.
Contour Lines	Shows or hides the contour lines.
Contour Fill	Fills the areas between the contour lines.
Select Points by Density	Selects points that fall in a user-specified quantile range.
Color by Density Quantile	Colors the points according to density.
Save Density Quantile	Creates a new column containing the density quantile each point is in.
Mesh Plot	Is a three-dimensional plot of the density over a grid of the two analysis variables. See Figure 4.18.
Model Clustering	Creates a new column in the current data table and fills it with cluster values. **Note:** If you save the modal clustering values first and then save the density grid, the grid table also contains the cluster values. The cluster values are useful for coloring and marking points in plots.
Save Density Grid	Saves the density estimates and the quantiles associated with them in a new data table. The grid data can be used to visualize the density in other ways, such as with the Scatterplot 3D or the Contour Plot platforms.

Figure 4.18 Example of a Mesh Plot

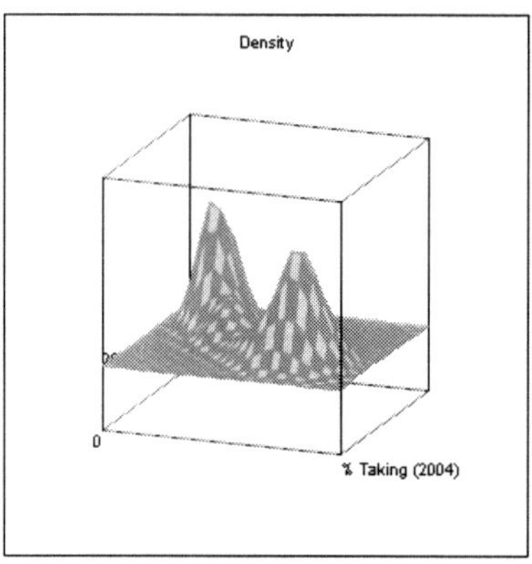

Diagnostics Plots

The **Plot Residuals** option creates residual plots and other plots to diagnose the model fit. The following plots are available:

Residual by Predicted Plot is a plot of the residuals vs. the predicted values. A histogram of the residuals is also created.

Actual by Predicted Plot is a plot of the actual values vs. the predicted values.

Residual by Row Plot is a plot of the residual values vs. the row number.

Residual Normal Quantile Plot is a Normal quantile plot of the residuals.

Additional Examples of the Bivariate Platform

This section contains additional examples using the fitting commands in the Bivariate platform.

Example of the Fit Special Command

To transform Y as log and X as square root, proceed as follows:

1. Open the SAT.jmp sample data table.
2. Select **Analyze > Fit Y by X**.
3. Select 2004 Verbal and click **Y, Response**.
4. Select % Taking (2004) and click **X, Factor.**
5. Click **OK**.

Figure 4.19 Example of SAT Scores by Percent Taking

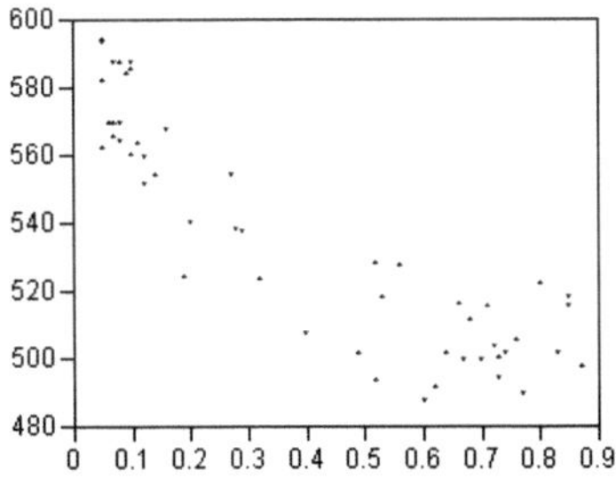

6. From the red triangle menu for Bivariate Fit, select **Fit Special**. The Specify Transformation or Constraint window appears. For a description of this window, see Table 4.8.

Figure 4.20 The Specify Transformation or Constraint Window

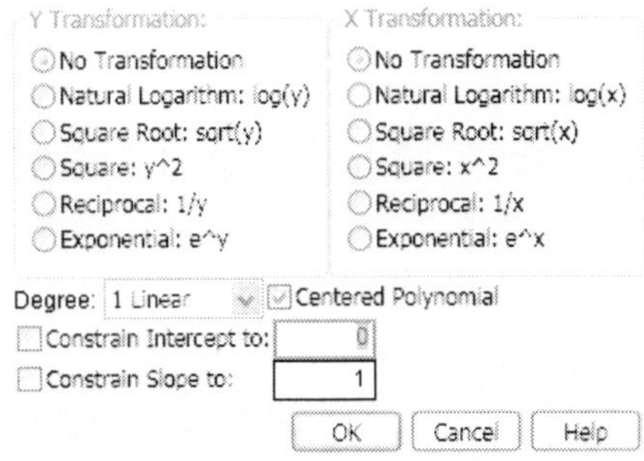

7. Within Y Transformation, select Natural Logarithm: log(y).
8. Within X Transformation, select Square Root: sqrt(x).
9. Click **OK**.

Figure 4.21 Example of Fit Special Report

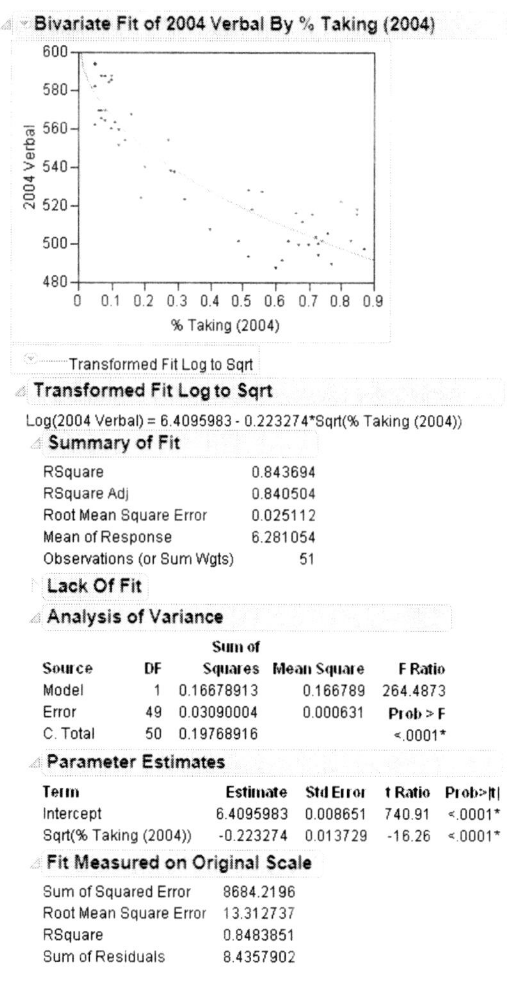

Figure 4.21 shows the fitted line plotted on the original scale. The model appears to fit the data well, as the plotted line goes through the cloud of points.

Example Using the Fit Orthogonal Command

This example involves two parts. First, standardize the variables using the Distribution platform. Then, use the standardized variables to fit the orthogonal model.

Standardize the Variables

1. Open the Big Class.jmp sample data table.
2. Select **Analyze > Distribution**.

3. Select height and weight and click **Y, Columns**.
4. Click **OK**.
5. Hold down the CTRL key. On the red triangle menu next to height, select **Save > Standardized.**

   Holding down the CTRL key broadcasts the operation to all variables in the report window. Notice that in the Big Class.jmp sample data table, two new columns have been added.
6. Close the Distribution report window.

Use the Standardized Variables to Fit the Orthogonal Model

1. From the Big Class.jmp sample data table, select **Analyze > Fit Y by X.**
2. Select Std weight and click **Y, Response.**
3. Select Std height and click **X, Factor**.
4. Click **OK.**
5. From the red triangle menu, select **Fit Line.**
6. From the red triangle menu, select **Fit Orthogonal**. Then select each of the following:
   - **Equal Variances**
   - **Fit X to Y**
   - **Specified Variance Ratio** and type 0.2.
   - **Specified Variance Ratio** and type 5.

Figure 4.22 Example of Orthogonal Fitting Options

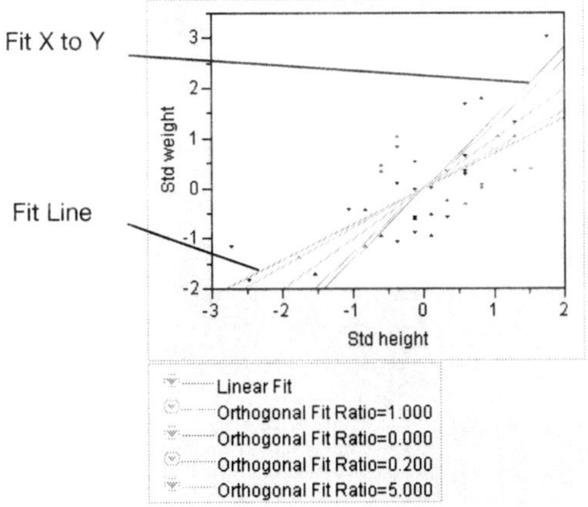

The scatterplot in Figure 4.22 shows the standardized height and weight values with various line fits that illustrate the behavior of the orthogonal variance ratio selections. The standard linear regression (**Fit Line**) occurs when the variance of the *X* variable is considered to be very small. **Fit X to Y** is the opposite extreme, when the variation of the *Y* variable is ignored. All other lines fall between these two extremes and shift as the variance ratio changes. As the variance ratio increases, the variation in the *Y* response dominates and the slope of the fitted line shifts closer to the *Y* by *X* fit. Likewise, when you decrease the ratio, the slope of the line shifts closer to the *X* by *Y* fit.

Example of Group By Using Density Ellipses

This example uses the Hot Dogs.jmp sample data table. The Type column identifies three different types of hot dogs: beef, meat, or poultry. You want to group the three types of hot dogs according to their cost variables.

1. Open the Hot Dogs.jmp sample data table.
2. Select **Analyze > Fit Y by X**.
3. Select $/oz and click **Y, Response**.
4. Select $/lb Protein and click **X, Factor**.
5. Click **OK**.
6. From the red triangle menu, select **Group By**.
7. From the list, select Type.
8. Click **OK**. If you look at the **Group By** option again, you see it has a check mark next to it.
9. From the red triangle menu, select **Density Ellipse > 0.90**.

To color the points according to Type, proceed as follows:

10. Right-click on the scatterplot and select **Row Legend**.
11. Select Type in the column list and click **OK**.

Figure 4.23 Example of Group By

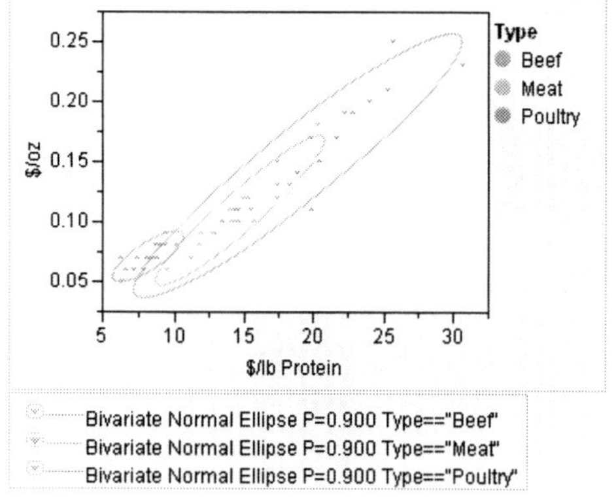

The ellipses in Figure 4.23 show clearly how the different types of hot dogs cluster with respect to the cost variables.

Example of Group By Using Regression Lines

Another use for grouped regression is overlaying lines to compare slopes of different groups.

1. Open the Big Class.jmp sample data table.
2. Select **Analyze > Fit Y by X**.
3. Select weight and click **Y, Response**.
4. Select height and click **X, Factor**.
5. Click **OK**.

To create the example on the left in Figure 4.24:

6. Select **Fit Line** from the red triangle menu.

To create the example on the right in Figure 4.24:

7. From the Linear Fit menu, select **Remove Fit**.
8. From the red triangle menu, select **Group By**.
9. From the list, select sex.
10. Click **OK**.
11. Select **Fit Line** from the red triangle menu.

Figure 4.24 Example of Regression Analysis for Whole Sample and Grouped Sample

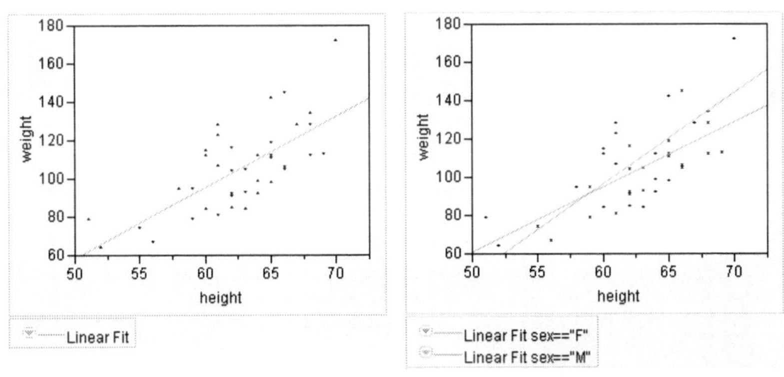

The scatterplot to the left in Figure 4.24 has a single regression line that relates weight to height. The scatterplot to the right shows separate regression lines for males and females.

Statistical Details for the Bivariate Platform

This section contains statistical details for selected commands and reports.

Statistical Details for Fit Line

The **Fit Line** command finds the parameters β_0 and β_1 for the straight line that fits the points to minimize the residual sum of squares. The model for the *i*th row is written $y_i = \beta_0 + \beta_1 x_i + \varepsilon_i$.

A polynomial of degree 2 is a parabola; a polynomial of degree 3 is a cubic curve. For degree *k*, the model for the *i*th observation is as follows:

$$y_i = \sum_{j=0}^{k} \beta_j x_i^j + \varepsilon_i$$

Statistical Details for Fit Spline

The cubic spline method uses a set of third-degree polynomials spliced together such that the resulting curve is continuous and smooth at the splices (knot points). The estimation is done by minimizing an objective function that is a combination of the sum of squares error and a penalty for curvature integrated over the curve extent. See the paper by Reinsch (1967) or the text by Eubank (1988) for a description of this method.

Statistical Details for Fit Orthogonal

Standard least square fitting assumes that the *X* variable is fixed and the *Y* variable is a function of *X* plus error. If there is random variation in the measurement of *X*, you should fit a line that minimizes the sum of the squared perpendicular differences. See Figure 4.25. However, the perpendicular distance depends on how *X* and *Y* are scaled, and the scaling for the perpendicular is reserved as a statistical issue, not a graphical one.

Figure 4.25 Line Perpendicular to the Line of Fit

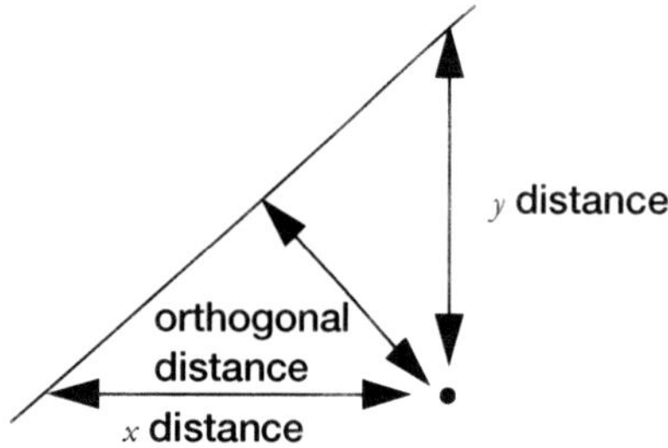

The fit requires that you specify the ratio of the variance of the error in *X* to the error in *Y*. This is the variance of the error, not the variance of the sample points, so you must choose carefully. The ratio $(\sigma_y^2)/(\sigma_x^2)$ is infinite in standard least squares because σ_x^2 is zero. If you do an orthogonal fit with a large error ratio, the fitted line approaches the standard least squares line of fit. If you specify a ratio of zero, the fit is equivalent to the regression of *X* on *Y*, instead of *Y* on *X*.

The most common use of this technique is in comparing two measurement systems that both have errors in measuring the same value. Thus, the *Y* response error and the *X* measurement error are both the same type of measurement error. Where do you get the measurement error variances? You cannot get them from bivariate data because you cannot tell which measurement system produces what proportion of the error. So, you either must blindly assume some ratio like 1, or you must rely on separate repeated measurements of the same unit by the two measurement systems.

An advantage to this approach is that the computations give you predicted values for both *Y* and *X*; the predicted values are the point on the line that is closest to the data point, where closeness is relative to the variance ratio.

Confidence limits are calculated as described in Tan and Iglewicz (1999).

Statistical Details for the Summary of Fit Report

Rsquare

Using quantities from the corresponding analysis of variance table, the **Rsquare** for any continuous response fit is calculated as follows:

$$\frac{\text{Sum of Squares for Model}}{\text{Sum of Squares for C. Total}}$$

RSquare Adj

The RSquare Adj is a ratio of mean squares instead of sums of squares and is calculated as follows:

$$1 - \frac{\text{Mean Square for Error}}{\text{Mean Square for C. Total}}$$

The mean square for **Error** is in the Analysis of Variance report. See Figure 4.12. You can compute the mean square for **C. Total** as the **Sum of Squares** for **C. Total** divided by its respective degrees of freedom.

Statistical Details for the Lack of Fit Report

Pure Error DF

For the Pure Error DF, consider the multiple instances in the **Big Class.jmp** sample data table where more than one subject has the same value of height. In general, if there are *g* groups having multiple rows with identical values for each effect, the pooled DF, denoted DF_p, is as follows:

$$DF_p = \sum_{i=1}^{g} (n_i - 1)$$

n_i is the number of subjects in the *i*th group.

Pure Error SS

For the Pure Error SS, in general, if there are *g* groups having multiple rows with the same *x* value, the pooled SS, denoted SS_p, is written as follows:

$$SS_p = \sum_{i=1}^{g} SS_i$$

where SS_i is the sum of squares for the *i*th group corrected for its mean.

Max RSq

Because Pure Error is invariant to the form of the model and is the minimum possible variance, **Max RSq** is calculated as follows:

$$1 - \frac{SS(\text{Pure error})}{SS(\text{Total for whole model})}$$

Statistical Details for the Parameter Estimates Report

Std Beta

Std Beta is calculated as follows:

$$\hat{\beta}(s_x / s_y)$$

where $\hat{\beta}$ is the estimated parameter, *sx* and *sy* are the standard deviations of the *X* and *Y* variables.

Design Std Error

Design Std Error is calculated as the standard error of the parameter estimate divided by the RMSE.

Statistical Details for the Smoothing Fit Reports

R-Square is equal to 1-(SSE/C.Total SS), where C.Total SS is available in the Fit Line ANOVA report.

Statistical Details for the Correlation Report

The Pearson correlation coefficient is denoted *r*, and is computed as follows:

$$r_{xy} = \frac{s_{xy}^2}{\sqrt{s_x^2 s_y^2}} \quad \text{where} \quad s_{xy}^2 = \frac{\sum w_i(x_i - \bar{x}_i)(y_i - \bar{y}_i)}{df}$$

Where w_i is either the weight of the *i*th observation if a weight column is specified, or 1 if no weight column is assigned.

Chapter 5

Performing Oneway Analysis

Using the Fit Y by X or Oneway Platform

Using the Oneway or Fit Y by X platform, you can explore how the distribution of a continuous Y variable differs across groups defined by a single categorical X variable. For example, you might want to find out how different categories of the same type of drug (X) affect patient pain levels on a numbered scale (Y).

The Oneway platform is the *continuous by nominal or ordinal* personality of the Fit Y by X platform. The analysis results appear in a plot, and you can interactively add additional analyses, such as the following:

- a one-way analysis of variance to fit means and to test that they are equal
- nonparametric tests
- a test for homogeneity of variance
- multiple-comparison tests on means, with means comparison circles
- outlier box plots overlaid on each group
- power details for the one-way layout

Figure 5.1 Oneway Analysis

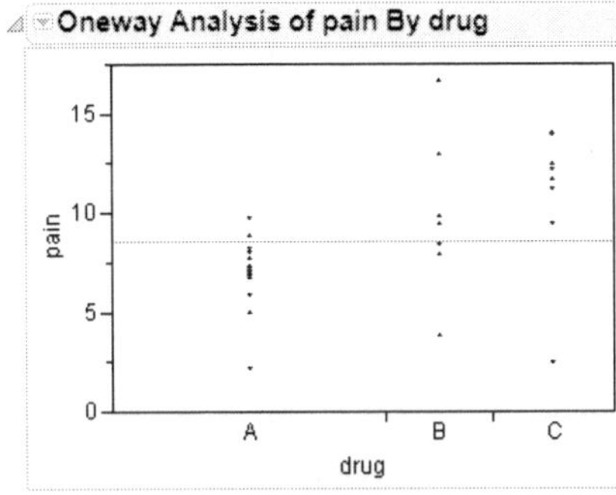

Contents

Overview of Oneway Analysis

A one-way analysis of variance tests for differences between group means. The total variability in the response is partitioned into two parts: within-group variability and between-group variability. If the between-group variability is large relative to the within-group variability, then the differences between the group means are considered to be significant.

Example of Oneway Analysis

This example uses the Analgesics.jmp sample data table. Thirty-three subjects were administered three different types of analgesics (A, B, and C). The subjects were asked to rate their pain levels on a sliding scale. You want to find out if the means for A, B, and C are significantly different.

1. Open the Analgesics.jmp sample data table.
2. Select **Analyze > Fit Y by X**.
3. Select pain and click **Y, Response**.
4. Select drug and click **X, Factor**.
5. Click **OK**.

Figure 5.2 Example of Oneway Analysis

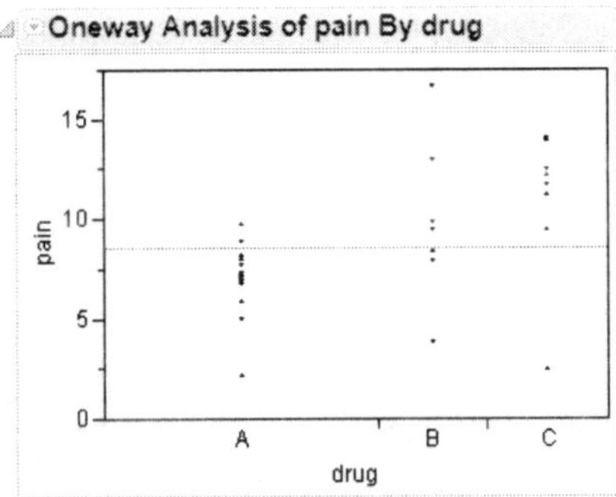

You notice that one drug (A) has consistently lower scores than the other drugs. You also notice that the *x*-axis ticks are unequally spaced. The length between the ticks is proportional to the number of scores (observations) for each drug.

Perform an analysis of variance on the data.

6. From the red triangle menu for Oneway Analysis, select **Means/Anova**.

Note: If the X factor has only two levels, the **Means/Anova** option appears as **Means/Anova/Pooled t,** and adds a pooled *t*-test report to the report window.

Figure 5.3 Example of the Means/Anova Option

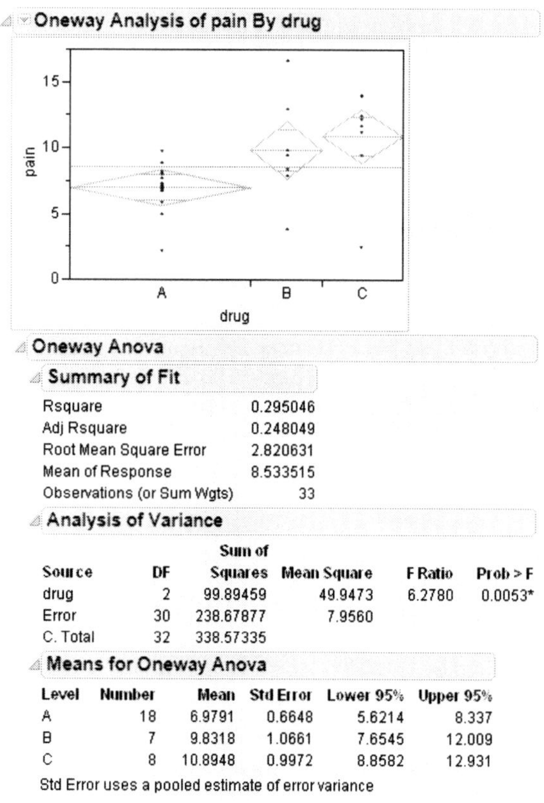

Oneway Anova

Summary of Fit

Rsquare	0.295046
Adj Rsquare	0.248049
Root Mean Square Error	2.820631
Mean of Response	8.533515
Observations (or Sum Wgts)	33

Analysis of Variance

Source	DF	Sum of Squares	Mean Square	F Ratio	Prob > F
drug	2	99.89459	49.9473	6.2780	0.0053*
Error	30	238.67877	7.9560		
C. Total	32	338.57335			

Means for Oneway Anova

Level	Number	Mean	Std Error	Lower 95%	Upper 95%
A	18	6.9791	0.6648	5.6214	8.337
B	7	9.8318	1.0661	7.6545	12.009
C	8	10.8948	0.9972	8.8582	12.931

Std Error uses a pooled estimate of error variance

Note the following observations:

- Mean diamonds representing confidence intervals appear.
  - The line near the center of each diamond represents the group mean. At a glance, you can see that the mean for each drug looks significantly different.
  - The vertical span of each diamond represents the 95% confidence interval for the mean of each group.

  See "Mean Diamonds and X-Axis Proportional" on page 143.
- The Summary of Fit table provides overall summary information about the analysis.
- The Analysis of Variance report shows the standard ANOVA information. You notice that the Prob > F (the *p*-value) is 0.0053, which supports your visual conclusion that there are significant differences between the drugs.
- The Means for Oneway Anova report shows the mean, sample size, and standard error for each level of the categorical factor.

Launch the Oneway Platform

You can perform a Oneway analysis using either the Fit Y by X platform or the Oneway platform. The two approaches are equivalent.

- To launch the Fit Y by X platform, select **Analyze > Fit Y by X**.

  or

- To launch the Oneway platform, from the JMP Starter window, click on the **Basic** category and click **Oneway**.

Figure 5.4 The Oneway Launch Window

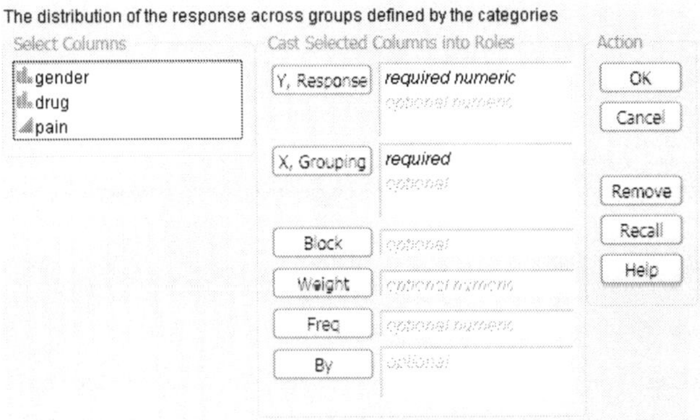

For more information about this launch window, see "Introduction to the Fit Y by X Platform" chapter on page 87.

The Oneway Plot

The Oneway plot shows the response points for each X factor value. You can compare the distribution of the response across the levels of the X factor. The distinct values of X are sometimes called levels.

Replace variables in the plot in one of two ways: swap existing variables by dragging and dropping a variable from one axis to the other axis; or, click on a variable in the Columns panel of the associated data table and drag it onto an axis.

You can add reports, additional plots, and tests to the report window using the options in the red triangle menu for Oneway Analysis. See "Oneway Platform Options" on page 132.

To produce the plot shown in Figure 5.5, follow the instructions in "Example of Oneway Analysis" on page 129.

Figure 5.5 The Oneway Plot

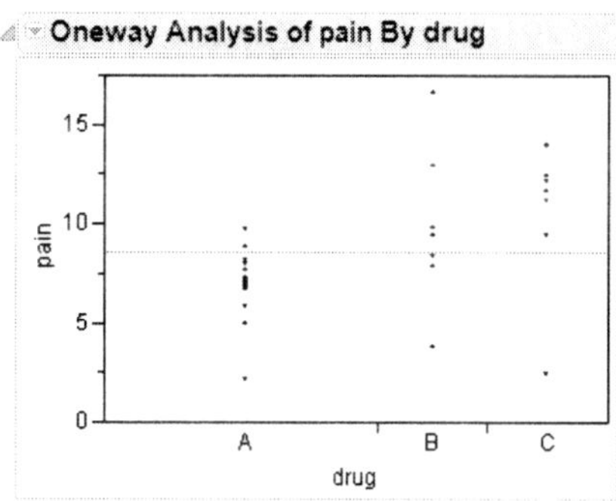

Oneway Platform Options

Figure 5.6 shows an example of the platform options in the red triangle menu for Oneway Analysis.

Figure 5.6 Example of Oneway Platform Options

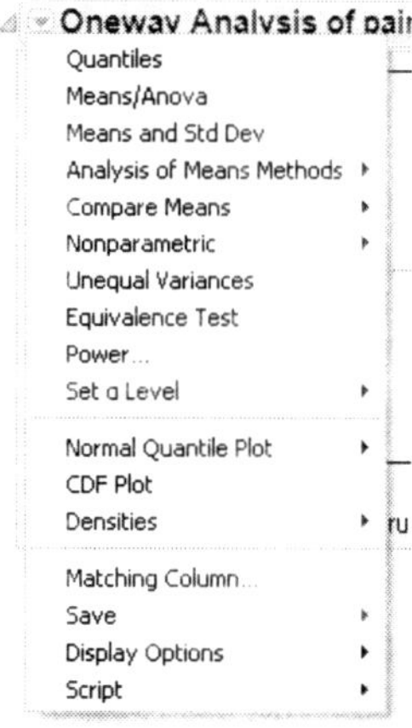

When you select a platform option, objects might be added to the plot, and a report is added to the report window.

Table 5.1 Examples of Options and Elements

Platform Option	Object Added to Plot	Report Added to Report Window
Quantiles	Box plots	Quantiles report
Means/Anova	Mean diamonds	Oneway ANOVA reports

Table 5.1 Examples of Options and Elements *(Continued)*

Platform Option	Object Added to Plot	Report Added to Report Window
Means and Std Dev	Mean lines, error bars, and standard deviation lines	Means and Std Deviations report
Compare Means	Comparison circles (except Nonparametric Multiple Comparisons option)	Means Comparison reports

The following table describes all of the platform options in the red triangle menu for Oneway Analysis. Some options might not appear unless specific conditions are met.

Table 5.2 Descriptions of Oneway Analysis Platform Options

Quantiles	Lists the following quantiles for each group: • 0% (Minimum) • 10% • 25% • 50% (Median) • 75% • 90% • 100% (Maximum) Activates **Box Plots** from the **Display Options** menu. See “Quantiles” on page 137.
Means/Anova or **Means/Anova/Pooled t**	Fits means for each group and performs a one-way analysis of variance to test if there are differences among the means. See “Means/Anova and Means/Anova/Pooled t” on page 138. The **Means/Anova/Pooled t** option appears only if the X factor has two levels.
Means and Std Dev	Gives summary statistics for each group. The standard errors for the means use individual group standard deviations rather than the pooled estimate of the standard deviation. The plot now contains mean lines, error bars, and standard deviation lines. For a brief description of these elements, see “Display Options” on page 136. For more details about these elements, see “Mean Lines, Error Bars, and Standard Deviation Lines” on page 144.

Table 5.2 Descriptions of Oneway Analysis Platform Options *(Continued)*

t test	Produces a *t*-test report assuming that the variances are not equal. See "The t-test Report" on page 140. This option appears only if the X factor has two levels.
Analysis of Means Methods	Provides four commands for performing Analysis of Means (ANOM) procedures. There are commands for comparing both means and variances. See "Analysis of Means Methods" on page 144.
Compare Means	Provides multiple-comparison methods for comparing sets of group means. See "Compare Means" on page 146.
Nonparametric	Provides nonparametric comparisons of group means. See "Nonparametric" on page 151.
Unequal Variances	Performs four tests for equality of group variances. Also gives the Welch test, which is an ANOVA test for comparing means when the variances within groups are not equal. See "Unequal Variances" on page 156.
Equivalence Test	Tests that a difference is less than a threshold value. See "Equivalence Test" on page 158.
Power	Provides calculations of statistical power and other details about a given hypothesis test. See "Power" on page 159. The Power Details window and reports also appear within the Fit Model platform. For further discussion and examples of power calculations, see the *Modeling and Multivariate Methods* book.
Set α Level	You can select an option from the most common alpha levels or specify any level with the **Other** selection. Changing the alpha level results in the following actions: • recalculates confidence limits • adjusts the mean diamonds on the plot (if they are showing) • modifies the upper and lower confidence level values in reports • changes the critical number and comparison circles for all Compare Means reports • changes the critical number for all Nonparametric Multiple Comparison reports

Table 5.2 Descriptions of Oneway Analysis Platform Options *(Continued)*

Normal Quantile Plot	Provides the following options for plotting the quantiles of the data in each group: • **Plot Actual by Quantile** generates a quantile plot with the response variable on the *y*-axis and quantiles on the *x*-axis. The plot shows quantiles computed within each level of the categorical X factor. • **Plot Quantile by Actual** reverses the *x*- and *y*-axes. • **Line of Fit** draws straight diagonal reference lines on the plot for each level of the X variable. This option is available only once you have created a plot (Actual by Quantile or Quantile by Actual).
CDF Plot	Plots the cumulative distribution function for all of the groups in the Oneway report. See "CDF Plot" on page 161.
Densities	Compares densities across groups. See "Densities" on page 161.
Matching Column	Specify a matching variable to perform a matching model analysis. Use this option when the data in your Oneway analysis comes from matched (paired) data, such as when observations in different groups come from the same subject. The plot now contains matching lines that connect the matching points. See "Matching Column" on page 161.
Save	Saves the following quantities as new columns in the current data table: • **Save Residuals** saves values computed as the response variable minus the mean of the response variable within each level of the factor variable. • **Save Standardized** saves standardized values of the response variable computed within each level of the factor variable. This is the centered response divided by the standard deviation within each level. • **Save Normal Quantiles** saves normal quantile values computed within each level of the categorical factor variable. • **Save Predicted** saves the predicted mean of the response variable for each level of the factor variable.
Display Options	Adds or removes elements from the plot. See "Display Options" on page 136.

Table 5.2 Descriptions of Oneway Analysis Platform Options *(Continued)*

Script	This menu contains options that are available to all platforms. They enable you to redo the analysis or save the JSL commands for the analysis to a window or a file. For more information, see *Using JMP*.

Display Options

Using Display Options, you can add or remove elements from a plot. Some options might not appear unless they are relevant.

Table 5.3 Descriptions of Display Options

All Graphs	Shows or hides all graphs.
Points	Shows or hides data points on the plot.
Box Plots	Shows or hides outlier box plots for each group.
Mean Diamonds	Draws a horizontal line through the mean of each group proportional to its *x*-axis. The top and bottom points of the mean diamond show the upper and lower 95% confidence points for each group. See "Mean Diamonds and X-Axis Proportional" on page 143.
Mean Lines	Draws a line at the mean of each group. See "Mean Lines, Error Bars, and Standard Deviation Lines" on page 144.
Mean CI Lines	Draws lines at the upper and lower 95% confidence levels for each group.
Mean Error Bars	Identifies the mean of each group and shows error bars one standard error above and below the mean. See "Mean Lines, Error Bars, and Standard Deviation Lines" on page 144.
Grand Mean	Draws the overall mean of the *Y* variable on the plot.
Std Dev Lines	Shows lines one standard deviation above and below the mean of each group. See "Mean Lines, Error Bars, and Standard Deviation Lines" on page 144.
Comparison Circles	Shows or hides comparison circles. This option is available only when one of the **Compare Means** options is selected. See "Statistical Details for Comparison Circles" on page 179.
Connect Means	Connects the group means with a straight line.
Mean of Means	Draws a line at the mean of the group means.

Table 5.3 Descriptions of Display Options *(Continued)*

X-Axis proportional	Makes the spacing on the *x*-axis proportional to the sample size of each level. See "Mean Diamonds and X-Axis Proportional" on page 143.
Points Spread	Spreads points over the width of the interval
Points Jittered	Adds small spaces between points that overlay on the same *y* value. The horizontal adjustment of points varies from 0.375 to 0.625 with a $4^*(\text{Uniform-}0.5)^5$ distribution.
Matching Lines	(Only appears when the **Matching Column** option is selected.) Connects matching points.
Matching Dotted Lines	Only appears when the **Matching Column** option is selected.) Draws dotted lines to connect cell means from missing cells in the table. The values used as the endpoints of the lines are obtained using a two-way ANOVA model.
Histograms	Draws side-by-side histograms to the right of the original plot.

Quantiles

The Quantiles report lists selected percentiles for each level of the X factor variable. The median is the 50th percentile, and the 25th and 75th percentiles are called the *quartiles*.

The **Quantiles** option adds the following elements to the plot:

- the grand mean representing the overall mean of the Y variable
- outlier box plots summarizing the distribution of points at each factor level

Figure 5.7 Outlier Box Plot and Grand Mean

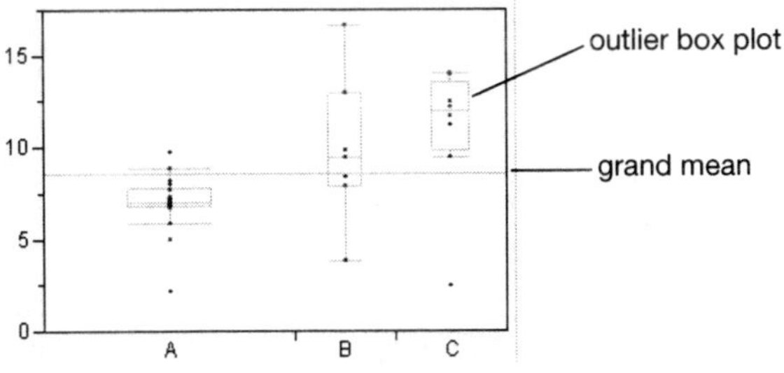

Note: To hide these elements, click the red triangle next to Oneway Analysis and select **Display Options > Box Plots** or **Grand Mean**.

Outlier Box Plots

The outlier box plot is a graphical summary of the distribution of data. Note the following aspects about outlier box plots (see Figure 5.8):

- The vertical line within the box represents the median sample value.
- The ends of the box represent the 75th and 25th quantiles, also expressed as the 3rd and 1st quartile, respectively.
- The difference between the 1st and 3rd quartiles is called the *interquartile range*.
- Each box has lines, sometimes called *whiskers*, that extend from each end. The whiskers extend from the ends of the box to the outermost data point that falls within the distances computed as follows:

  ```
  3rd quartile + 1.5*(interquartile range)
  1st quartile - 1.5*(interquartile range)
  ```

  If the data points do not reach the computed ranges, then the whiskers are determined by the upper and lower data point values (not including outliers).

Figure 5.8 Examples of Outlier Box Plots

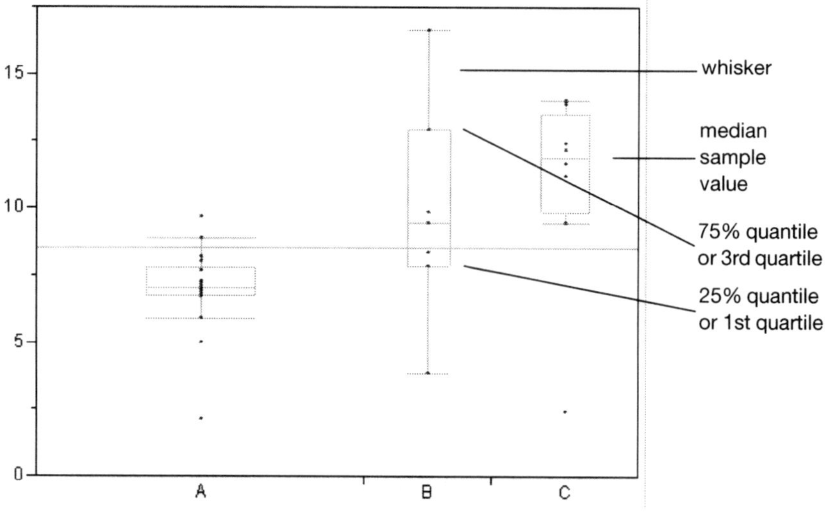

Means/Anova and Means/Anova/Pooled t

The **Means/Anova** option performs an analysis of variance. If the X factor contains exactly two levels, this option appears as **Means/Anova/Pooled t**. In addition to the other reports, a *t*-test report assuming pooled (or equal) variances appears.

Table 5.4 Description of Report Window Elements

Element	Reference
Mean diamonds are added to the Oneway plot	See "Display Options" on page 136 and "Mean Diamonds and X-Axis Proportional" on page 143.
Reports	See "The Summary of Fit Report" on page 139.
	See "The Analysis of Variance Report" on page 141.
	See "The Means for Oneway Anova Report" on page 142.
	See "The t-test Report" on page 140. **Note:** This report appears only if the **Means/Anova/Pooled t** option is selected.
	See "The Block Means Report" on page 143. **Note:** This report appears only if you have specified a Block variable in the launch window.

The Summary of Fit Report

The Summary of Fit report shows a summary for a one-way analysis of variance.

The report lists the following quantities:

Table 5.5 Description of the Summary of Fit Report

Rsquare	Measures the proportion of the variation accounted for by fitting means to each factor level. The remaining variation is attributed to random error. The R^2 value is 1 if fitting the group means accounts for all the variation with no error. An R^2 of 0 indicates that the fit serves no better as a prediction model than the overall response mean. For more information, see "Statistical Details for the Summary of Fit Report" on page 181. R^2 is also called the *coefficient of determination*.
Adj Rsquare	Adjusts R^2 to make it more comparable over models with different numbers of parameters by using the degrees of freedom in its computation. For more information, see "Statistical Details for the Summary of Fit Report" on page 181.
Root Mean Square Error	Estimates the standard deviation of the random error. It is the square root of the mean square for Error found in the Analysis of Variance report.
Mean of Response	Overall mean (arithmetic average) of the response variable.

Table 5.5 Description of the Summary of Fit Report *(Continued)*

Observations (or Sum Wgts)	Number of observations used in estimating the fit. If weights are used, this is the sum of the weights. See "Statistical Details for the Summary of Fit Report" on page 181.

The t-test Report

Note: This option is applicable only for the **Means/Anova/Pooled t** option.

There are two types of *t*-Tests:

- Equal variances. If you select the **Means/Anova/Pooled t** option, a *t*-Test report appears. This *t*-Test assumes equal variances.
- Unequal variances. If you select the **t-Test** option from the red triangle menu, a *t*-Test report appears. This *t*-Test assumes unequal variances.

The report shows the following information:

Table 5.6 Description of the t-Test Report

t Test plot	Shows the sampling distribution of the difference in the means, assuming the null hypothesis is true. The vertical red line is the actual difference in the means. The shaded areas correspond to the p-values.
Difference	Shows the estimated difference between the two X levels. In the plots, the Difference value appears as a red line that compares the two levels.
Std Err Dif	Shows the standard error of the difference.
Upper CL Dif	Shows the upper confidence limit for the difference.
Lower CL Dif	Shows the lower confidence limit for the difference.
Confidence	Shows the level of confidence (1-alpha). To change the level of confidence, select a new alpha level from the **Set α Level** command from the platform red triangle menu.
t Ratio	Value of the *t*-statistic.
DF	The degrees of freedom used in the *t*-test.
Prob > |t|	The *p*-value associated with a two-tailed test.
Prob > t	The *p*-value associated with a lower-tailed test.

Table 5.6 Description of the t-Test Report *(Continued)*

Prob < t	The *p*-value associated with an upper-tailed test.

The Analysis of Variance Report

The Analysis of Variance report partitions the total variation of a sample into two components. The ratio of the two mean squares forms the *F* ratio. If the probability associated with the *F* ratio is small, then the model is a better fit statistically than the overall response mean.

Note: If you specified a **Block** column, then the Analysis of Variance report includes the **Block** variable.

The Analysis of Variance report shows the following information:

Table 5.7 Description of the Analysis of Variance Report

Source	Lists the three sources of variation, which are the model source, **Error**, and **C. Total** (corrected total).
DF	Records an associated degrees of freedom (DF for short) for each source of variation: • The degrees of freedom for **C. Total** are N - 1, where N is the total number of observations used in the analysis. • If the X factor has k levels, then the model has k - 1 degrees of freedom. The **Error** degrees of freedom is the difference between the **C. Total** degrees of freedom and the **Model** degrees of freedom (in other words, N - k).
Sum of Squares	Records a sum of squares (SS for short) for each source of variation: • The total (**C. Total**) sum of squares of each response from the overall response mean. The **C. Total** sum of squares is the base model used for comparison with all other models. • The sum of squared distances from each point to its respective group mean. This is the remaining unexplained **Error** (residual) SS after fitting the analysis of variance model. The total SS minus the error SS gives the sum of squares attributed to the model. This tells you how much of the total variation is explained by the model.

Table 5.7 Description of the Analysis of Variance Report *(Continued)*

Mean Square	Is a sum of squares divided by its associated degrees of freedom: • The **Model** mean square estimates the variance of the error, but only under the hypothesis that the group means are equal. • The **Error** mean square estimates the variance of the error term independently of the model mean square and is unconditioned by any model hypothesis.
F Ratio	Model mean square divided by the error mean square. If the hypothesis that the group means are equal (there is no real difference between them) is true, then both the mean square for error and the mean square for model estimate the error variance. Their ratio has an *F* distribution. If the analysis of variance model results in a significant reduction of variation from the total, the *F* ratio is higher than expected.
Prob>F	Probability of obtaining (by chance alone) an *F* value greater than the one calculated if, in reality, there is no difference in the population group means. Observed significance probabilities of 0.05 or less are often considered evidence that there are differences in the group means.

The Means for Oneway Anova Report

The Means for Oneway Anova report summarizes response information for each level of the nominal or ordinal factor.

The Means for Oneway Anova report shows the following information:

Table 5.8 Description of the Means for Oneway Anova Report

Level	Lists the levels of the X variable.
Number	Lists the number of observations in each group.
Mean	Lists the mean of each group.
Std Error	Lists the estimates of the standard deviations for the group means. This standard error is estimated assuming that the variance of the response is the same in each level. It is the root mean square error found in the Summary of Fit report divided by the square root of the number of values used to compute the group mean.
Lower 95% and Upper 95%	Lists the lower and upper 95% confidence interval for the group means.

The Block Means Report

If you have specified a Block variable on the launch window, the **Means/Anova** and **Means/Anova/Pooled t** commands produce a Block Means report. This report shows the means for each block and the number of observations in each block.

Mean Diamonds and X-Axis Proportional

A mean diamond illustrates a sample mean and confidence interval.

Figure 5.9 Examples of Mean Diamonds and X-Axis Proportional Options

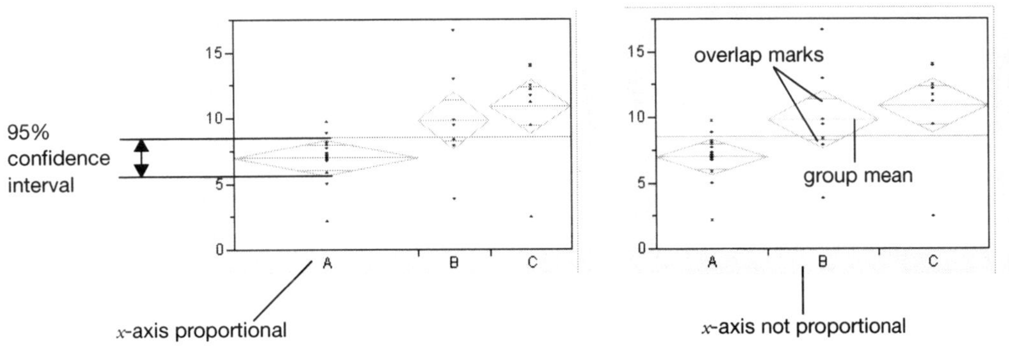

Note the following observations:

- The top and bottom of each diamond represent the (1-alpha)x100 confidence interval for each group. The confidence interval computation assumes that the variances are equal across observations. Therefore, the height of the diamond is proportional to the reciprocal of the square root of the number of observations in the group.
- If the **X-Axis proportional** option is selected, the horizontal extent of each group along the *x*-axis (the horizontal size of the diamond) is proportional to the sample size for each level of the X variable. Therefore, the narrower diamonds are usually taller, because fewer data points results in a wider confidence interval.
- The mean line across the middle of each diamond represents the group mean.
- Overlap marks appear as lines above and below the group mean. For groups with equal sample sizes, overlapping marks indicate that the two group means are not significantly different at the given confidence level. Overlap marks are computed as group mean $\pm (\sqrt{2})/2 \times \mathrm{CI}/2$. Overlap marks in one diamond that are closer to the mean of another diamond than that diamond's overlap marks indicate that those two groups are not different at the given confidence level.
- The mean diamonds automatically appear when you select the **Means/Anova/Pooled t** or **Means/Anova** option from the platform menu. However, you can show or hide them at any time by selecting **Display Options > Mean Diamonds** from the red triangle menu.

Mean Lines, Error Bars, and Standard Deviation Lines

Show mean lines by selecting **Display Options > Mean Lines**. Mean lines indicate the mean of the response for each level of the X variable.

Mean error bars and standard deviation lines appear when you select the **Means and Std Dev** option from the red triangle menu. See Figure 5.10. To turn each option on or off singly, select **Display Options > Mean Error Bars** or **Std Dev Lines**.

Figure 5.10 Mean Lines, Mean Error Bars, and Std Dev Lines

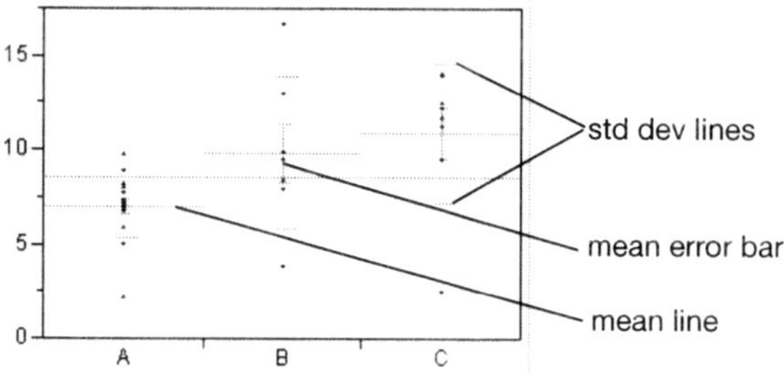

Analysis of Means Methods

Note: For a description of ANOM methods, see the document by Nelson, Wludyka, and Copeland (2005). For a description of the specific ANOM for Variances method, see the paper by Wludyka and Nelson (1997).

Analysis of means (ANOM) methods compare means and variances across several groups. You might want to use these methods under these circumstances:

- if you want to test whether any of the group means are statistically different from the overall mean
- if you want to test whether any of the group standard deviations are statistically different from the root mean square error (RMSE)

Note: Within the Contingency platform, you can use the **Analysis of Means for Proportions** when the response has two categories. For details, see the "Performing Contingency Analysis" chapter on page 183.

Compare Means

Use the **ANOM** and **ANOM with Transformed Ranks** options to compare group means to the overall mean.

Table 5.9 Descriptions of Methods for Comparing Means

ANOM	Compares group means to the overall mean. This method assumes that your data is approximately normally distributed. See "Example of an Analysis of Means Chart" on page 162.
ANOM with Transformed Ranks	This is the nonparametric version of the **ANOM** analysis. Use this method if your data is clearly non-normal and cannot be transformed to normality. Compares each group mean transformed rank to the overall mean transformed rank.

Compare Standard Deviations (or Variances)

Use the **ANOM for Variances** and **ANOM for Variances with Levene (ADM)** options to compare group standard deviations to the root mean square error. This is a type of variance heterogeneity test.

Table 5.10 Descriptions of Methods for Comparing Standard Deviations (or Variances)

ANOM for Variances	Compares group standard deviations to the root mean square error. This method assumes that your data is approximately normally distributed. To use this method, each group must have at least four observations. See "Example of an Analysis of Means for Variances Chart" on page 163.
ANOM for Variances with Levene (ADM)	This is the nonparametric version of the **ANOM for Variances** analysis. Use this method if you suspect your data is non-normal and cannot be transformed to normality. Compares the group means of the absolute deviation from the median (ADM) to the overall mean ADM.

Analysis of Means Charts

Each Analysis of Means Methods option adds a chart to the report window, that shows the following:

- a center line indicating the overall mean or root mean square error (or MSE when in variance scale)
- upper decision limits (UDL)
- lower decision limits (LDL)

If a group mean falls outside of the decision limits, then that mean is significantly different from the overall mean. If a group standard deviation falls outside of the decision limits, then that standard deviation is significantly different from the root mean square error.

Analysis of Means Options

Each Analysis of Means Methods option adds an Analysis of Means red triangle menu to the report window.

Table 5.11 Descriptions of the Analysis of Means Methods Options

Set Alpha Level	Select an option from the most common alpha levels or specify any level with the **Other** selection. Changing the alpha level modifies the upper and lower decision limits.
Show Summary Report	For **ANOM**, creates a report showing group means and decision limits. • For **ANOM with Transformed Ranks**, creates a report showing group mean transformed ranks and decision limits. • For **ANOM for Variances**, creates a report showing group standard deviations (or variances) and decision limits. • For **ANOM for Variances with Levene (ADM)**, creates a report showing group mean ADMs and decision limits.
Graph in Variance Scale	(Only for **ANOM for Variances**) Changes the scale of the *y*-axis from standard deviations to variances.
Display Options	Display options include the following: • **Show Decision Limits** shows or hides decision limit lines. • **Show Decision Limit Shading** shows or hides decision limit shading. • **Show Center Line** shows or hides the center line statistic. • **Show Needles** shows or hides the needles.

Compare Means

Note: Another method for comparing means is ANOM. See "Analysis of Means Methods" on page 144.

Use the Compare Means options to perform multiple comparisons of group means. All of these methods use pooled variance estimates for the means. Each Compare Means option adds comparison circles next to the plot and specific reports to the report window. For details about comparison circles, see "Using Comparison Circles" on page 148.

Option	Description	Reference	Nonparametric Menu Option
Each Pair, Student's t	Computes individual pairwise comparisons using Student's *t*-tests. If you make many pairwise tests, there is no protection across the inferences. Therefore, the alpha-size (Type I error rate) across the hypothesis tests is higher than that for individual tests.	See "Each Pair, Student's t" on page 149.	**Nonparametric > Nonparametric Multiple Comparisons > Wilcoxon Each Pair**
All Pairs, Tukey HSD	Shows a test that is sized for all differences among the means. This is the *Tukey* or *Tukey-Kramer* HSD (honestly significant difference) test. (Tukey 1953, Kramer 1956). This test is an exact alpha-level test if the sample sizes are the same, and conservative if the sample sizes are different (Hayter 1984).	See "All Pairs, Tukey HSD" on page 149.	**Nonparametric > Nonparametric Multiple Comparisons > Steel-Dwass All Pairs**
With Best, Hsu MCB	Tests whether the means are less than the unknown maximum or greater than the unknown minimum. This is the Hsu MCB test (Hsu 1981).	See "With Best, Hsu MCB" on page 149.	none
With Control, Dunnett's	Tests whether the means are different from the mean of a control group. This is Dunnett's test (Dunnett 1955).	See "With Control, Dunnett's" on page 150.	**Nonparametric > Nonparametric Multiple Comparisons > Steel With Control**

Note: If you have specified a **Block** column, then the multiple comparison methods are performed on data that has been adjusted for the Block means.

Related Information

- "Example Contrasting All of the Compare Means Tests" on page 169

Using Comparison Circles

Each multiple comparison test begins with a *comparison circles* plot, which is a visual representation of group mean comparisons. Figure 5.11 shows the comparison circles for the All Pairs, Tukey HSD method. Other comparison tests lengthen or shorten the radii of the circles.

Figure 5.11 Visual Comparison of Group Means

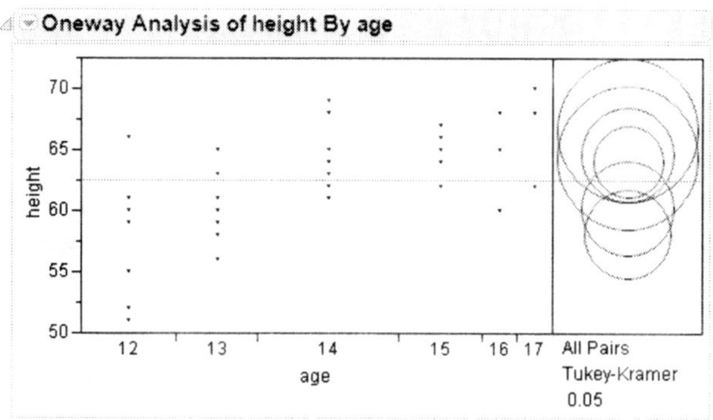

Compare each pair of group means visually by examining the intersection of the comparison circles. The outside angle of intersection tells you whether the group means are significantly different. See Figure 5.12.

- Circles for means that are significantly different either do not intersect, or intersect slightly, so that the outside angle of intersection is less than 90 degrees.
- If the circles intersect by an angle of more than 90 degrees, or if they are nested, the means are not significantly different.

Figure 5.12 Angles of Intersection and Significance

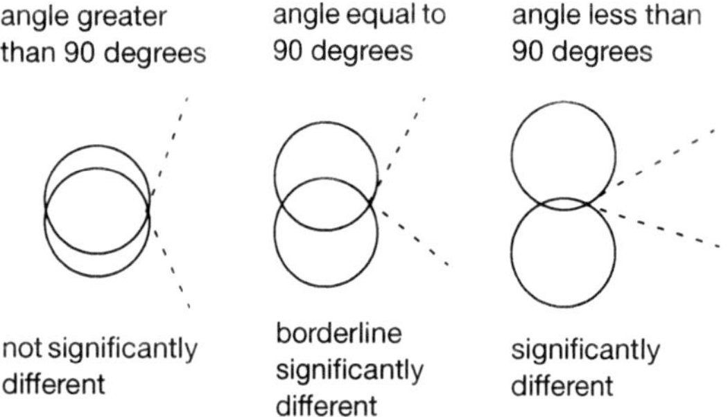

If the intersection angle is close to 90 degrees, you can verify whether the means are significantly different by clicking on the comparison circle to select it. See Figure 5.13. To deselect circles, click in the white space outside the circles.

Figure 5.13 Highlighting Comparison Circles

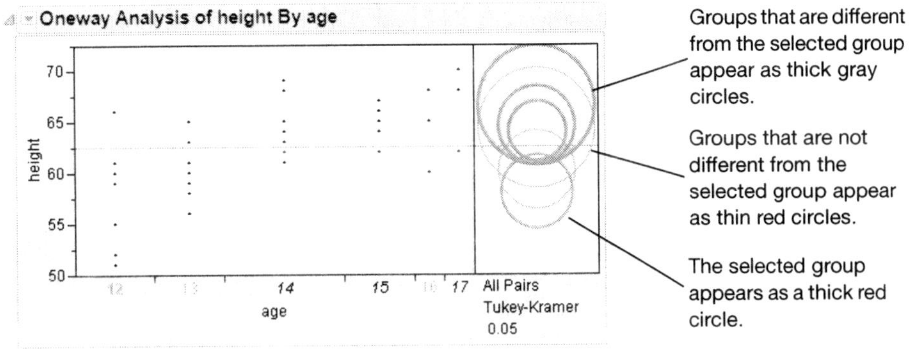
Oneway Analysis of height By age
70
65
60
55
50
height
12
13
14
15
16
17
age
All Pairs
Tukey-Kramer
0.05
Groups that are different from the selected group appear as thick gray circles.
Groups that are not different from the selected group appear as thin red circles.
The selected group appears as a thick red circle.

Related Information

- "Statistical Details for Comparison Circles" on page 179

Each Pair, Student's t

The **Each Pair, Student's t** test shows the Student's *t*-test for each pair of group levels and tests only individual comparisons.

Related Information

- "Example of the Each Pair, Student's t Test" on page 163

All Pairs, Tukey HSD

The **All Pairs, Tukey HSD** test (also called Tukey-Kramer) protects the significance tests of all combinations of pairs, and the HSD intervals become greater than the Student's *t* pairwise LSDs. Graphically, the comparison circles become larger and differences are less significant.

Related Information

- "Example of the All Pairs, Tukey HSD Test" on page 165

With Best, Hsu MCB

The **With Best, Hsu MCB** test determines whether a mean can be rejected as the maximum or minimum among the means. The Hsu's MCB test is given for each level versus the maximum, and each level versus the minimum.

The quantile that scales the LSD is not the same for each level. Therefore, the comparison circles technique does not work as well, because the circles must be rescaled according to the level being tested (unless the sample sizes are equal). The comparison circles plot uses the largest quantiles value shown to make the circles. Use the p-values of the tests to obtain precise assessments of significant differences.

Note the following:

- If a mean has means significantly separated *above* it, it is not regarded as the maximum.
- If a mean has means significantly separated *below* it, it is not regarded as the minimum.
- If a mean is significantly separated *above* all other means, it is regarded as the maximum.
- If a mean is significantly separated *below* all other means, it is regarded as the minimum.

Note: Means that are not regarded as the maximum or the minimum by MCB are also the means that are not contained in the selected subset of Gupta (1965) of potential maximums or minimum means.

For the *maximum* report, a column shows the row mean minus the column mean minus the LSD. If a value is positive, the row mean is significantly higher than the mean for the column, and the mean for the column is not the maximum.

For the *minimum* report, a column shows the row mean minus the column mean plus the LSD. If a value is negative, the row mean is significantly less than the mean for the column, and the mean for the column is not the minimum.

Related Information

- "Example of the With Best, Hsu MCB Test" on page 167

With Control, Dunnett's

The **With Control, Dunnett's** test compares a set of means against the mean of a control group. The LSDs that it produces are between the Student's *t* and Tukey-Kramer LSDs, because they are sized to guard against an intermediate number of comparisons.

In the Dunnett's report, the $|d|$ quantile appears, and can be used in a manner similar to a Student's *t*-statistic. The LSD threshold matrix shows the absolute value of the difference minus the LSD. If a value is positive, its mean is more than the LSD apart from the control group mean and is therefore significantly different.

Related Information

- "Example of the With Control, Dunnett's Test" on page 168

Compare Means Options

The Means Comparisons reports for all four tests contain a red triangle menu with customization options.

Table 5.12 Descriptions of Compare Means Options

Difference Matrix	Shows a table of all differences of means.
Confidence Quantile	Shows the *t*-value or other corresponding quantiles used for confidence intervals.
LSD Threshold Matrix	Shows a matrix showing if a difference exceeds the least significant difference for all comparisons. **Note:** For Hsu's MCB and Dunnett's test, only Difference Matrix, Confidence Quantile, and LSD Threshold Matrix are applicable.
Connecting Letters Report	Shows the traditional letter-coded report where means that are not sharing a letter are significantly different.
Ordered Differences Report	Shows all the positive-side differences with their confidence interval in sorted order. For the Student's t and Tukey-Kramer comparisons, an Ordered Difference report appears below the text reports. This report shows the ranked differences, from highest to lowest, with a confidence interval band overlaid on the plot. Confidence intervals that do not fully contain their corresponding bar are significantly different from each other.
Detailed Comparisons Report	Shows a detailed report for each comparison. Each section shows the difference between the levels, standard error and confidence intervals, *t*-ratios, *p*-values, and degrees of freedom. A plot illustrating the comparison appears on the right of each report. This option is not available for All Pairs, Tukey's HSD, and Nonparametric Multiple Comparisons.

Nonparametric

Nonparametric tests are useful for testing whether group means or medians are located the same across groups. However, the usual analysis of variance assumption of normality is not made. Nonparametric tests use functions of the response ranks, called rank scores (Hajek 1969).

Note: If you have specified a Block column, then the nonparametric tests are performed on data that has been adjusted for the Block means.

Table 5.13 Descriptions of Nonparametric Tests

Wilcoxon Test[a]	Performs the test based on Wilcoxon rank scores. The Wilcoxon rank scores are the simple ranks of the data. The Wilcoxon test is the most powerful rank test for errors with logistic distributions. If the factor has two or more levels, the Kruskal-Wallis test is performed. The Wilcoxon test is also called the Mann-Whitney test.
Median Test	Performs the test based on Median rank scores. The Median rank scores are either 1 or 0, depending on whether a rank is above or below the median rank. The Median test is the most powerful rank test for errors with double-exponential distributions.
van der Waerden Test	Performs the test based on Van der Waerden rank scores. The Van der Waerden rank scores are the ranks of the data divided by one plus the number of observations transformed to a normal score by applying the inverse of the normal distribution function. The Van der Waerden test is the most powerful rank test for errors with normal distributions.
Kolmogorov-Smirnov Test	Performs the test based on the empirical distribution function, which tests whether the distribution of the response is the same across the groups. Both an approximate and an exact test are given. This test is available only when the X factor has two levels.
Exact Test	Provides options for performing exact versions of the Wilcoxon, Median, van der Waerden, and Kolmogorov-Smirnov tests. These options are available only when the X factor has two levels, and after the approximate test is requested.

Table 5.13 Descriptions of Nonparametric Tests *(Continued)*

Nonparametric Multiple Comparisons	Provides several options for performing nonparametric multiple comparisons. These tests are based on ranks, and control for the overall alpha level, except for the Wilcoxon Each Pair test. The following tests are available: **Wilcoxon Each Pair** performs the Wilcoxon test on each pair, and does not control for the overall alpha level. This is the nonparametric version of the **Each Pair, Student's t** option found on the Compare Means menu. **Steel-Dwass All Pairs** performs the Steel-Dwass test on each pair. This is the nonparametric version of the **All Pairs, Tukey HSD** option found on the Compare Means menu. **Steel With Control** compares each level to a control level. This is the nonparametric version of the **With Control, Dunnett's** option found on the Compare Means menu. **Dunn With Control for Joint Ranks** compares each level to a control level, similar to the Steel With Control option. The Dunn method is different in that it computes ranks on all the data, not just the pair being compared. **Dunn All Pairs for Joint Ranks** performs a comparison of each pair, similar to the Steel-Dwass All Pairs option. The Dunn method is different in that it computes ranks on all the data, not just the pair being compared. See Dunn (1964) and Hsu (1996).

a. For the Wilcoxon, Median, and Van der Waerden tests, if the X factor has more than two levels, a chi-square approximation to the one-way test is performed. If the X factor has two levels, a normal approximation to the two-sample test is performed, in addition to the chi-square approximation to the one-way test.

Nonparametric Report Descriptions

All nonparametric reports are described in the following tables:

- Table 5.14 on page 153
- Table 5.15 on page 154
- Table 5.16 on page 155

Table 5.14 Descriptions of the Wilcoxon, Median, and Van der Waerden Tests

Level	Lists the factor levels occurring in the data.

Table 5.14 Descriptions of the Wilcoxon, Median, and Van der Waerden Tests *(Continued)*

Count	Records the frequencies of each level.
Score Sum	Records the sum of the rank score for each level.
Expected Score	Records the expected score under the null hypothesis that there is no difference among class levels.
Score Mean	Records the mean rank score for each level.
(Mean-Mean0)/Std0	Records the standardized score. Mean0 is the mean score expected under the null hypothesis. Std0 is the standard deviation of the score sum expected under the null hypothesis. The null hypothesis is that the group means or medians are in the same location across groups.
ChiSquare	Gives the values of the chi-square test statistic.
DF	Gives the degrees of freedom for the test.
Prob>ChiSq	Gives the p-value for the test.
S	Gives the sum of the rank scores. This is reported only when the X factor has two levels.
Z	Gives the test statistic for the normal approximation test. This is reported only when the X factor has two levels.
Prob>|Z|	Gives the p-value for the normal approximation test. This is reported only when the X factor has two levels.
Prob≥S	Gives a one-sided p-value for the test. This is reported only when the X factor has two levels, and the exact version of the test is requested. JMP PRO Exact tests are available only in JMP Pro.
Prob≥|S-Mean|	Gives a two-sided p-value for the test. This is reported only when the X factor has two levels, and the exact version of the test is requested. JMP PRO Exact tests are available only in JMP Pro.

Table 5.15 Description of the Kolmogorov-Smirnov Test

Level	Lists the factor levels occurring in the data.
Count	Records the frequencies of each level.
EDF at Maximum	Lists the value at which the maximum deviation from the empirical distribution function (EDF) of each level and the overall EDF occurs.
Deviation from Mean at Maximum	Lists the value of the EDF of a sample at the maximum deviation from the mean of the EDF for the overall sample.

Table 5.15 Description of the Kolmogorov-Smirnov Test *(Continued)*

KS	A Kolmogorov-Smirnov statistic.
KSa	An asymptotic Kolmogorov-Smirnov statistic.
D=max|F1-F2|	Lists the maximum absolute deviation between the EDF of two class levels.
Prob > D	Lists the p-value for the test. In other words, the probability that D is greater than the observed value d, under the null hypothesis of no difference between class levels or samples.
D+=max(F1-F2)	Lists a one-sided test statistic that max deviation between the EDF of two class levels is positive.
Prob > D+	Lists the probability that D+ is greater than the observed value d+, under the null hypothesis of no difference between the two class levels.
D-=max(F2-F1)	Lists a one-sided test statistic that max deviation between the EDF of two class levels is negative.
Prob > D-	Lists the probability that D- is greater than the observed value for d-.

Table 5.16 Descriptions of the Nonparametric Multiple Comparisons Tests

q*	Gives the quantile value used in the confidence intervals.
Alpha	Gives the alpha level used in the confidence intervals
Level	Gives the pair used in the current comparison
Score Mean Diff	Gives the difference of the score means.
Std Err Dif	Gives the standard error of the difference between the score means.
Z	Gives the standardized test statistic, which has an asymptotic standard normal deviation under the null hypothesis.
p-Value	Gives the asymptotic two-sided p-value for Z.
Hodges-Lehmann	Gives the Hodges-Lehmann estimator of location shift. It is the median of all paired differences between observations in the two samples.
Lower CL	Gives the lower confidence limit for the Hodges-Lehmann statistic.
Upper CL	Gives the upper confidence limit for the Hodges-Lehmann statistic.

Unequal Variances

When the variances across groups are not equal, the usual analysis of variance assumptions are not satisfied and the ANOVA *F* test is not valid. JMP provides four tests for equality of group variances and an ANOVA that is valid when the group sample variances are unequal. The concept behind the first three tests of equal variances is to perform an analysis of variance on a new response variable constructed to measure the spread in each group. The fourth test is Bartlett's test, which is similar to the likelihood-ratio test under normal distributions.

Note: Another method to test for unequal variances is ANOMV. See "Analysis of Means Methods" on page 144.

Table 5.17 Descriptions of Tests for Equal Variances

O'Brien	Constructs a dependent variable so that the group means of the new variable equal the group sample variances of the original response. An ANOVA on the O'Brien variable is actually an ANOVA on the group sample variances (O'Brien 1979, Olejnik, and Algina 1987).
Brown-Forsythe	Shows the *F* test from an ANOVA where the response is the absolute value of the difference of each observation and the group median (Brown and Forsythe 1974a).
Levene	Shows the *F* test from an ANOVA where the response is the absolute value of the difference of each observation and the group mean (Levene 1960). The spread is measured as $z_{ij} = \lvert y_{ij} - \bar{y}_i \rvert$ (as opposed to the SAS default $z_{ij}^2 = (y_{ij} - \bar{y}_i)^2$).
Bartlett	Compares the weighted arithmetic average of the sample variances to the weighted geometric average of the sample variances. The geometric average is always less than or equal to the arithmetic average with equality holding only when all sample variances are equal. The more variation there is among the group variances, the more these two averages differ. A function of these two averages is created, which approximates a χ^2-distribution (or, in fact, an *F* distribution under a certain formulation). Large values correspond to large values of the arithmetic or geometric ratio, and therefore to widely varying group variances. Dividing the Bartlett Chi-square test statistic by the degrees of freedom gives the *F* value shown in the table. Bartlett's test is not very robust to violations of the normality assumption (Bartlett and Kendall 1946).

If there are only two groups tested, then a standard *F* test for unequal variances is also performed. The *F* test is the ratio of the larger to the smaller variance estimate. The *p*-value from the *F* distribution is doubled to make it a two-sided test.

Note: If you have specified a **Block** column, then the variance tests are performed on data after it has been adjusted for the Block means.

Tests That the Variances Are Equal Report

The Tests That the Variances Are Equal report shows the differences between group means to the grand mean and to the median, and gives a summary of testing procedures.

If the equal variances test reveals that the group variances are significantly different, use Welch's test instead of the regular ANOVA test. The Welch statistic is based on the usual ANOVA *F* test. However, the means are weighted by the reciprocal of the group mean variances (Welch 1951; Brown and Forsythe 1974b; Asiribo, Osebekwin, and Gurland 1990). If there are only two levels, the Welch ANOVA is equivalent to an unequal variance *t*-test.

Table 5.18 Description of the Tests That the Variances Are Equal Report

Level	Lists the factor levels occurring in the data.
Count	Records the frequencies of each level.
Std Dev	Records the standard deviations of the response for each factor level. The standard deviations are equal to the means of the O'Brien variable. If a level occurs only once in the data, no standard deviation is calculated.
MeanAbsDif to Mean	Records the mean absolute difference of the response and group mean. The mean absolute differences are equal to the group means of the Levene variable.
MeanAbsDif to Median	Records the absolute difference of the response and group median. The mean absolute differences are equal to the group means of the Brown-Forsythe variable.
Test	Lists the names of the tests performed.
F Ratio	Records a calculated *F* statistic for each test. See "Statistical Details for the Tests That the Variances Are Equal Report" on page 181.
DFNum	Records the degrees of freedom in the numerator for each test. If a factor has *k* levels, the numerator has k - 1 degrees of freedom. Levels occurring only once in the data are not used in calculating test statistics for O'Brien, Brown-Forsythe, or Levene. The numerator degrees of freedom in this situation is the number of levels used in calculations minus one.
DFDen	Records the degrees of freedom used in the denominator for each test. For O'Brien, Brown-Forsythe, and Levene, a degree of freedom is subtracted for each factor level used in calculating the test statistic. One more degree of freedom is subtracted for the overall mean. If a factor has k levels, the denominator degrees of freedom is $n - k - 1$.
p-Value	Probability of obtaining, by chance alone, an *F* value larger than the one calculated if in reality the variances are equal across all levels.

Table 5.19 Description of the Welch's Test Report

F Ratio	Shows the F test statistic for the equal variance test. See "Statistical Details for the Tests That the Variances Are Equal Report" on page 181.
DFNum	Records the degrees of freedom in the numerator of the test. If a factor has *k* levels, the numerator has *k* - 1 degrees of freedom. Levels occurring only once in the data are not used in calculating the Welch ANOVA. The numerator degrees of freedom in this situation is the number of levels used in calculations minus one.
DFDen	Records the degrees of freedom in the denominator of the test. See "Statistical Details for the Tests That the Variances Are Equal Report" on page 181.
Prob>F	Probability of obtaining, by chance alone, an *F* value larger than the one calculated if in reality the means are equal across all levels. Observed significance probabilities of 0.05 or less are considered evidence of unequal means across the levels.
t Test	Shows the relationship between the F ratio and the t Test. Calculated as the square root of the F ratio. Appears only if the *X* factor has two levels.

Related Information

- "Example of the Unequal Variances Option" on page 170
- "Statistical Details for the Tests That the Variances Are Equal Report" on page 181

Equivalence Test

Equivalence tests assess whether there is a practical difference in means. You must pick a threshold difference for which smaller differences are considered practically equivalent. The most straightforward test to construct uses two one-sided *t*-tests from both sides of the difference interval. If both tests reject (or conclude that the difference in the means differs significantly from the threshold), then the groups are practically equivalent. The **Equivalence Test** option uses the Two One-Sided Tests (TOST) approach.

Related Information

- "Example of an Equivalence Test" on page 171

Power

The **Power** option calculates statistical power and other details about a given hypothesis test.

- *LSV* (the Least Significant Value) is the value of some parameter or function of parameters that would produce a certain *p*-value alpha. Said another way, you want to know how small an effect would be declared significant at some *p*-value alpha. The LSV provides a measuring stick for significance on the scale of the parameter, rather than on a probability scale. It shows how sensitive the design and data are.
- *LSN* (the Least Significant Number) is the total number of observations that would produce a specified *p*-value alpha given that the data has the same form. The LSN is defined as the number of observations needed to reduce the variance of the estimates enough to achieve a significant result with the given values of alpha, sigma, and delta (the significance level, the standard deviation of the error, and the effect size). If you need more data to achieve significance, the LSN helps tell you how many more. The LSN is the total number of observations that yields approximately 50% power.
- *Power* is the probability of getting significance (p-value < alpha) when a real difference exists between groups. It is a function of the sample size, the effect size, the standard deviation of the error, and the significance level. The power tells you how likely your experiment is to detect a difference (effect size), at a given alpha level.

Note: When there are only two groups in a one-way layout, the LSV computed by the power facility is the same as the least significant difference (LSD) shown in the multiple-comparison tables.

Power Details Window and Reports

The Power Details window and reports are the same as those in the general fitting platform launched by the Fit Model platform. For more details about power calculation, see the *Modeling and Multivariate Methods* book.

For each of four columns Alpha, Sigma, Delta, and Number, fill in a single value, two values, or the start, stop, and increment for a sequence of values. See Figure 5.26. Power calculations are performed on all possible combinations of the values that you specify.

Table 5.20 Description of the Power Details Report

Alpha (α)	Significance level, between 0 and 1 (usually 0.05, 0.01, or 0.10). Initially, a value of 0.05 shows.
Sigma (σ)	Standard error of the residual error in the model. Initially, RMSE, the estimate from the square root of the mean square error is supplied here.
Delta (δ)	Raw effect size. For details about effect size computations, see the *Modeling and Multivariate Methods* book. The first position is initially set to the square root of the sums of squares for the hypothesis divided by *n*; that is, $\delta = \sqrt{SS/n}$.

Table 5.20 Description of the Power Details Report *(Continued)*

Number (n)	Total sample size across all groups. Initially, the actual sample size is put in the first position.
Solve for Power	Solves for the power (the probability of a significant result) as a function of all four values: α, σ, δ, and n.
Solve for Least Significant Number	Solves for the number of observations needed to achieve approximately 50% power given α, σ, and δ.
Solve for Least Significant Value	Solves for the value of the parameter or linear test that produces a *p*-value of α. This is a function of α, σ, n, and the standard error of the estimate. This feature is available only when the X factor has two levels and is usually used for individual parameters.
Adjusted Power and Confidence Interval	When you look at power retrospectively, you use estimates of the standard error and the test parameters. • Adjusted power is the power calculated from a more unbiased estimate of the non-centrality parameter. • The confidence interval for the adjusted power is based on the confidence interval for the non-centrality estimate. Adjusted power and confidence limits are computed only for the original **Delta**, because that is where the random variation is.

Related Information

- "Example of the Power Option" on page 172
- "Statistical Details for Power" on page 180

Normal Quantile Plot

You can create two types of normal quantile plots:

- **Plot Actual by Quantile** creates a plot of the response values versus the normal quantile values. The quantiles are computed and plotted separately for each level of the X variable.
- **Plot Quantile by Actual** creates a plot of the normal quantile values versus the response values. The quantiles are computed and plotted separately for each level of the X variable.

The **Line of Fit** option shows or hides the lines of fit on the quantile plots.

Related Information

- "Example of a Normal Quantile Plot" on page 174

CDF Plot

A CDF plot shows the cumulative distribution function for all of the groups in the Oneway report. CDF plots are useful if you want to compare the distributions of the response across levels of the X factor.

Related Information

- "Example of a CDF Plot" on page 175

Densities

The **Densities** options provide several ways to compare the distribution and composition of the response across the levels of the X factor. There are three density options:

- **Compare Densities** shows a smooth curve estimating the density of each group. The smooth curve is the kernel density estimate for each group.
- **Composition of Densities** shows the summed densities, weighted by each group's counts. At each X value, the Composition of Densities plot shows how each group contributes to the total.
- **Proportion of Densities** shows the contribution of the group as a proportion of the total at each X level.

Related Information

- "Example of the Densities Options" on page 176

Matching Column

Use the **Matching Column** option to specify a matching (ID) variable for a matching model analysis. The **Matching Column** option addresses the case when the data in a one-way analysis come from matched (paired) data, such as when observations in different groups come from the same subject.

Note: A special case of matching leads to the paired *t*-test. The **Matched Pairs** platform handles this type of data, but the data must be organized with the pairs in different columns, not in different rows.

The **Matching Column** option performs two primary actions:

- It fits an additive model (using an iterative proportional fitting algorithm) that includes both the grouping variable (the X variable in the Fit Y by X analysis) and the matching variable that you select. The iterative proportional fitting algorithm makes a difference if there are hundreds of subjects, because the equivalent linear model would be very slow and would require huge memory resources.
- It draws lines between the points that match across the groups. If there are multiple observations with the same matching ID value, lines are drawn from the mean of the group of observations.

The **Matching Column** option automatically activates the **Matching Lines** option connecting the matching points. To turn the lines off, select **Display Options > Matching Lines**.

The Matching Fit report shows the effects with *F* tests. These are equivalent to the tests that you get with the Fit Model platform if you run two models, one with the interaction term and one without. If there are only two levels, then the *F* test is equivalent to the paired *t*-test.

Note: For details about the Fit Model platform, see the *Modeling and Multivariate Methods* book.

Related Information

- "Example of the Matching Column Option" on page 177

Additional Examples of the Oneway Platform

This section contains additional examples of selected options and reports in the Oneway platform.

Example of an Analysis of Means Chart

1. Open the Analgesics.jmp sample data table.
2. Select **Analyze > Fit Y by X**.
3. Select pain and click **Y, Response**.
4. Select drug and click **X, Factor**.
5. Click **OK**.
6. From the red triangle menu, select **Analysis of Means Methods > ANOM**.

Figure 5.14 Example of Analysis of Means Chart

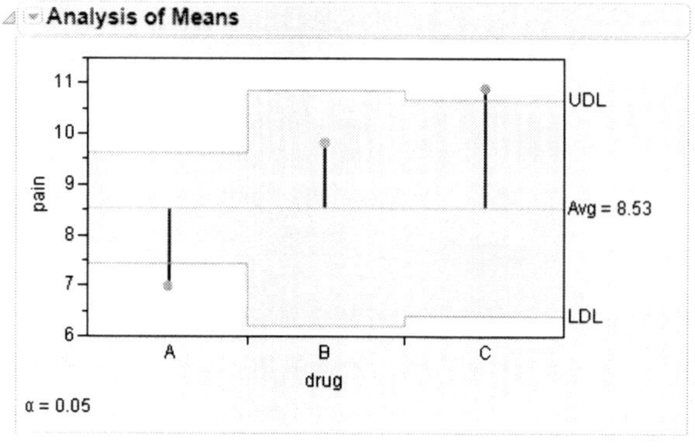

For the example in Figure 5.14, the means for drug A and C are statistically different from the overall mean. The drug A mean is lower and the drug C mean is higher. Note the decision limits for the drug types are not the same, due to different sample sizes.

Example of an Analysis of Means for Variances Chart

This example uses the Spring Data.jmp sample data table. Four different brands of springs were tested to see what weight is required to extend a spring 0.10 inches. Six springs of each brand were tested. The data was checked for normality, since the ANOMV test is not robust to non-normality. Examine the brands to determine whether the variability is significantly different between brands.

1. Open the Spring Data.jmp sample data table.
2. Select **Analyze > Fit Y by X.**
3. Select Weight and click **Y, Response.**
4. Select Brand and click **X, Factor.**
5. Click **OK.**
6. From the red triangle menu, select **Analysis of Means Methods > ANOM for Variances.**
7. From the red triangle menu next to Analysis of Means for Variances, select **Show Summary Report.**

Figure 5.15 Example of Analysis of Means for Variances Chart

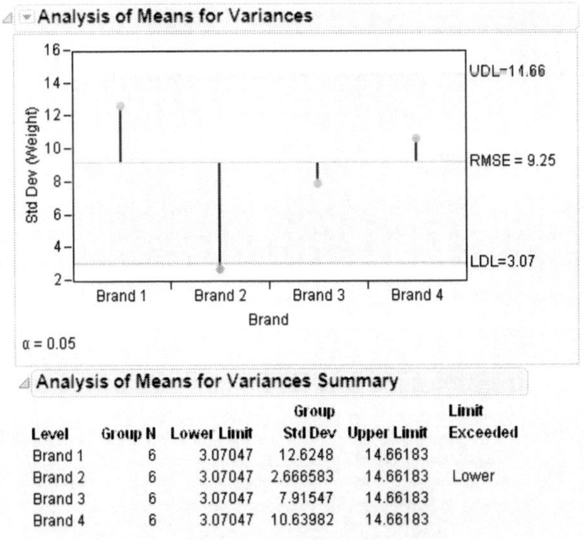

Analysis of Means for Variances Summary

Level	Group N	Lower Limit	Group Std Dev	Upper Limit	Limit Exceeded
Brand 1	6	3.07047	12.6248	14.66183	
Brand 2	6	3.07047	2.666583	14.66183	Lower
Brand 3	6	3.07047	7.91547	14.66183	
Brand 4	6	3.07047	10.63982	14.66183	

From Figure 5.15, notice that the standard deviation for Brand 2 exceeds the lower decision limit. Therefore, Brand 2 has significantly lower variance than the other brands.

Example of the Each Pair, Student's t Test

This example uses the Big Class.jmp sample data table. It shows a one-way layout of weight by age, and shows the group comparison using comparison circles that illustrate all possible *t*-tests.

1. Open the Big Class.jmp sample data table.

2. Select **Analyze > Fit Y by X**.
3. Select weight and click **Y, Response.**
4. Select age and click **X, Factor**.
5. Click **OK.**
6. From the red triangle menu, select **Compare Means > Each Pair, Student's t.**

Figure 5.16 Example of Each Pair, Student's t Comparison Circles

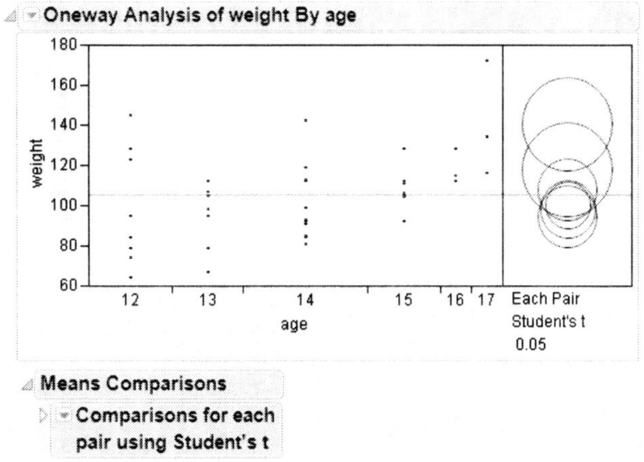

The means comparison method can be thought of as seeing if the actual difference in the means is greater than the difference that would be significant. This difference is called the LSD (least significant difference). The LSD term is used for Student's *t* intervals and in context with intervals for other tests. In the comparison circles graph, the distance between the circles' centers represent the actual difference. The LSD is what the distance would be if the circles intersected at right angles.

Figure 5.17 Example of Means Comparisons Report for Each Pair, Student's *t*

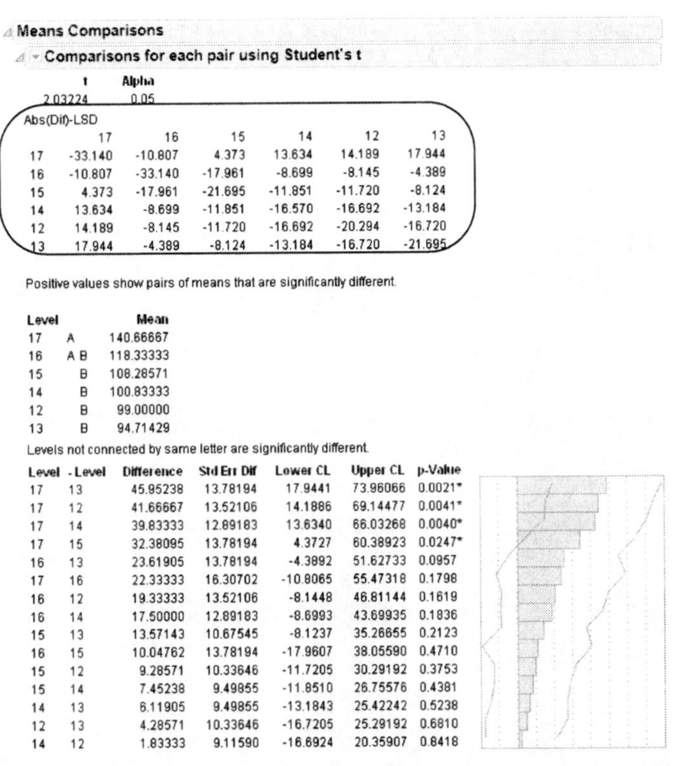

Means Comparisons

Comparisons for each pair using Student's t

t	Alpha
2.03224	0.05

Abs(Dif)-LSD

	17	16	15	14	12	13
17	-33.140	-10.807	4.373	13.634	14.189	17.944
16	-10.807	-33.140	-17.961	-8.699	-8.145	-4.389
15	4.373	-17.961	-21.695	-11.851	-11.720	-8.124
14	13.634	-8.699	-11.851	-16.570	-16.692	-13.184
12	14.189	-8.145	-11.720	-16.692	-20.294	-16.720
13	17.944	-4.389	-8.124	-13.184	-16.720	-21.695

Positive values show pairs of means that are significantly different.

Level		Mean
17	A	140.66667
16	A B	118.33333
15	B	108.28571
14	B	100.83333
12	B	99.00000
13	B	94.71429

Levels not connected by same letter are significantly different.

Level	- Level	Difference	Std Err Dif	Lower CL	Upper CL	p-Value
17	13	45.95238	13.78194	17.9441	73.96066	0.0021*
17	12	41.66667	13.52106	14.1886	69.14477	0.0041*
17	14	39.83333	12.89183	13.6340	66.03268	0.0040*
17	15	32.38095	13.78194	4.3727	60.38923	0.0247*
16	13	23.61905	13.78194	-4.3892	51.62733	0.0957
17	16	22.33333	16.30702	-10.8065	55.47318	0.1798
16	12	19.33333	13.52106	-8.1448	46.81144	0.1619
16	14	17.50000	12.89183	-8.6993	43.69935	0.1836
15	13	13.57143	10.67545	-8.1237	35.26655	0.2123
16	15	10.04762	13.78194	-17.9607	38.05590	0.4710
15	12	9.28571	10.33646	-11.7205	30.29192	0.3753
15	14	7.45238	9.49855	-11.8510	26.75576	0.4381
14	13	6.11905	9.49855	-13.1843	25.42242	0.5238
12	13	4.28571	10.33646	-16.7205	25.29192	0.6810
14	12	1.83333	9.11590	-16.6924	20.35907	0.8418

In Figure 5.17, the LSD threshold table shows the difference between the absolute difference in the means and the LSD (least significant difference). If the values are positive, the difference in the two means is larger than the LSD, and the two groups are significantly different.

Example of the All Pairs, Tukey HSD Test

1. Open the Big Class.jmp sample data table.
2. Select **Analyze > Fit Y by X.**
3. Select weight and click **Y, Response.**
4. Select age and click **X, Factor.**
5. Click **OK.**
6. From the red triangle menu, select **Compare Means > All Pairs, Tukey HSD.**

Figure 5.18 Example of All Pairs, Tukey HSD Comparison Circles

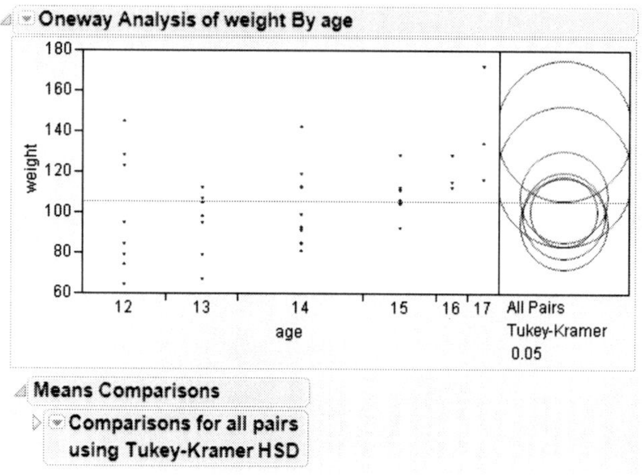

Figure 5.19 Example of Means Comparisons Report for All Pairs, Tukey HSD

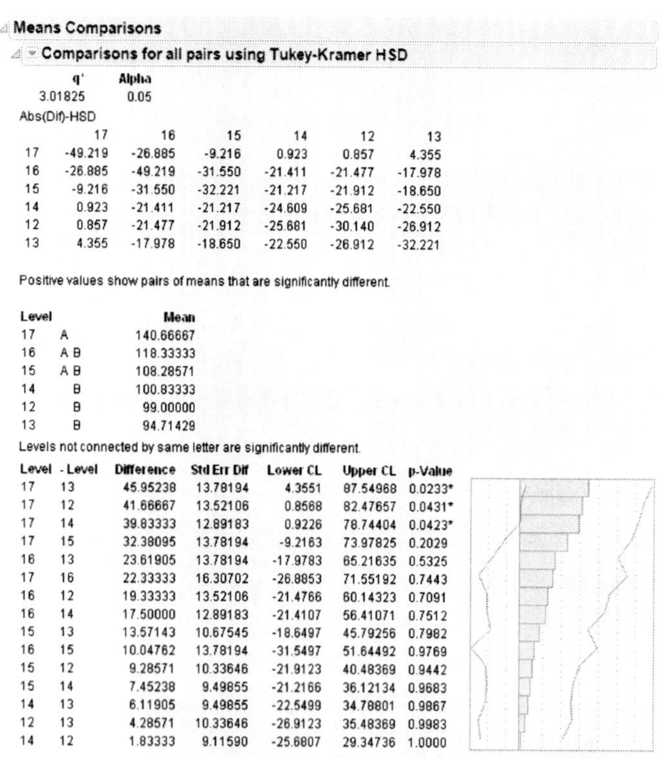

Means Comparisons

Comparisons for all pairs using Tukey-Kramer HSD

q*	Alpha
3.01825	0.05

Abs(Dif)-HSD

	17	16	15	14	12	13
17	-49.219	-26.885	-9.216	0.923	0.857	4.355
16	-26.885	-49.219	-31.550	-21.411	-21.477	-17.978
15	-9.216	-31.550	-32.221	-21.217	-21.912	-18.650
14	0.923	-21.411	-21.217	-24.609	-25.681	-22.550
12	0.857	-21.477	-21.912	-25.681	-30.140	-26.912
13	4.355	-17.978	-18.650	-22.550	-26.912	-32.221

Positive values show pairs of means that are significantly different.

Level			Mean
17	A		140.66667
16	A	B	118.33333
15	A	B	108.28571
14		B	100.83333
12		B	99.00000
13		B	94.71429

Levels not connected by same letter are significantly different.

Level	- Level	Difference	Std Err Dif	Lower CL	Upper CL	p-Value
17	13	45.95238	13.78194	4.3551	87.54968	0.0233*
17	12	41.66667	13.52106	0.8568	82.47657	0.0431*
17	14	39.83333	12.89183	0.9226	78.74404	0.0423*
17	15	32.38095	13.78194	-9.2163	73.97825	0.2029
16	13	23.61905	13.78194	-17.9783	65.21635	0.5325
17	16	22.33333	16.30702	-26.8853	71.55192	0.7443
16	12	19.33333	13.52106	-21.4766	60.14323	0.7091
16	14	17.50000	12.89183	-21.4107	56.41071	0.7512
15	13	13.57143	10.67545	-18.6497	45.79256	0.7982
16	15	10.04762	13.78194	-31.5497	51.64492	0.9769
15	12	9.28571	10.33646	-21.9123	40.48369	0.9442
15	14	7.45238	9.49855	-21.2166	36.12134	0.9683
14	13	6.11905	9.49855	-22.5499	34.78801	0.9867
12	13	4.28571	10.33646	-26.9123	35.48369	0.9983
14	12	1.83333	9.11590	-25.6807	29.34736	1.0000

In Figure 5.19, the Tukey-Kramer HSD Threshold matrix shows the actual absolute difference in the means minus the HSD, which is the difference that would be significant. Pairs with a positive value are significantly different. The q* (appearing above the HSD Threshold Matrix table) is the quantile that is used to scale the HSDs. It has a computational role comparable to a Student's *t*.

Example of the With Best, Hsu MCB Test

1. Open the Big Class.jmp sample data table.
2. Select **Analyze > Fit Y by X.**
3. Select weight and click **Y, Response.**
4. Select age and click **X, Factor**.
5. Click **OK**.
6. From the red triangle menu, select **Compare Means > With Best, Hsu MCB.**

Figure 5.20 Examples of With Best, Hsu MCB Comparison Circles

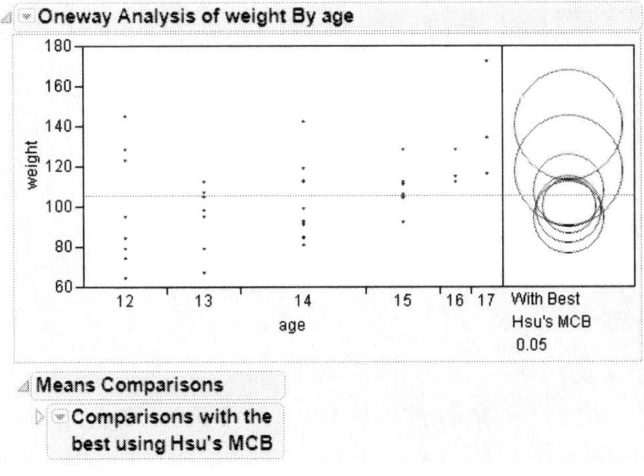

Figure 5.21 Example of Means Comparisons Report for With Best, Hsu MCB

Comparisons with the best using Hsu's MCB

Confidence Quantile

d	Alpha
2.23157	0.05
2.23157	
2.34056	
2.38159	
2.35278	
2.34056	

Comparison vs. Min/Max

Level	vs. Max p-Value	vs. Min p-Value
17	.	0.0039*
16	0.2162	0.1268
15	0.0441*	0.2978
14	0.0075*	0.6416
12	0.0080*	0.7071
13	0.0036*	0.9365

LSD Threshold Matrix

Mean[i]-Mean[j]-LSD

	17	16	15	14	12	13
17	-36.390	-14.057	0.124	9.130	9.855	13.695
16	-58.724	-36.390	-22.210	-13.203	-12.479	-8.638
15	-63.136	-40.803	-24.987	-15.169	-15.034	-11.415
14	-68.602	-46.269	-29.684	-19.418	-19.614	-16.113
12	-71.840	-49.507	-33.479	-23.544	-23.495	-19.907
13	-76.708	-54.374	-38.558	-28.741	-28.605	-24.987

If a column has any positive values, the mean is significantly less than the max.

Mean[i]-Mean[j]+LSD

	17	16	15	14	12	13
17	36.390	58.724	64.638	70.536	73.479	78.210
16	14.057	36.390	42.305	48.203	51.145	55.876
15	-1.626	20.708	24.987	30.074	33.605	38.558
14	-11.064	11.269	14.780	19.418	23.281	28.351
12	-11.493	10.840	14.907	19.877	23.495	28.479
13	-15.197	7.136	11.415	16.503	20.034	24.987

If a column has any negative values, the mean is significantly greater than the min.

The Comparison vs. Min/Max report compares each level to the minimum and the maximum level. For example, level 17 is the only level that is significantly different from the minimum level.

Example of the With Control, Dunnett's Test

1. Open the Big Class.jmp sample data table.
2. Select **Analyze > Fit Y by X**.
3. Select weight and click **Y, Response.**
4. Select age and click **X, Factor.**
5. Click **OK.**
6. From the red triangle menu, select **Compare Means > With Control, Dunnett's.**
7. Select the group to use as the control group. In this example, select age 12.

   Alternatively, click on a row to highlight it in the scatterplot before selecting the **Compare Means > With Control, Dunnett's** option. The test uses the selected row as the control group.
8. Click **OK.**

Figure 5.22 Example of With Control, Dunnett's Comparison Circles

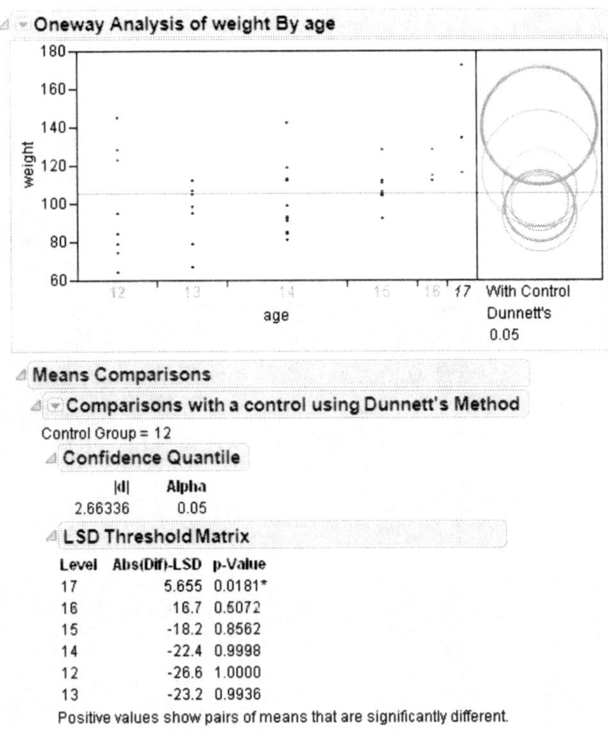

|d|	Alpha
2.66336	0.05

Level	Abs(Dif)-LSD	p-Value
17	5.655	0.0181*
16	16.7	0.5072
15	-18.2	0.8562
14	-22.4	0.9998
12	-26.6	1.0000
13	-23.2	0.9936

Using the comparison circles in Figure 5.22, you can conclude that level 17 is the only level that is significantly different from the control level of 12.

Example Contrasting All of the Compare Means Tests

1. Open the Big Class.jmp sample data table.
2. Select **Analyze > Fit Y by X.**
3. Select weight and click **Y, Response.**
4. Select age and click **X, Factor.**
5. Click **OK.**
6. From the red triangle menu, select each one of the **Compare Means** options.

Although the four methods all test differences between group means, different results can occur. Figure 5.23 shows the comparison circles for all four tests, with the age 17 group as the control group.

Figure 5.23 Comparison Circles for Four Multiple Comparison Tests

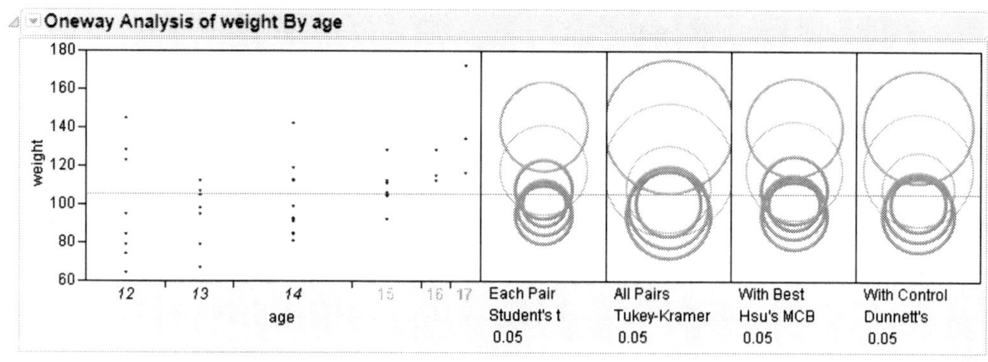

From Figure 5.23, notice that for the Student's t and Hsu methods, age group 15 (the third circle from the top) is significantly different from the control group and appears gray. But, for the Tukey and Dunnett method, age group 15 is not significantly different, and appears red.

Example of the Unequal Variances Option

Suppose you want to test whether two variances (males and females) are equal, instead of two means.

1. Open the Big Class.jmp sample data table.
2. Select **Analyze > Fit Y by X**.
3. Select height and click **Y, Response.**
4. Select sex and click **X, Factor**.
5. Click **OK**.
6. From the red triangle menu, select **Unequal Variances**.

Figure 5.24 Example of the Unequal Variances Report

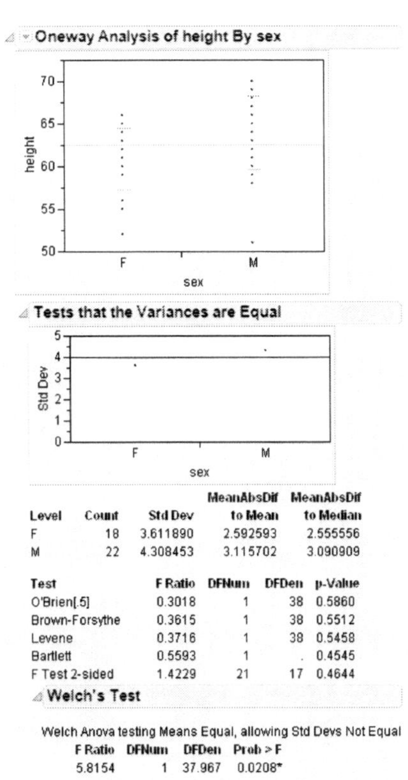

Since the p-value from the 2-sided F-Test is large, you can conclude that the variances are equal.

Example of an Equivalence Test

This example uses the Big Class.jmp sample data table. Examine if the difference in height between males and females is less than 6 inches.

1. Open the Big Class.jmp sample data table.
2. Select **Analyze > Fit Y by X.**
3. Select height and click **Y, Response.**
4. Select sex and click **X, Factor.**
5. Click **OK.**
6. From the red triangle menu, select **Equivalence Test.**
7. Type 6 as the difference considered practically zero.
8. Click **OK.**

Figure 5.25 Example of an Equivalence Test

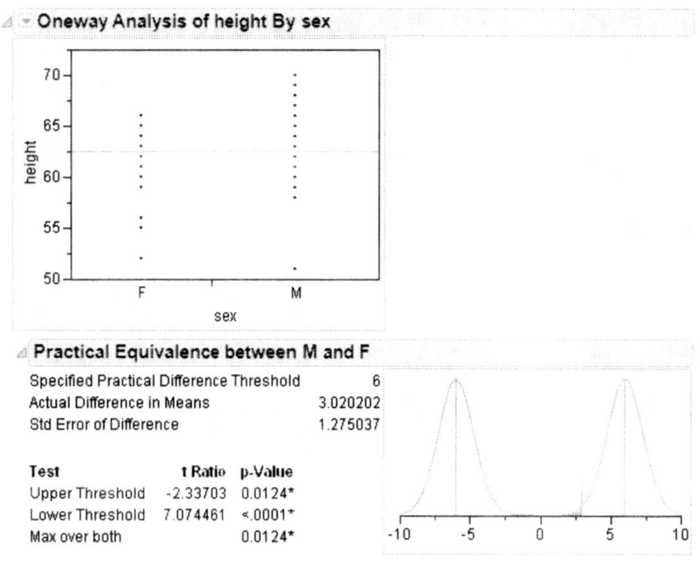

From Figure 5.25, notice the following:

- The Upper Threshold test compares the actual difference to 6.
- The Lower Threshold test compares the actual difference to -6.
- For both tests, the p-value is small. Therefore, you can conclude that the actual difference in means (3.02) is significantly different from 6 and -6. For your purposes, you can declare the means to be practically equivalent.

Example of the Power Option

1. Open the Typing Data.jmp sample data table.
2. Select **Analyze > Fit Y by X.**
3. Select speed and click **Y, Response.**
4. Select brand and click **X, Factor.**
5. Click **OK.**
6. From the red triangle menu, select **Power.**
7. Within the From row, type 2 for Delta (the third box) and type 11 for Number.
8. Within the To row, type 6 for Delta, and type 17 in the Number box.
9. Within the By row, type 2 for both Delta and Number.
10. Select the **Solve for Power** check box.

Figure 5.26 Example of the Power Details Window

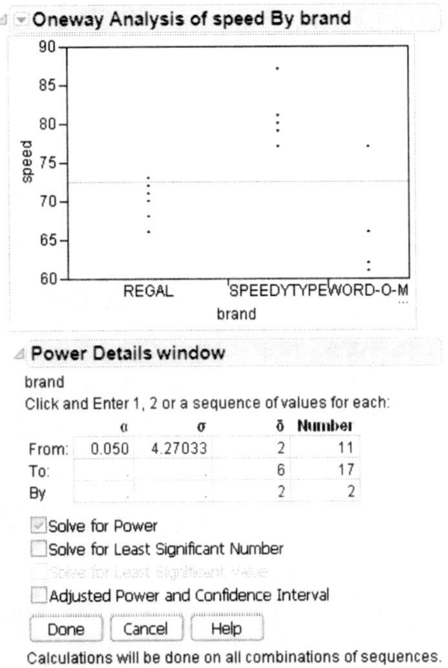

11. Click **Done**.

Power is computed for each combination of Delta and Number, and appears in the Power report.

To plot the Power values:

12. From the red triangle menu at the bottom of the report, select **Power Plot**.

Figure 5.27 Example of the Power Report

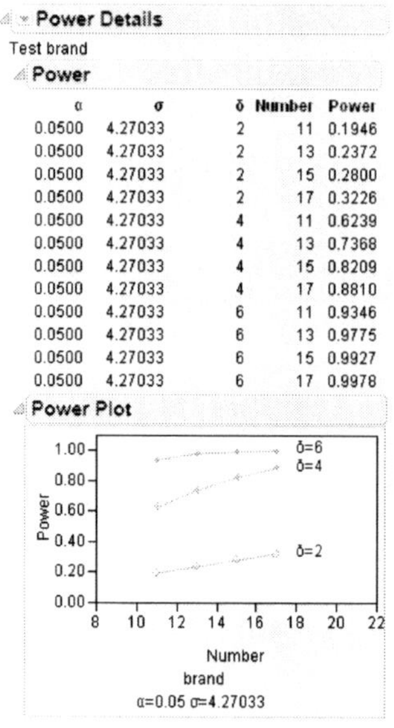
Power Details
Test brand
Power

α	σ	δ	Number	Power
0.0500	4.27033	2	11	0.1946
0.0500	4.27033	2	13	0.2372
0.0500	4.27033	2	15	0.2800
0.0500	4.27033	2	17	0.3226
0.0500	4.27033	4	11	0.6239
0.0500	4.27033	4	13	0.7368
0.0500	4.27033	4	15	0.8209
0.0500	4.27033	4	17	0.8810
0.0500	4.27033	6	11	0.9346
0.0500	4.27033	6	13	0.9775
0.0500	4.27033	6	15	0.9927
0.0500	4.27033	6	17	0.9978

13. You might need to click and drag vertically on the Power axis to see all of the data in the plot.

Power is plotted for each combination of Delta and Number. As you might expect, the power rises for larger Number (sample sizes) values and for larger Delta values (difference in means).

Example of a Normal Quantile Plot

1. Open the Big Class.jmp sample data table.
2. Select **Analyze > Fit Y by X**.
3. Select height and click **Y, Response.**
4. Select sex and click **X, Factor**.
5. Click **OK**.
6. From the red triangle menu, select **Normal Quantile Plot > Plot Actual by Quantile**.

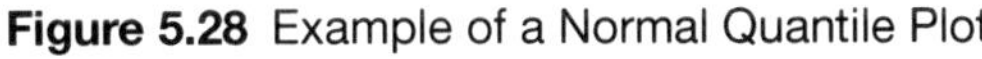
Figure 5.28 Example of a Normal Quantile Plot

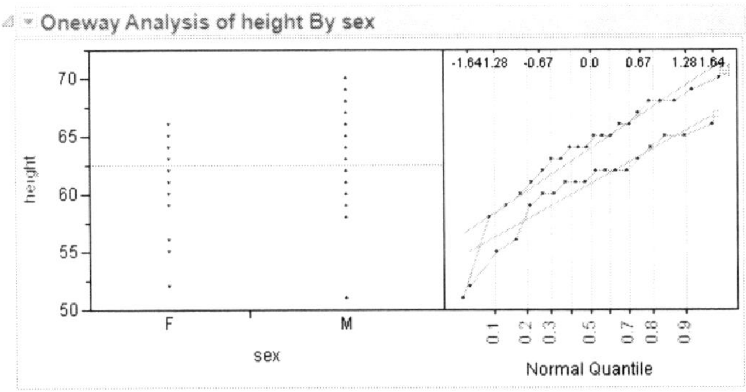

From Figure 5.28, notice the following:

- The Line of Fit appears by default.
- The data points track very closely to the line of fit, indicating a normal distribution.

Example of a CDF Plot

1. Open the Analgesics.jmp sample data table.
2. Select **Analyze > Fit Y by X**.
3. Select pain and click **Y, Response**.
4. Select drug and click **X, Factor**.
5. Click **OK**.
6. From the red triangle menu, select **CDF Plot**.

Figure 5.29 Example of a CDF Plot

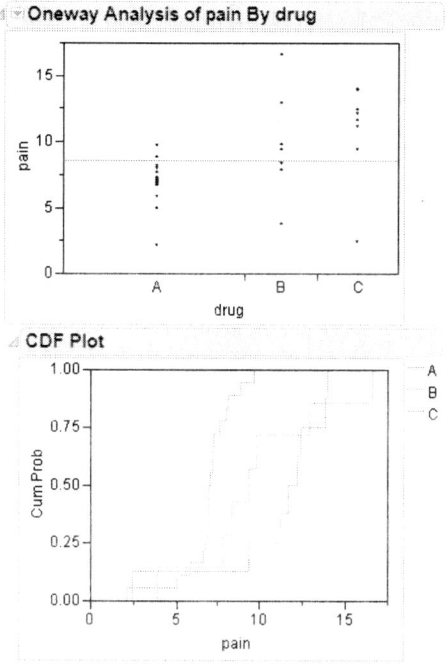

The levels of the X variables in the initial Oneway analysis appear in the CDF plot as different curves. The horizontal axis of the CDF plot uses the *y* value in the initial Oneway analysis.

Example of the Densities Options

1. Open the Big Class.jmp sample data table.
2. Select **Analyze > Fit Y by X**.
3. Select height and click **Y, Response.**
4. Select sex and click **X, Factor.**
5. Click **OK.**
6. From the red triangle menu, select **Densities > Compare Densities**, **Densities > Composition of Densities**, and **Densities > Proportion of Densities**.

Figure 5.30 Example of the Densities Options

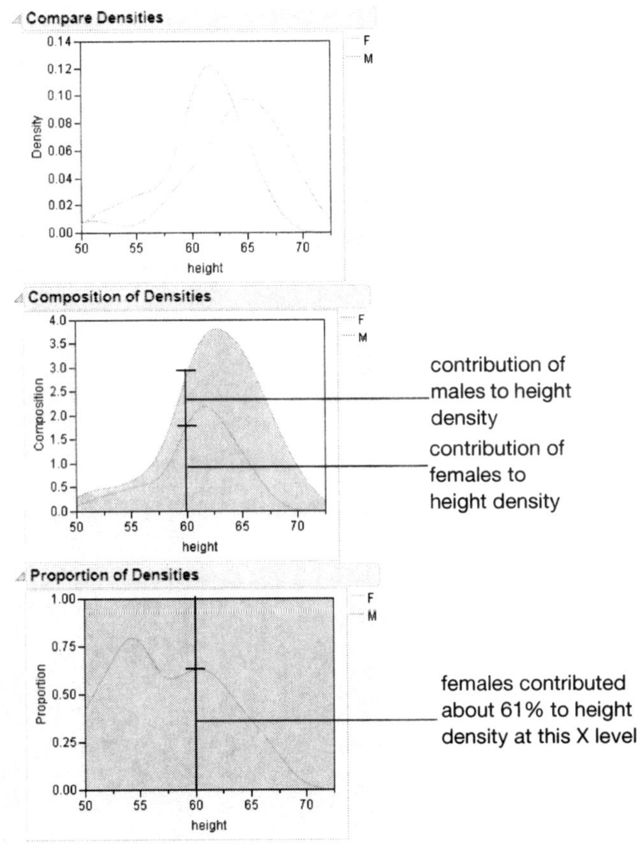

Example of the Matching Column Option

This example uses the Matching.jmp sample data table, which contains data on six animals and the miles that they travel during different seasons.

1. Open the Matching.jmp sample data table.
2. Select **Analyze > Fit Y by X.**
3. Select miles and click **Y, Response.**
4. Select season and click **X, Factor.**
5. Click **OK.**
6. From the red triangle menu, select **Matching Column.**
7. Select subject as the matching column.
8. Click **OK.**

Figure 5.31 Example of the Matching Column Report

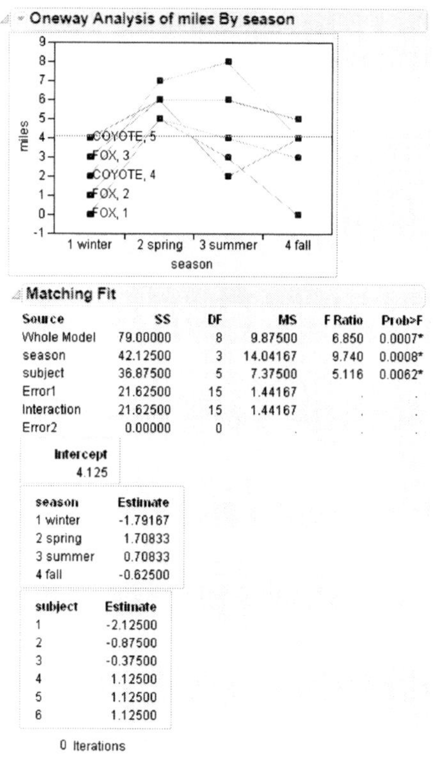

Matching Fit

Source	SS	DF	MS	F Ratio	Prob>F
Whole Model	79.00000	8	9.87500	6.850	0.0007*
season	42.12500	3	14.04167	9.740	0.0008*
subject	36.87500	5	7.37500	5.116	0.0062*
Error1	21.62500	15	1.44167	.	.
Interaction	21.62500	15	1.44167	.	.
Error2	0.00000	0	.	.	.

Intercept
4.125

season	Estimate
1 winter	-1.79167
2 spring	1.70833
3 summer	0.70833
4 fall	-0.62500

subject	Estimate
1	-2.12500
2	-0.87500
3	-0.37500
4	1.12500
5	1.12500
6	1.12500

0 Iterations

The plot graphs the miles traveled by season, with subject as the matching variable. The labels next to the first measurement for each subject on the graph are determined by the species and subject variables.

The Matching Fit report shows the season and subject effects with F tests. These are equivalent to the tests that you get with the Fit Model platform if you run two models, one with the interaction term and one without. If there are only two levels, then the F test is equivalent to the paired t-test.

Note: For details about the Fit Model platform, see the *Modeling and Multivariate Methods* book.

Statistical Details for the Oneway Platform

The following sections provide statistical details for selected options and reports.

Statistical Details for Comparison Circles

One approach to comparing two means is to determine whether their actual difference is greater than their *least significant difference* (LSD). This least significant difference is a Student's *t*-statistic multiplied by the standard error of the difference of the two means and is written as follows:

$$\text{LSD} = t_{\alpha/2}\text{std}(\hat{\mu}_1 - \hat{\mu}_2)$$

The standard error of the difference of two independent means is calculated from the following relationship:

$$[\text{std}(\hat{\mu}_1 - \hat{\mu}_2)]^2 = [\text{std}(\hat{\mu}_1)]^2 + [\text{std}(\hat{\mu}_2)]^2$$

When the means are un correlated, these quantities have the following relationship:

$$\text{LSD}^2 = \left[t_{\alpha/2}\text{std}((\hat{\mu}_1 - \hat{\mu}_2))\right]^2 = \left[t_{\alpha/2}\text{std}(\hat{\mu}_1)\right]^2 + \left[t_{\alpha/2}\text{std}\hat{\mu}_2\right]^2$$

These squared values form a Pythagorean relationship, illustrated graphically by the right triangle shown in Figure 5.32.

Figure 5.32 Relationship of the Difference between Two Means

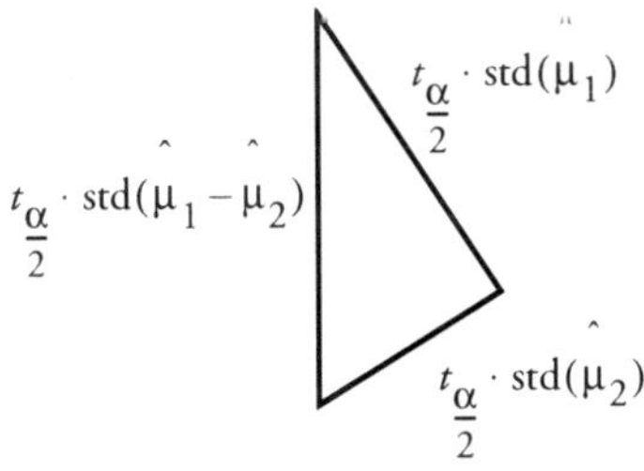

The hypotenuse of this triangle is a measuring stick for comparing means. The means are significantly different if and only if the actual difference is greater than the hypotenuse (LSD).

Suppose that you have two means that are exactly on the borderline, where the actual difference is the same as the least significant difference. Draw the triangle with vertices at the means measured on a vertical scale. Also, draw circles around each mean so that the diameter of each is equal to the confidence interval for that mean.

Figure 5.33 Geometric Relationship of *t*-test Statistics

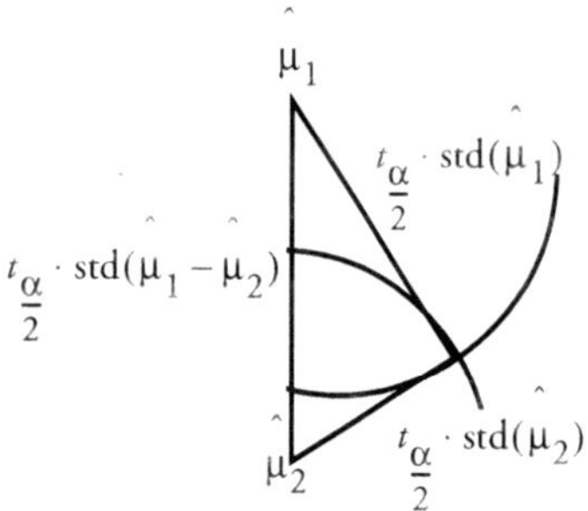

The radius of each circle is the length of the corresponding leg of the triangle, which is $t_{\alpha/2}\text{std}(\hat{\mu}_i)$.

The circles must intersect at the same right angle as the triangle legs, giving the following relationship:

- If the means differ exactly by their least significant difference, then the confidence interval circles around each mean intersect at a right angle. That is, the angle of the tangents is a right angle.

Now, consider the way that these circles must intersect if the means are different by greater than or less than the least significant difference:

- If the circles intersect so that the outside angle is greater than a right angle, then the means *are not* significantly different. If the circles intersect so that the outside angle is less than a right angle, then the means *are* significantly different. An outside angle of less than 90 degrees indicates that the means are farther apart than the least significant difference.
- If the circles do not intersect, then they are significantly different. If they nest, they are not significantly different. See Figure 5.12.

The same graphical technique works for many multiple-comparison tests, substituting a different probability quantile value for the Student's *t*.

Statistical Details for Power

To compute power, you make use of the noncentral *F* distribution. The formula (O'Brien and Lohr 1984) is given as follows:

Power = Prob($F > F_{crit}$, ν_1, ν_2, nc)

where:

- F is distributed as the noncentral F(nc, ν_1, ν_2)and $F_{crit} = F_{(1 - \alpha, \nu 1, \nu 2)}$ is the 1 - α quantile of the F distribution with ν_1 and ν_2 degrees of freedom.
- $\nu_1 = r - 1$ is the numerator df.
- $\nu_2 = r(n - 1)$ is the denominator df.
- n is the number per group.
- r is the number of groups.

- $nc = n(CSS)/\sigma^2$ is the non-centrality parameter.

  $$CSS = \sum_{g=1}^{r} (\mu_g - \mu)^2$$ is the corrected sum of squares.

- μ_g is the mean of the gth group.
- μ is the overall mean.
- σ^2 is estimated by the mean squared error (MSE).

Statistical Details for the Summary of Fit Report

Rsquare

Using quantities from the Analysis of Variance report for the model, the R^2 for any continuous response fit is always calculated as follows:

$$\frac{\text{Sum of Squares (Model)}}{\text{Sum of Squares (C Total)}}$$

Adj Rsquare

Adj Rsquare is a ratio of mean squares instead of sums of squares and is calculated as follows:

$$1 - \frac{\text{Mean Square (Error)}}{\text{Mean Square (C Total)}}$$

The mean square for Error is found in the Analysis of Variance report and the mean square for **C. Total** can be computed as the **C. Total** Sum of Squares divided by its respective degrees of freedom. See "The Analysis of Variance Report" on page 141.

Statistical Details for the Tests That the Variances Are Equal Report

F Ratio

O'Brien's test constructs a dependent variable so that the group means of the new variable equal the group sample variances of the original response. The O'Brien variable is computed as follows:

$$r_{ijk} = \frac{(n_{ij} - 1.5)n_{ij}(y_{ijk} - \bar{y}_{ij})^2 - 0.5s_{ij}^2(n_{ij} - 1)}{(n_{ij} - 1)(n_{ij} - 2)}$$

where n represents the number of y_{ijk} observations.

Brown-Forsythe is the model F statistic from an ANOVA on $z_{ij} = |y_{ij} - \tilde{y}_i|$ where $\tilde{y}_i$ is the median response for the ith level. If any z_{ij} is zero, then it is replaced with the next smallest z_{ij} in the same level. This corrects for the artificial zeros occurring in levels with odd numbers of observations.

The Levene F is the model F statistic from an ANOVA on $z_{ij} = |y_{ij} - \bar{y}_{i.}|$ where $\bar{y}_{i.}$ is the mean response for the ith level.

Bartlett's test is calculated as follows:

$$T = \frac{v\log\left(\sum_i \frac{v_i}{v} s_i^2\right) - \sum_i v_i \log(s_i^2)}{1 + \left(\frac{\sum_i \frac{1}{v_i} - \frac{1}{v}}{3(k-1)}\right)} \quad \text{where } v_i = n_i - 1 \text{ and } v = \sum_i v_i$$

and n_i is the count on the ith level and s_i^2 is the response sample variance on the ith level. The Bartlett statistic has a χ^2-distribution. Dividing the Chi-square test statistic by the degrees of freedom results in the reported F value.

Welch's Test F Ratio

The Welch's Test F Ratio is computed as follows:

$$F = \frac{\left[\frac{\sum_i w_i(\bar{y}_i - \tilde{y}_{..})^2}{k-1}\right]}{\left\{1 + \frac{2(k-2)}{k^2-1}\left[\sum_i \frac{\left(1 - \frac{w_i}{u}\right)^2}{n_i - 1}\right]\right\}} \quad \text{where } w_i = \frac{n_i}{s_i^2},\ u = \sum_i w_i,\ \tilde{y}_{..} = \sum_i \frac{w_i \bar{y}_{i.}}{u},$$

and n_i is the count on the ith level, $\bar{y}_{i.}$ is the mean response for the ith level, and s_i^2 is the response sample variance for the ith level.

Welch's Test DF Den

The Welch approximation for the denominator degrees of freedom is as follows:

$$df = \frac{1}{\left(\frac{3}{k^2-1}\right)\left[\sum_i \frac{\left(1 - \frac{w_i}{u}\right)^2}{n_i - 1}\right]}$$

where w_i, n_i, and u are defined as in the F ratio formula.

Chapter 6

Performing Contingency Analysis

Using the Fit Y by X or Contingency Platform

The Contingency or Fit Y by X platform lets you explore the distribution of a categorical (nominal or ordinal) variable Y across the levels of a second categorical variable *X*. The Contingency platform is the *categorical by categorical* personality of the Fit Y by X platform. The analysis results include a mosaic plot, frequency counts, and proportions. You can interactively perform additional analyses and tests on your data, such as an Analysis of Means for Proportions, a correspondence analysis plot, and so on.

Figure 6.1 Example of Contingency Analysis

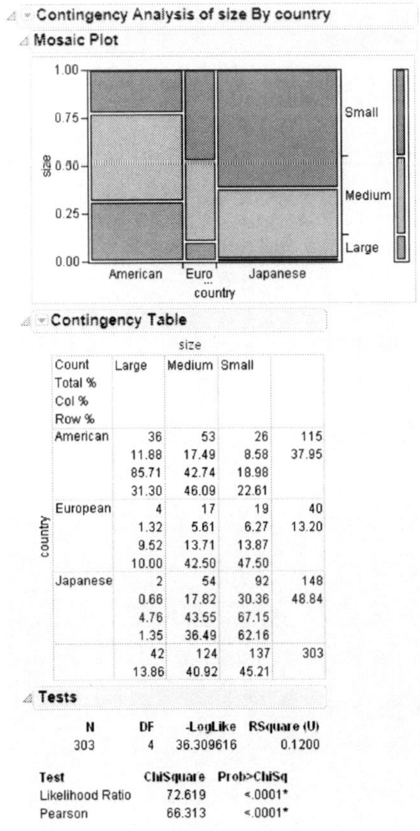

Contingency Table

Count / Total % / Col % / Row % (country)	Large	Medium	Small	
American	36	53	26	115
	11.88	17.49	8.58	37.95
	85.71	42.74	18.98	
	31.30	46.09	22.61	
European	4	17	19	40
	1.32	5.61	6.27	13.20
	9.52	13.71	13.87	
	10.00	42.50	47.50	
Japanese	2	54	92	148
	0.66	17.82	30.36	48.84
	4.76	43.55	67.15	
	1.35	36.49	62.16	
	42	124	137	303
	13.86	40.92	45.21	

Tests

N	DF	-LogLike	RSquare (U)
303	4	36.309616	0.1200

Test	ChiSquare	Prob>ChiSq
Likelihood Ratio	72.619	<.0001*
Pearson	66.313	<.0001*

Contents

Example of Contingency Analysis

This example uses the Car Poll.jmp sample data table, which contains data collected from car polls. The data includes aspects about the individual polled, such as their sex, marital status, and age. The data also includes aspects about the car that they own, such as the country of origin, the size, and the type. Examine the relationship between car sizes (small, medium, and large) and the cars' country of origin.

1. Open the Car Poll.jmp sample data table.
2. Select **Analyze > Fit Y by X**.
3. Select size and click **Y, Response**.
4. Select country and click **X, Factor**.
5. Click **OK**.

Figure 6.2 Example of Contingency Analysis

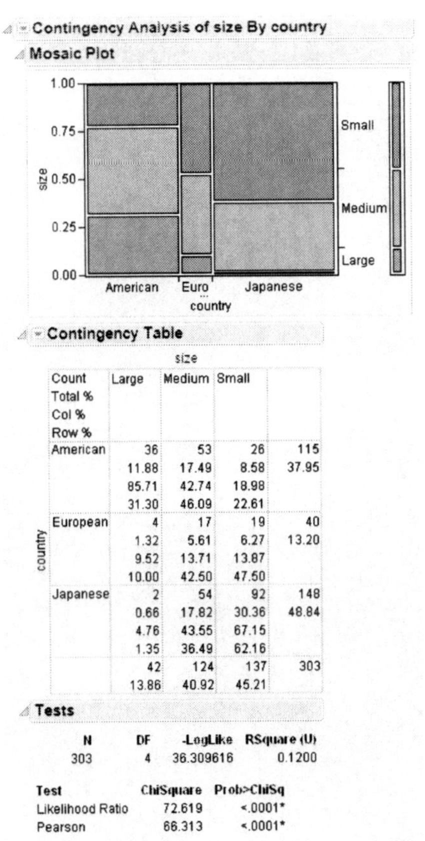

size

Count / Total % / Col % / Row %	Large	Medium	Small	
American	36	53	26	115
	11.88	17.49	8.58	37.95
	85.71	42.74	18.98	
	31.30	46.09	22.61	
European	4	17	19	40
	1.32	5.61	6.27	13.20
	9.52	13.71	13.87	
	10.00	42.50	47.50	
Japanese	2	54	92	148
	0.66	17.82	30.36	48.84
	4.76	43.55	67.15	
	1.35	36.49	62.16	
	42	124	137	303
	13.86	40.92	45.21	

N	DF	-LogLike	RSquare (U)
303	4	36.309616	0.1200

Test	ChiSquare	Prob>ChiSq
Likelihood Ratio	72.619	<.0001*
Pearson	66.313	<.0001*

From the mosaic plot and legend in Figure 6.2, notice the following:

- Very few Japanese cars fall into the Large size category.
- The majority of the European cars fall into the Small and Medium size categories.
- The majority of the American cars fall into the Large and Medium size categories.

Launch the Contingency Platform

You can perform a contingency analysis using either the Fit Y by X platform or the Contingency platform. The two approaches are equivalent.

- To launch the Fit Y by X platform, select **Analyze > Fit Y by X**.

  or

- To launch the Contingency platform, from the JMP Starter window, click on the **Basic** category and click **Contingency**.

Figure 6.3 The Contingency Launch Window

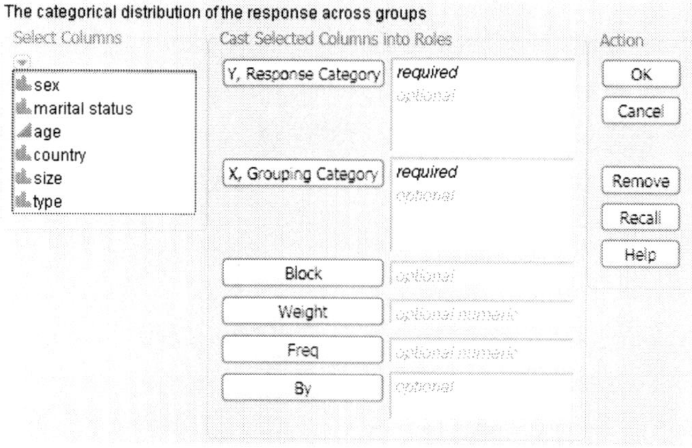

For more information about this launch window, see "Introduction to the Fit Y by X Platform" chapter on page 87.

The Contingency Report

To produce the plot shown in Figure 6.4, follow the instructions in "Example of Contingency Analysis" on page 185.

Figure 6.4 Example of a Contingency Report

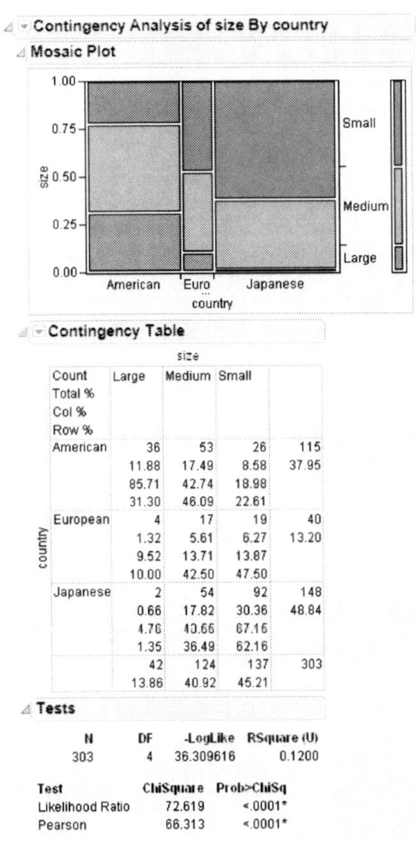

size	Large	Medium	Small	
Count Total % Col % Row %				
American	36 11.88 85.71 31.30	53 17.49 42.74 46.09	26 8.58 18.98 22.61	115 37.95
European	4 1.32 9.52 10.00	17 5.61 13.71 42.50	19 6.27 13.87 47.50	40 13.20
Japanese	2 0.66 4.76 1.35	54 17.82 43.55 36.49	92 30.36 67.16 62.16	148 48.84
	42 13.86	124 40.92	137 45.21	303

N	DF	-LogLike	RSquare (U)
303	4	36.309616	0.1200

Test	ChiSquare	Prob>ChiSq
Likelihood Ratio	72.619	<.0001*
Pearson	66.313	<.0001*

The Contingency report initially shows a Mosaic Plot, a Contingency Table, and a Tests report. You can add other analyses and tests using the options that are located within the red triangle menu. For details about all of these reports and options, see "Contingency Platform Options" on page 187.

Contingency Platform Options

Use the platform options within the red triangle menu next to Contingency Analysis to perform additional analyses and tests on your data.

Mosaic Plot	A graphical representation of the data in the Contingency Table. See "Mosaic Plot" on page 189.
Contingency Table	A two-way frequency table. There is a row for each factor level and a column for each response level. See "Contingency Table" on page 191.

Tests	Analogous to the Analysis of Variance table for continuous data. The tests show that the response level rates are the same across X levels. See "Tests" on page 193.
Set α level	Changes the alpha level used in confidence intervals. Select one of the common values (0.10, 0.05, 0.01) or select a specific value using the **Other** option.
Analysis of Means for Proportions	Only appears if the response has exactly two levels. Compares response proportions for the X levels to the overall response proportion. See "Analysis of Means for Proportions" on page 194.
Correspondence Analysis	Shows which rows or columns of a frequency table have similar patterns of counts. In the correspondence analysis plot, there is a point for each row and for each column of the contingency table. See "Correspondence Analysis" on page 195.
Cochran Mantel Haenszel	Tests if there is a relationship between two categorical variables after blocking across a third classification. See "Cochran-Mantel-Haenszel Test" on page 196.
Agreement Statistic	Only appears when both the X and Y variables have the same levels. Displays the Kappa statistic (Agresti 1990), its standard error, confidence interval, hypothesis test, and Bowker's test of symmetry, also know as McNemar's test. See "Agreement Statistic" on page 197.
Relative Risk	Calculates risk ratios. Appears only when both the X and Y variables have only two levels. See "Relative Risk" on page 197.
Odds Ratio	Appears only when there are exactly two levels for each variable. Produces a report of the odds ratio. For more information, see "Statistical Details for the Odds Ratio Option" on page 213. The report also gives a confidence interval for this ratio. You can change the alpha level using the **Set α Level** option.
Two Sample Test for Proportions	Performs a two-sample test for proportions. This test compares the proportions of the Y variable between the two levels of the X variable. Appears only when both the X and Y variables have only two levels. See "Two Sample Test for Proportions" on page 198.
Measures of Association	Describes the association between the variables in the contingency table. See "Measures of Association" on page 198.
Cochran Armitage Trend Test	Tests for trends in binomial proportions across levels of a single variable. This test is appropriate only when one variable has two levels and the other variable is ordinal. See "Cochran Armitage Trend Test" on page 200.

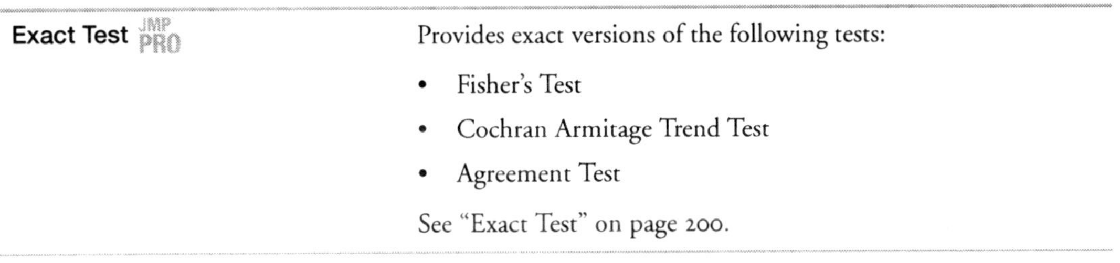

Exact Test JMP PRO	Provides exact versions of the following tests: • Fisher's Test • Cochran Armitage Trend Test • Agreement Test See "Exact Test" on page 200.

Mosaic Plot

The mosaic plot is a graphical representation of the two-way frequency table or Contingency Table. A mosaic plot is divided into rectangles, so that the area of each rectangle is proportional to the proportions of the Y variable in each level of the X variable. The mosaic plot was introduced by Hartigan and Kleiner in 1981 and refined by Friendly (1994).

To produce the plot shown in Figure 6.5, follow the instructions in "Example of Contingency Analysis" on page 185.

Figure 6.5 Example of a Mosaic Plot

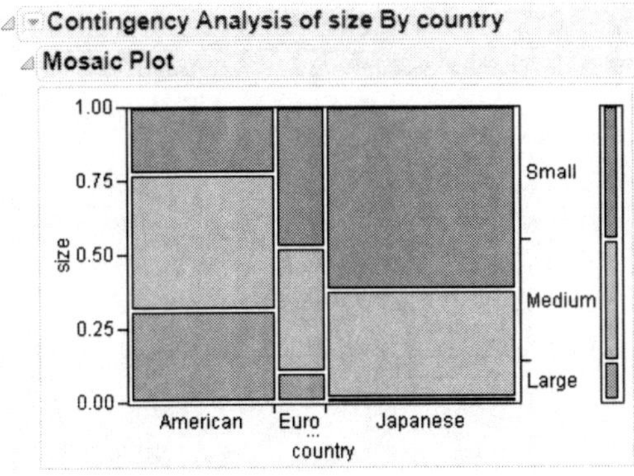

Note the following about the mosaic plot in Figure 6.5:

- The proportions on the *x*-axis represent the number of observations for each level of the X variable, which is country.
- The proportions on the *y*-axis at right represent the overall proportions of Small, Medium, and Large cars for the combined levels (American, European, and Japanese).
- The scale of the *y*-axis at left shows the response probability, with the whole axis being a probability of one (representing the total sample).

Clicking on a rectangle in the mosaic plot highlights the selection and highlights the corresponding data in the associated data table.

Replace variables in the mosaic plot by dragging and dropping a variable, in one of two ways: swap existing variables by dragging and dropping a variable from one axis to the other axis; or, click on a variable in the Columns panel of the associated data table and drag it onto an axis.

Context Menu

Right-click on the mosaic plot to change colors and label the cells.

Table 6.1 Context Menu Options

Set Colors	Shows the current assignment of colors to levels. See "Set Colors" on page 190.
Cell Labeling	Specify a label to be drawn in the mosaic plot. Select one of the following options: **Unlabeled** Shows no labels, and removes any of the other options. **Show Counts** Shows the number of observations in each cell. **Show Percents** Shows the percent of observations in each cell. **Show Labels** Shows the levels of the Y variable corresponding to each cell. **Show Row Labels** Shows the row labels for all of the rows represented by the cell.

Note: For descriptions of the remainder of the right-click options, see the *Using JMP* book.

Set Colors

When you select the **Set Colors** option, the Select Colors for Values window appears.

Figure 6.6 Select Colors for Values Window

The default mosaic colors depend on whether the response column is ordinal or nominal, and whether there is an existing Value Colors column property. To change the color for any level, click on the oval in the second column of colors and pick a new color.

Table 6.2 Description of the Select Colors for Values Window

Macros	Computes a color gradient between any two levels, as follows: • If you select a range of levels (by dragging the mouse over the levels that you want to select, or pressing the SHIFT key and clicking the first and last level), the **Gradient Between Selected Points** option applies a color gradient to the levels that you have selected. • The **Gradient Between Ends** option applies a gradient to all levels of the variable. • Undo any of your changes by selecting **Revert to Old Colors**.
Color Theme	Changes the colors for each value based on a color theme.
Save Colors to Column	If you select this check box, a new column property (**Value Colors**) is added to the column in the associated data table. To edit this property from the data table, select **Cols > Column Info**.

Contingency Table

The Contingency Table is a two-way frequency table. There is a row for each factor level and a column for each response level.

To produce the plot shown in Figure 6.7, follow the instructions in "Example of Contingency Analysis" on page 185.

Figure 6.7 Example of a Contingency Table

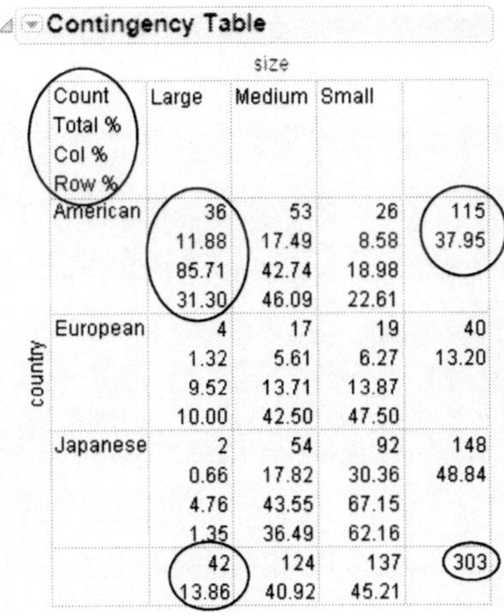

Contingency Table

country	Count Total % Col % Row %	size: Large	Medium	Small	
	American	36 11.88 85.71 31.30	53 17.49 42.74 46.09	26 8.58 18.98 22.61	115 37.95
	European	4 1.32 9.52 10.00	17 5.61 13.71 42.50	19 6.27 13.87 47.50	40 13.20
	Japanese	2 0.66 4.76 1.35	54 17.82 43.55 36.49	92 30.36 67.15 62.16	148 48.84
		42 13.86	124 40.92	137 45.21	303

Note the following about Contingency tables:

- The Count, Total%, Col%, and Row% correspond to the data within each cell that has row and column headings (such as the cell under American and Large).
- The last column contains the total counts for each row and percentages for each row.
- The bottom row contains total counts for each column and percentages for each column.

For example, in Figure 6.7, focus on the cars that are large and come from America. The following table explains the conclusions that you can make about these cars using the Contingency Table.

Number	Description	Label in Table
36	Number of cars that are both large and come from America	Count
11.88%	Percentage of all cars that are both large and come from America (36/303)[a].	Total%
85.71%	Percentage of large cars that come from America (36/42)[b]	Col%
31.30%	Percentage of American cars that are large (36/115)[c].	Row%
37.95%	Percentage of all cars that come from America (115/303).	(none)
13.86%	Percentage of all cars that are large (42/303).	(none)

a. 303 is the total number of cars in the poll.
b. 42 is the total number of large cars in the poll.
c. 115 is the total number of American cars in the poll.

Tip: To show or hide data in the Contingency Table, from the red triangle menu next to Contingency Table, select the option that you want to show or hide.

Table 6.3 Description of the Contingency Table

Count	Cell frequency, margin total frequencies, and grand total (total sample size).
Total%	Percent of cell counts and margin totals to the grand total.
Row%	Percent of each cell count to its row total.
Col%	Percent of each cell count to its column total.
Expected	Expected frequency (E) of each cell under the assumption of independence. Computed as the product of the corresponding row total and column total divided by the grand total.
Deviation	Observed cell frequency (O) minus the expected cell frequency (E).
Cell Chi Square	Chi-square values computed for each cell as $(O - E)^2 / E$.
Col Cum	Cumulative column total.

Table 6.3 Description of the Contingency Table *(Continued)*

Col Cum%	Cumulative column percentage.
Row Cum	Cumulative row total.
Row Cum%	Cumulative row percentage.

Tests

The Tests report shows the results for two tests to determine whether the response level rates are the same across X levels.

To produce the report shown in Figure 6.8, follow the instructions in “Example of Contingency Analysis” on page 185.

Figure 6.8 Example of a Tests Report

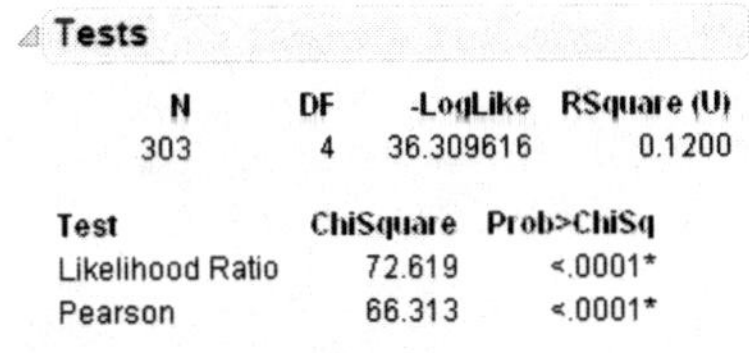

Tests

N	DF	-LogLike	RSquare (U)
303	4	36.309616	0.1200

Test	ChiSquare	Prob>ChiSq
Likelihood Ratio	72.619	<.0001*
Pearson	66.313	<.0001*

Note the following about the Chi-square statistics:

- When both categorical variables are responses (Y variables), the Chi-square statistics test that they are independent.
- You might have a Y variable with a fixed X variable. In this case, the Chi-square statistics test that the distribution of the Y variable is the same across each X level.

Table 6.4 Description of the Tests Report

N	Total number of observations.
DF	Records the degrees of freedom associated with the test. The degrees of freedom are equal to $(c - 1)(r - 1)$, where c is the number of columns and r is the number of rows.
-LogLike	Negative log-likelihood, which measures fit and uncertainty (much like sums of squares in continuous response situations).

Table 6.4 Description of the Tests Report *(Continued)*

Rsquare (U)	Portion of the total uncertainty attributed to the model fit. • An R^2 of 1 means that the factors completely predict the categorical response. • An R^2 of 0 means that there is no gain from using the model instead of fixed background response rates. For more information, see "Statistical Details for the Tests Report" on page 213.
Test	Lists two Chi-square statistical tests of the hypothesis that the response rates are the same in each sample category. For more information, see "Statistical Details for the Tests Report" on page 213.
Prob>ChiSq	Lists the probability of obtaining, by chance alone, a Chi-square value greater than the one computed if no relationship exists between the response and factor. If both variables have only two levels, Fisher's exact probabilities for the one-tailed tests and the two-tailed test also appear.

Fisher's Exact Test

This report gives the results of Fisher's exact test for a 2x2 table. The results appear automatically for 2x2 tables. For more details about Fisher's exact test, and for details about the test for *r* x *c* tables, see "Exact Test" on page 200.

Analysis of Means for Proportions

If the response has two levels, you can use this option to compare response proportions for the X levels to the overall response proportion. This method uses the normal approximation to the binomial. Therefore, if the sample sizes are too small, a warning appears in the results.

Note: For a description of Analysis of Means methods, see the document by Nelson, Wludyka, and Copeland (2005). For a description of the specific Analysis of Means for Variances method, see the paper by Wludyka and Nelson (1997).

Table 6.5 Descriptions of the Analysis of Means for Proportions Options

Set Alpha Level	Selects the alpha level used in the analysis.
Show Summary Report	Produces a report that shows the response proportions with decision limits for each level of the X variable. The report indicates whether a limit has been exceeded.
Switch Response Level for Proportion	Changes the response category used in the analysis.

Table 6.5 Descriptions of the Analysis of Means for Proportions Options *(Continued)*

Display Options	Shows or hides the decision limits, decision limit shading, center line, and needles.

Related Information

- "Example of Analysis of Means for Proportions" on page 201

Correspondence Analysis

Correspondence analysis is a graphical technique to show which rows or columns of a frequency table have similar patterns of counts. In the correspondence analysis plot, there is a point for each row and for each column. Use Correspondence Analysis when you have many levels, making it difficult to derive useful information from the mosaic plot.

Understanding Correspondence Analysis Plots

The *row profile* can be defined as the set of rowwise rates, or in other words, the counts in a row divided by the total count for that row. If two rows have very similar row profiles, their points in the correspondence analysis plot are close together. Squared distances between row points are approximately proportional to Chi-square distances that test the homogeneity between the pair of rows.

Column and row profiles are alike because the problem is defined symmetrically. The distance between a row point and a column point has no meaning. However, the directions of columns and rows from the origin are meaningful, and the relationships help interpret the plot.

Correspondence Analysis Options

Use the options in the red triangle menu next to Correspondence Analysis to produce a 3-D scatterplot and add column properties to the data table.

Table 6.6 Descriptions of the Correspondence Analysis Options

3D Correspondence Analysis	Produces a 3-D scatterplot.
Save Value Ordering	Takes the order of the levels sorted by the first correspondence score coefficient and makes a column property for both the X and Y columns.

The Details Report

The Details report contains statistical information about the correspondence analysis and shows the values used in the plot.

Table 6.7 Description of the Details Report

Singular Value	Provides the singular value decomposition of the contingency table. For the formula, see "Statistical Details for the Details Report" on page 214.
Inertia	Lists the square of the singular values, reflecting the relative variation accounted for in the canonical dimensions.
Portion	Portion of inertia with respect to the total inertia.
Cumulative	Shows the cumulative portion of inertia. If the first two singular values capture the bulk of the inertia, then the 2-D correspondence analysis plot is sufficient to show the relationships in the table.
X variable (Cheese) c1, c2, c3	The values plotted on the Correspondence Analysis plot.
Y variable (Response) c1, c2, c3	The values plotted on the Correspondence Analysis plot.

Related Information

- "Example of Correspondence Analysis" on page 202

Cochran-Mantel-Haenszel Test

The Cochran-Mantel-Haenszel test discovers if there is a relationship between two categorical variables after blocking across a third classification.

Table 6.8 Description of the Cochran-Mantel-Haenszel Test

Correlation of Scores	Applicable when either the Y or X is ordinal. The alternative hypothesis is that there is a linear association between Y and X in at least one level of the blocking variable.
Row Score by Col Categories	Applicable when Y is ordinal or interval. The alternative hypothesis is that, for at least one level of the blocking variable, the mean scores of the *r* rows are unequal.
Col Score by Row Categories	Applicable when X is ordinal or interval. The alternative hypothesis is that, for at least one level of the blocking variable, the mean scores of the *c* columns are unequal.
General Assoc. of Categories	Tests that for at least one level of the blocking variable, there is some type of association between X and Y.

Related Information

- "Example of a Cochran Mantel Haenszel Test" on page 206

Agreement Statistic

When the two variables have the same levels, the **Agreement Statistic** option is available. This option shows the Kappa statistic (Agresti 1990), its standard error, confidence interval, hypothesis test, and Bowker's test of symmetry.

The Kappa statistic and associated p-value given in this section are approximate. An exact version of the agreement test is available. See "Exact Test" on page 200.

Table 6.9 Description of the Agreement Statistic Report

Kappa	Shows the Kappa statistic.
Std Err	Shows the standard error of the Kappa statistic.
Lower 95%	Shows the lower endpoint of the confidence interval for Kappa.
Upper 95%	Shows the upper endpoint of the confidence interval for Kappa.
Prob>Z	Shows the p-value for a one-sided test for Kappa. The null hypothesis tests if Kappa equals zero.
Prob>|Z|	Shows the p-value for a two-sided test for Kappa.
ChiSquare	Shows the test statistic for Bowker's test. For Bowker's test of symmetry, the null hypothesis is that the probabilities in the square table satisfy symmetry, or that $p_{ij}=p_{ji}$ for all pairs of table cells. When both X and Y have two levels, this test is equal to McNemar's test.
Prob>ChiSq	Shows the p-value for the Bowker's test.

Related Information

- "Example of the Agreement Statistic Option" on page 207
- "Statistical Details for the Agreement Statistic Option" on page 212

Relative Risk

Calculate risk ratios for 2x2 contingency tables using the **Relative Risk** option. Confidence intervals also appear in the report. You can find more information about this method in Agresti (1990) section 3.4.2.

The Choose Relative Risk Categories window appears when you select the **Relative Risk** option. You can select a single response and factor combination, or you can calculate the risk ratios for all combinations of response and factor levels.

Related Information

- "Example of the Relative Risk Option" on page 208

Two Sample Test for Proportions

When both the X and Y variables have two levels, you can request a two-sample test for proportions. This test determines whether the chosen response level has equal proportions across the levels of the X variable.

Table 6.10 Description of the Two Sample Test for Proportions Report

Description	Shows the test being performed.
Proportion Difference	Shows the difference in the proportions between the levels of the X variable.
Lower 95%	Shows the lower endpoint of the confidence interval for the difference. Based on the adjusted Wald confidence interval.
Upper 95%	Shows the upper endpoint of the confidence interval for the difference. Based on the adjusted Wald confidence interval.
Adjusted Wald Test	Shows two-tailed and one-tailed tests.
Prob	Shows the p-values for the tests.
Response <variable> category of interest	Select which response level to use in the test.

Related Information

- "Example of a Two Sample Test for Proportions" on page 210

Measures of Association

You can request several statistics that describe the association between the variables in the contingency table by selecting the **Measures of Association** option.

For details about measures of association, see the following references:

- Brown and Benedetti (1977)
- Goodman and Kruskal (1979)
- Kendall and Stuart (1979)
- Snedecor and Cochran (1980)
- Somers (1962)

Table 6.11 Description of the Measures of Association Report

Gamma	Based on the number of concordant and discordant pairs and ignores tied pairs. Takes values in the range -1 to 1.

Table 6.11 Description of the Measures of Association Report *(Continued)*

Kendall's Tau-b	Similar to Gamma and uses a correction for ties. Takes values in the range -1 to 1.
Stuart's Tau-c	Similar to Gamma and uses an adjustment for table size and a correction for ties. Takes values in the range -1 to 1.
Somer's D	An asymmetric modification of Tau-b. • The C|R denotes that the row variable X is regarded as an independent variable and the column variable Y is regarded as dependent. • Similarly, the R|C denotes that the column variable Y is regarded as an independent variable and the row variable X is dependent. Somer's D differs from Tau-b in that it uses a correction for ties only when the pair is tied on the independent variable. It takes values in the range -1 to 1.
Lambda Asymmetric	• For C|R, is interpreted as the probable improvement in predicting the column variable Y given knowledge of the row variable X. • For R|C, is interpreted as the probable improvement in predicting the row variable X given knowledge about the column variable Y. Takes values in the range 0 to 1.
Lambda Symmetric	Loosely interpreted as the average of the two Lambda Asymmetric measures. Takes values in the range 0 to 1.
Uncertainty Coef	• For C|R, is the proportion of uncertainty in the column variable Y that is explained by the row variable X. • For R|C, is interpreted as the proportion of uncertainty in the row variable X that is explained by the column variable Y. Takes values in the range 0 to 1.
Uncertainty Coef Symmetric	Symmetric version of the two Uncertainty Coef measures. Takes values in the range 0 to 1.

Each statistic appears with its standard error and confidence interval. Note the following:

- Gamma, Kendall's Tau-b, Stuart's Tau-c, and Somer's D are measures of ordinal association that consider whether the variable Y tends to increase as X increases. They classify pairs of observations as concordant or discordant. A pair is concordant if an observation with a larger value of X also has a larger value of Y. A pair is discordant if an observation with a larger value of X has a smaller value of Y. These measures are appropriate only when both variables are ordinal.
- The Lambda and Uncertainty measures are appropriate for ordinal and nominal variables.

Related Information

- "Example of the Measures of Association Option" on page 211

Cochran Armitage Trend Test

This Cochran Armitage Trend tests for trends in binomial proportions across the levels of a single variable. This test is appropriate only when one variable has two levels and the other variable is ordinal. The two-level variable represents the response, and the other represents an explanatory variable with ordered levels. The null hypothesis is the hypothesis of no trend, which means that the binomial proportion is the same for all levels of the explanatory variable.

The test statistic and p-values given in this section are approximate. An exact version of the trend test is available. See "Exact Test" on page 200.

Related Information

- "Example of the Cochran Armitage Trend Test" on page 211

Exact Test

The following table describes the exact versions of three of the tests available in the Contingency platform.

Fisher's Exact Test	Performs Fisher's Exact test for an r x c table. This is a test for association between two variables. Fisher's exact test assumes that the row and column totals are fixed, and uses the hypergeometric distribution to compute probabilities. This test does not depend on any large-sample distribution assumptions. This means it is appropriate for situations where the Likelihood Ratio and Pearson tests become less reliable, like for small sample sizes or sparse tables. The report includes the following information: **Table Probability (P)** gives the probability for the observed table. This is not the p-value for the test. **Two-sided Prob ≤ P** gives the p-value for the two-sided test. For 2x2 tables, the Fisher's Exact test is automatically performed. See "Tests" on page 193.

Exact Cochran Armitage Trend Test	Performs the exact version of the Cochran Armitage Trend Test. This test is available only when one of the variables has two levels. For more details about the trend test, see "Cochran Armitage Trend Test" on page 200.
Exact Agreement Test	Performs an exact test for testing agreement between variables. This is an exact test for the Kappa statistic. This is available only when the two variables have the same levels. For more details about agreement testing, see "Agreement Statistic" on page 197.

Additional Examples of the Contingency Platform

This section contains additional examples using the options in the Contingency platform.

Example of Analysis of Means for Proportions

This example uses the Office Visits.jmp sample data table, which records late and on-time appointments for six clinics in a geographic region. 60 random appointments were selected from 1 week of records for each of the six clinics. To be considered on-time, the patient must be taken to an exam room within five minutes of their scheduled appointment time. Examine the proportion of patients that arrived on-time to their appointment.

1. Open the Office Visits.jmp sample data table.
2. Select **Analyze > Fit Y by X**.
3. Select On Time and click **Y, Response**.
4. Select Clinic and click **X, Factor**.
5. Select Frequency and click **Freq**.
6. Click **OK**.
7. From the red triangle menu next to Contingency Analysis, select **Analysis of Means for Proportions**.
8. From the red triangle menu next to Analysis of Means for Proportions, select **Show Summary Report**.

Figure 6.9 Example of Analysis of Means for Proportions

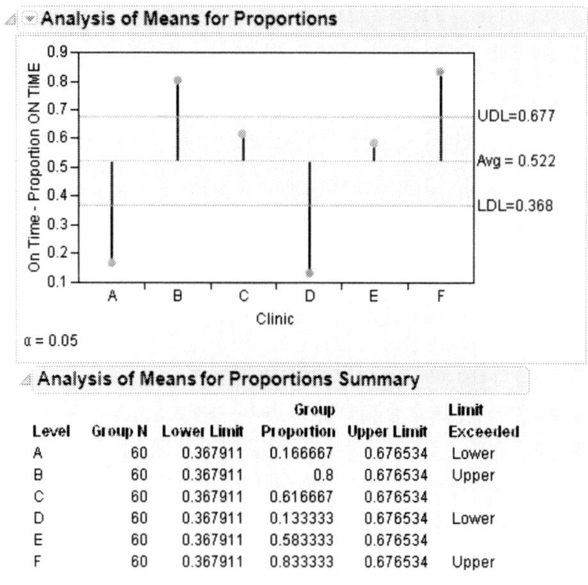

Analysis of Means for Proportions Summary

Level	Group N	Lower Limit	Group Proportion	Upper Limit	Limit Exceeded
A	60	0.367911	0.166667	0.676534	Lower
B	60	0.367911	0.8	0.676534	Upper
C	60	0.367911	0.616667	0.676534	
D	60	0.367911	0.133333	0.676534	Lower
E	60	0.367911	0.583333	0.676534	
F	60	0.367911	0.833333	0.676534	Upper

Figure 6.9 shows the proportion of patients who were on-time from each clinic. From Figure 6.9, notice the following:

- The proportion of on-time arrivals is the highest for clinic F, followed by clinic B.
- Clinic D has the lowest proportion of on-time arrivals, followed by clinic A.
- Clinic E and clinic C are close to the average, and do not exceed the decision limits.

Example of Correspondence Analysis

This example uses the Cheese.jmp sample data table, which is taken from the Newell cheese tasting experiment, reported in McCullagh and Nelder (1989). The experiment records counts more than nine different response levels across four different cheese additives.

1. Open the Cheese.jmp sample data table.
2. Select **Analyze > Fit Y by X**.
3. Select Response and click **Y, Response**.

   The Response values range from one to nine, where one is the least liked, and nine is the best liked.
4. Select Cheese and click **X, Factor**.

   A, B, C, and D represent four different cheese additives.
5. Select Count and click **Freq**.
6. Click **OK**.

Figure 6.10 Mosaic Plot for the Cheese Data

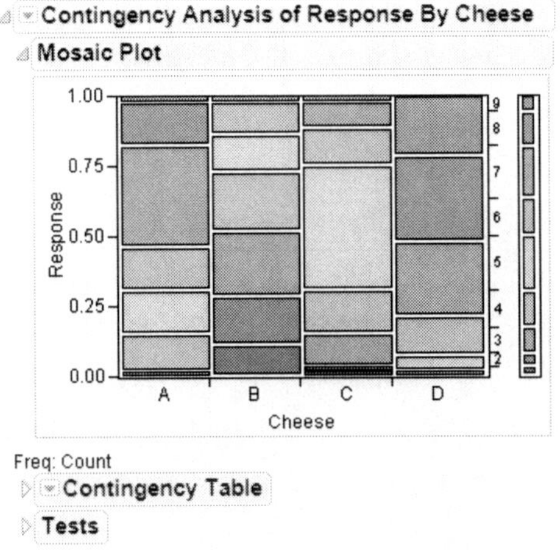

From the mosaic plot in Figure 6.10, you notice that the distributions do not appear alike. However, it is challenging to make sense of the mosaic plot across nine levels. A correspondence analysis can help define relationships in this type of situation.

7. To see the correspondence analysis plot, from the red triangle menu next to Contingency Analysis, select **Correspondence Analysis**.

Figure 6.11 Example of a Correspondence Analysis Plot

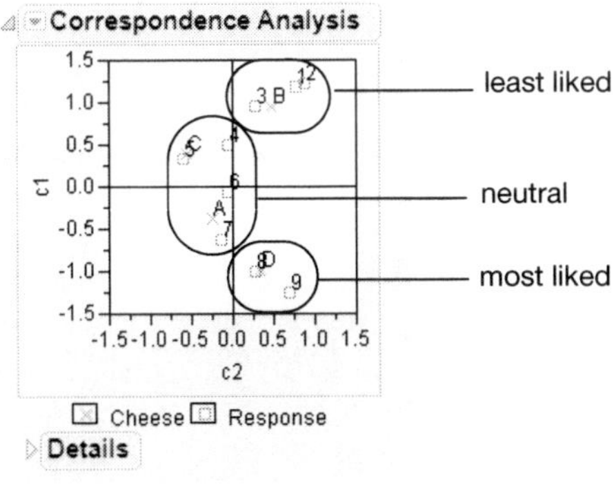

Figure 6.11 shows the correspondence analysis graphically, with the plot axes labeled c1 and c2. Notice the following:

- c1 seems to correspond to a general satisfaction level. The cheeses on the c1 axis go from least liked at the top to most liked at the bottom.
- c2 seems to capture some quality that makes B and D different from A and C.
- Cheese D is the most liked cheese, with responses of 8 and 9.

- Cheese B is the least liked cheese, with responses of 1,2, and 3.
- Cheeses C and A are in the middle, with responses of 4,5,6, and 7.

8. From the red triangle menu next to Correspondence Analysis, select **3D Correspondence Analysis**.

Figure 6.12 Example of a 3-D Scatterplot

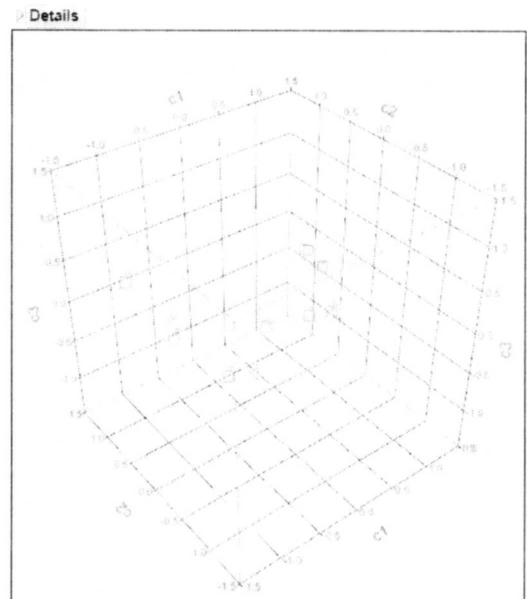

From Figure 6.12, notice the following:

- Looking at the c1 axis, responses 1 through 5 appear to the right of 0 (positive). Responses 6 through 9 appear to the left of 0 (negative).
- Looking at the c2 axis, A and C appear to the right of 0 (positive). B and D appear to the left of 0 (negative).
- You can have two conclusions: c1 corresponds to the general satisfaction (from least to most liked). c2 corresponds to a quality that makes B and D different from A and C.

Additional Example of Correspondence Analysis

This example uses the Mail Messages.jmp sample data table, which contains data about e-mail messages that were sent and received. The data includes the time, sender, and receiver. Examine the pattern of e-mail senders and receivers.

1. Open the Mail Messages.jmp sample data table.
2. Select **Analyze > Fit Y by X**.
3. Select To and click **Y, Response**.

4. Select From and click **X, Factor.**
5. Click **OK.**
6. From the red triangle menu next to Contingency Table, deselect everything except **Count**.

Figure 6.13 Contingency Analysis for E-mail Data

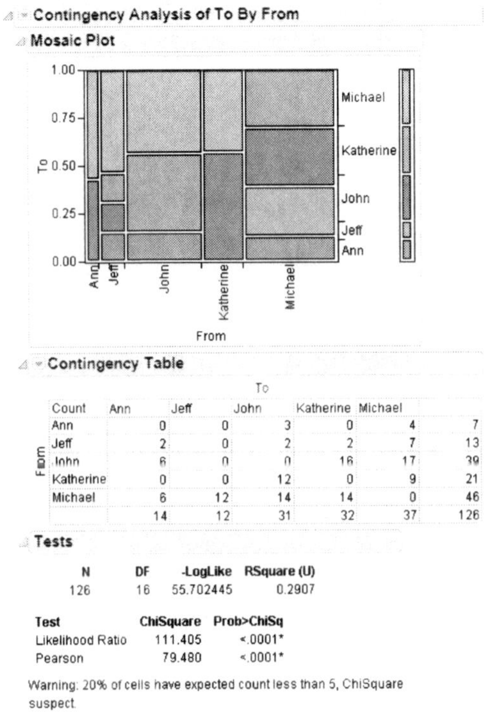

Count (From \ To)	Ann	Jeff	John	Katherine	Michael	
Ann	0	0	3	0	4	7
Jeff	2	0	2	2	7	13
John	6	0	0	16	17	39
Katherine	0	0	12	0	9	21
Michael	6	12	14	14	0	46
	14	12	31	32	37	126

N	DF	-LogLike	RSquare (U)
126	16	55.702445	0.2907

Test	ChiSquare	Prob>ChiSq
Likelihood Ratio	111.405	<.0001*
Pearson	79.480	<.0001*

Looking at the frequency table, you notice the following differences:

- Jeff sends messages to everyone but receives messages only from Michael.
- Michael and John send many more messages than the others.
- Michael sends messages to everyone.
- John sends messages to everyone except Jeff.
- Katherine and Ann send messages only to Michael and John.

Further visualize the results of the contingency table with a correspondence analysis. From the red triangle menu next to Contingency Analysis, select **Correspondence Analysis**.

Figure 6.14 Correspondence Analysis for E-mail Data

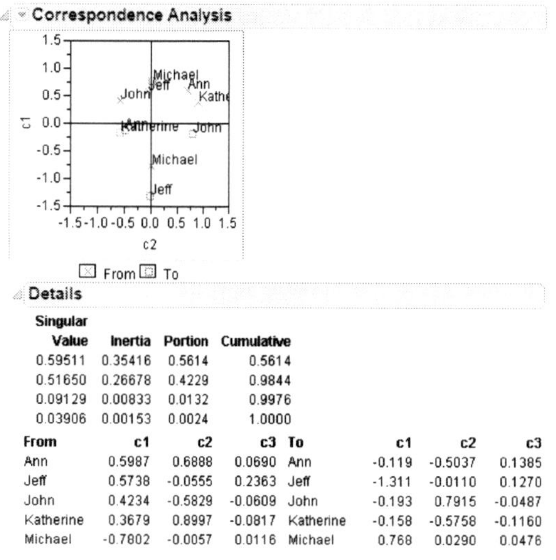

Singular Value	Inertia	Portion	Cumulative
0.59511	0.35416	0.5614	0.5614
0.51650	0.26678	0.4229	0.9844
0.09129	0.00833	0.0132	0.9976
0.03906	0.00153	0.0024	1.0000

From	c1	c2	c3	To	c1	c2	c3
Ann	0.5987	0.6888	0.0690	Ann	-0.119	-0.5037	0.1385
Jeff	0.5738	-0.0555	0.2363	Jeff	-1.311	-0.0110	0.1270
John	0.4234	-0.5829	-0.0609	John	-0.193	0.7915	-0.0487
Katherine	0.3679	0.8997	-0.0817	Katherine	-0.158	-0.5758	-0.1160
Michael	-0.7802	-0.0057	0.0116	Michael	0.768	0.0290	0.0476

From the Details report in Figure 6.14, you notice the following:

- The Portion column shows that the bulk of the variation (56% + 42%) of the mail sending pattern is summarized by c1 and c2, for the To and From groups.
- The Correspondence Analysis plot of c1 and c2 shows the pattern of mail distribution among the mail group, as follows:
  - Katherine and Ann have similar sending and receiving patterns; they both send e-mails to Michael and John and receive e-mails from Michael, John, and Jeff.
  - Jeff and Michael lie in a single dimension, but have opposite sending and receiving patterns. Jeff sends e-mails to everyone and receives e-mails only from Michael. Michael sends e-mail to everyone and receives e-mail from everyone.
  - John's patterns differ from the others. He sends e-mail to Ann, Katherine, and Michael, and receives e-mail from everyone.

Example of a Cochran Mantel Haenszel Test

This example uses the Hot Dogs.jmp sample data table. Examine the relationship between hot dog type and taste.

1. Open the Hot Dogs.jmp sample data table.
2. Select **Analyze > Fit Y by X**.
3. Select Type and click **Y, Response**.
4. Select Taste and click **X, Factor**.
5. Click **OK**.

6. From the red triangle menu next to Contingency Analysis, select **Cochran Mantel Haenszel.**
7. Select Protein/Fat as the grouping variable and click **OK**.

Figure 6.15 Example of a Cochran-Mantel-Haenszel Test

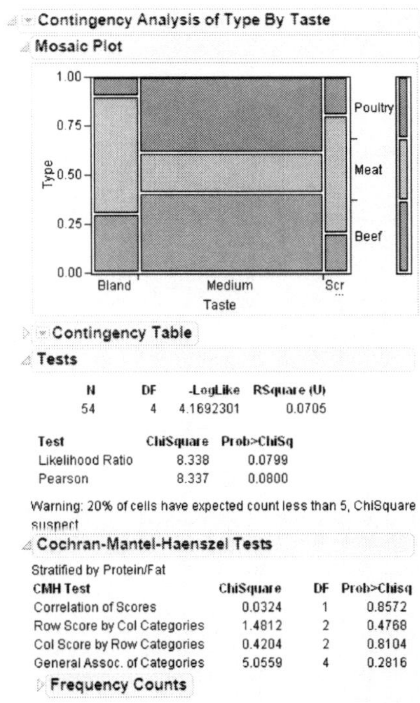

From Figure 6.15, you notice the following:

- The Tests report shows a marginally significant Chi-square probability of about 0.0799, indicating some significance in the relationship between hot dog taste and type.
- However, the Cochran Mantel Haenszel report shows no relationship at all between hot dog taste and type after adjusting for the protein-fat ratio. The Chi-square test for the adjusted correlation has a probability of 0.857, and the Chi-square probability associated with the general association of categories is 0.282.

Example of the Agreement Statistic Option

This example uses the Attribute Gauge.jmp sample data table. The data gives results from three people (raters) rating fifty parts three times each. Examine the relationship between raters A and B.

1. Open the Attribute Gauge.jmp sample data table.
2. Select **Analyze > Fit Y by X.**
3. Select A and click **Y, Response.**

4. Select B and click **X, Factor**.
5. Click **OK**.
6. From the red triangle menu next to Contingency Analysis, select **Agreement Statistic**.

Figure 6.16 Example of the Agreement Statistic Report

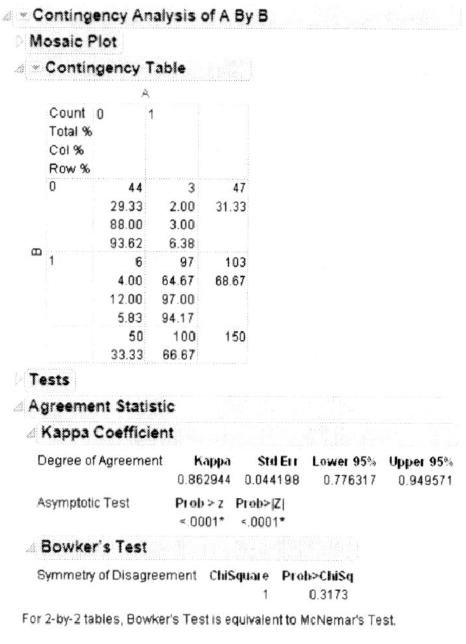

From Figure 6.16, you notice that the agreement statistic of 0.86 is high (close to 1) and the p-value of <.0001 is small. This reinforces the high agreement seen by looking at the diagonal of the contingency table. Agreement between the raters occurs when both raters give a rating of 0 or both give a rating of 1.

Example of the Relative Risk Option

This example uses the Car Poll.jmp sample data table. Examine the relative probabilities of being married and single for the participants in the poll.

1. Open the Car Poll.jmp sample data table.
2. Select **Analyze > Fit Y by X**.
3. Select marital status and click **Y, Response**.
4. Select sex and click **X, Factor**.
5. Click **OK**.
6. From the red triangle menu next to Contingency Analysis, select **Relative Risk**.

   The Choose Relative Risk Categories window appears.

Figure 6.17 The Choose Relative Risk Categories Window

Risk Ratio is P(Y=yj | X=xi) / P(Y=yj | X=xk)

Response marital status category of interest: Y=yj

○ Married
○ Single

Sample sex category for numerator: X=xi

○ Female
○ Male

☐ Calculate all combinations

OK Cancel Help

Note the following about the Choose Relative Risk Categories window:

- If you are interested in only a single response and factor combination, you can select that here. For example, if you clicked **OK** in the window in Figure 6.17, the calculation would be as follows:

  $$\frac{P(Y = \text{Married} \mid X = \text{Female})}{P(Y = \text{Married} \mid X = \text{Male})}$$

- If you would like to calculate the risk ratios for all ($2 \times 2 = 4$) combinations of response and factor levels, select the **Calculate All Combinations** check box. See Figure 6.18.

7. Ask for all combinations by selecting the **Calculate All Combinations** check box. Leave all other default selections as is.

Figure 6.18 Example of the Risk Ratio Report

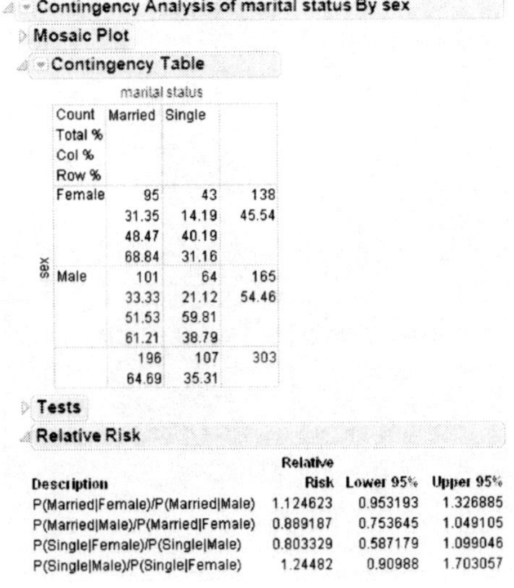

Contingency Analysis of marital status By sex

Mosaic Plot

Contingency Table

sex	Count / Total % / Col % / Row %	Married	Single	
	Female	95	43	138
		31.35	14.19	45.54
		48.47	40.19	
		68.84	31.16	
	Male	101	64	165
		33.33	21.12	54.46
		51.53	59.81	
		61.21	38.79	
		196	107	303
		64.69	35.31	

Tests

Relative Risk

Description	Relative Risk	Lower 95%	Upper 95%
P(Married|Female)/P(Married|Male)	1.124623	0.953193	1.326885
P(Married|Male)/P(Married|Female)	0.889187	0.753645	1.049105
P(Single|Female)/P(Single|Male)	0.803329	0.587179	1.099046
P(Single|Male)/P(Single|Female)	1.24482	0.90988	1.703057

To see how the relative risk is calculated, proceed as follows:

1. Examine the first entry in the Relative Risk report, which is P(Married|Female)/P(Married|Male).

2. You can find these probabilities in the Contingency Table. Since the probabilities are computed based on two levels of sex, which differs across the rows of the table, use the Row% to read the probabilities, as follows:

   P(Married|Female)=0.6884

   P(Married|Male) = 0.6121

Therefore, the calculations are as follows:

$$P(\text{Married}|\text{Female})/P(\text{Married}|\text{Male}) = \frac{0.6884}{0.6121} = 1.1247$$

Example of a Two Sample Test for Proportions

This example uses the Car Poll.jmp sample data table. Examine the probability of being married for males and females.

1. Open the Car Poll.jmp sample data table.
2. Select **Analyze > Fit Y by X**.
3. Select marital status and click **Y, Response**.
4. Select sex and click **X, Factor**.
5. Click **OK**.
6. From the red triangle menu next to Contingency Analysis, select **Two Sample Test for Proportions**.

Figure 6.19 Example of the Two Sample Test for Proportions Report

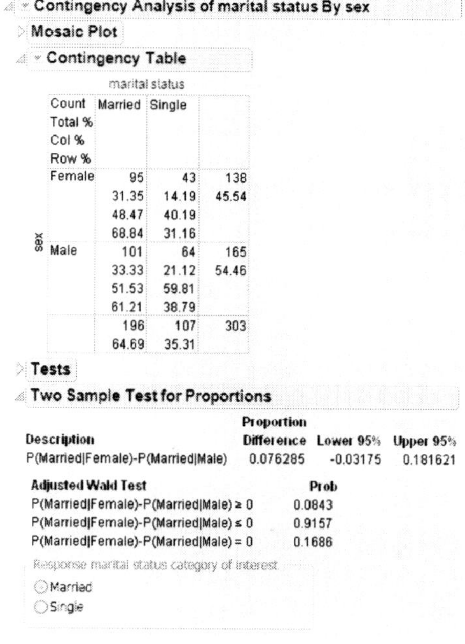

In this example, you are comparing the probability of being married between females and males. See the Row% in the Contingency Table to obtain the following:

P(Married|Female)=0.6884

P(Married|Male) = 0.6121

The difference between these two numbers, 0.0763, is the Proportion Difference shown in the report. The two-sided p-value (0.1686) is large, indicating that there is no significant difference between the proportions.

Example of the Measures of Association Option

This example uses the Car Poll.jmp sample data table. Examine the probability of being married for males and females.

1. Open the Car Poll.jmp sample data table.
2. Select **Analyze > Fit Y by X**.
3. Select marital status and click **Y, Response**.
4. Select sex and click **X, Factor**.
5. Click **OK**.
6. From the red triangle menu next to Contingency Analysis, select **Measures of Association.**

Figure 6.20 Example of the Measures of Association Report

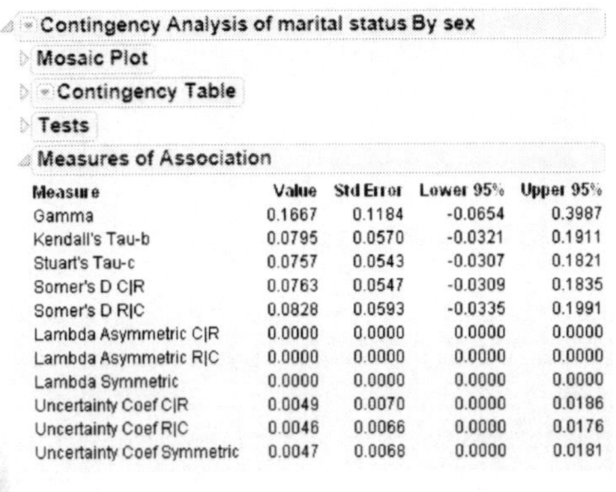

Contingency Analysis of marital status By sex
Mosaic Plot
Contingency Table
Tests
Measures of Association

Measure	Value	Std Error	Lower 95%	Upper 95%
Gamma	0.1667	0.1184	-0.0654	0.3987
Kendall's Tau-b	0.0795	0.0570	-0.0321	0.1911
Stuart's Tau-c	0.0757	0.0543	-0.0307	0.1821
Somer's D C|R	0.0763	0.0547	-0.0309	0.1835
Somer's D R|C	0.0828	0.0593	-0.0335	0.1991
Lambda Asymmetric C|R	0.0000	0.0000	0.0000	0.0000
Lambda Asymmetric R|C	0.0000	0.0000	0.0000	0.0000
Lambda Symmetric	0.0000	0.0000	0.0000	0.0000
Uncertainty Coef C|R	0.0049	0.0070	0.0000	0.0186
Uncertainty Coef R|C	0.0046	0.0066	0.0000	0.0176
Uncertainty Coef Symmetric	0.0047	0.0068	0.0000	0.0181

Since the variables that you want to examine (sex and marital status) are nominal, use the Lambda and Uncertainty measures. The confidence intervals include zero, so you conclude that there is no association between sex and marital status.

Example of the Cochran Armitage Trend Test

1. Open the Car Poll.jmp sample data table.

For the purposes of this test, change size to an ordinal variable:

2. In the Columns panel, right-click on the icon next to size and select **Ordinal.**
3. Select **Analyze > Fit Y by X.**
4. Select sex and click **Y, Response.**
5. Select size and click **X, Factor.**
6. Click **OK.**
7. From the red triangle menu next to Contingency Analysis, select **Cochran Armitage Trend Test.**

Figure 6.21 Example of the Cochran Armitage Trend Test Report

Contingency Analysis of size By sex
Mosaic Plot
Contingency Table
Tests
Cochran Armitage Trend Test

Asymptotic Test	Z	Prob<Z	Prob>|Z|
	-0.37271	0.3547	0.7094

The two-sided p-value (0.7094) is large. From this, you can conclude that there is no trend in the proportion of male and females that purchase different sizes of cars.

Statistical Details for the Contingency Platform

This section contains statistical details for selected options and reports in the Contingency platform.

Statistical Details for the Agreement Statistic Option

Viewing the two response variables as two independent ratings of the *n* subjects, the Kappa coefficient equals +1 when there is complete agreement of the raters. When the observed agreement exceeds chance agreement, the Kappa coefficient is positive, with its magnitude reflecting the strength of agreement. Although unusual in practice, Kappa is negative when the observed agreement is less than chance agreement. The minimum value of Kappa is between -1 and 0, depending on the marginal proportions.

Quantities associated with the Kappa statistic are computed as follows:

$$\hat{\kappa} = \frac{P_0 - P_e}{1 - P_e} \text{ where } P_0 = \sum_i p_{ii} \text{ and } P_e = \sum_i p_{i.} p_{.i}$$

The asymptotic variance of the simple kappa coefficient is estimated by the following:

$$\text{var} = \frac{A + B - C}{(1 - P_e)^2 n} \text{ where } A = \sum_i p_{ii}[1 - (p_{i.} + p_{.i})(1 - \hat{\kappa})]^2,\ B = (1 - \hat{\kappa})^2 \sum_{i \neq j} \sum p_{ij}(p_{.i} + p_{j.})^2 \text{ and}$$

$$C = \left[\hat{\kappa} - P_e(1 - \hat{\kappa})\right]^2$$

See Fleiss, Cohen, and Everitt (1969).

For Bowker's test of symmetry, the null hypothesis is that the probabilities in the two-by-two table satisfy symmetry ($p_{ij}=p_{ji}$).

Statistical Details for the Odds Ratio Option

The Odds Ratio is calculated as follows:

$$\frac{p_{11} \times p_{22}}{p_{12} \times p_{21}}$$

where p_{ij} is the count in the ith row and j^{th} column of the 2x2 table.

Statistical Details for the Tests Report

Rsquare (U)

Rsquare (U) is computed as follows:

$$\frac{-\text{log likelihood for Model}}{-\text{log likelihood for Corrected Total}}$$

The total negative log-likelihood is found by fitting fixed response rates across the total sample.

Test

The two Chi-square tests are as follows:

The **Likelihood Ratio** Chi-square test is computed as twice the negative log-likelihood for Model in the Tests table. Some books use the notation G^2 for this statistic. The difference of two negative log-likelihoods, one with *whole-population* response probabilities and one with *each-population* response rates, is written as follows:

$$G^2 = 2\left[\sum_{ij}(-n_{ij})\ln(p_j) - \sum_{ij}-n_{ij}\ln(p_{ij})\right] \text{ where } p_{ij} = \frac{n_{ij}}{N} \text{ and } p_j = \frac{N_j}{N}$$

This formula can be more compactly written as follows:

$$G^2 = 2\sum_i\sum_j n_{ij}\ln\left(\frac{n_{ij}}{e_{ij}}\right)$$

The **Pearson** Chi-square is calculated by summing the squares of the differences between the observed and expected cell counts. The Pearson Chi-square exploits the property that frequency counts tend to a normal distribution in very large samples. The familiar form of this Chi-square statistic is as follows:

$$\chi^2 = \sum \frac{(O - E)^2}{E}$$

where O is the observed cell counts and E is the expected cell counts. The summation is over all cells. There is no continuity correction done here, as is sometimes done in 2-by-2 tables.

Statistical Details for the Details Report

Lists the singular values of the following equation:

$$D_r^{-0.5}(P - rc')D_c^{-0.5}$$

where:

- P is the matrix of counts divided by the total frequency
- r and c are row and column sums of P
- the Ds are diagonal matrices of the values of r and c

Chapter 7

Performing Simple Logistic Regression

Using the Fit Y by X or Logistic Platform

The Logistic platform fits the probabilities for response categories to a continuous *x* predictor. The fitted model estimates probabilities for each *x* value. The Logistic platform is the *nominal* or *ordinal* by *continuous* personality of the Fit Y by X platform. There is a distinction between nominal and ordinal responses on this platform:

- Nominal logistic regression estimates a set of curves to partition the probability among the responses.
- Ordinal logistic regression models the probability of being less than or equal to a given response. This has the effect of estimating a single logistic curve, which is shifted horizontally to produce probabilities for the ordered categories. This model is less complex and is recommended for ordered responses.

Figure 7.1 Examples of Logistic Regression

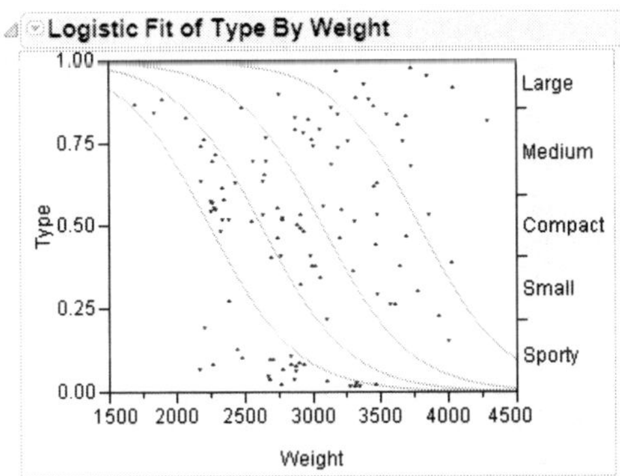

Ordinal Logistic Regression

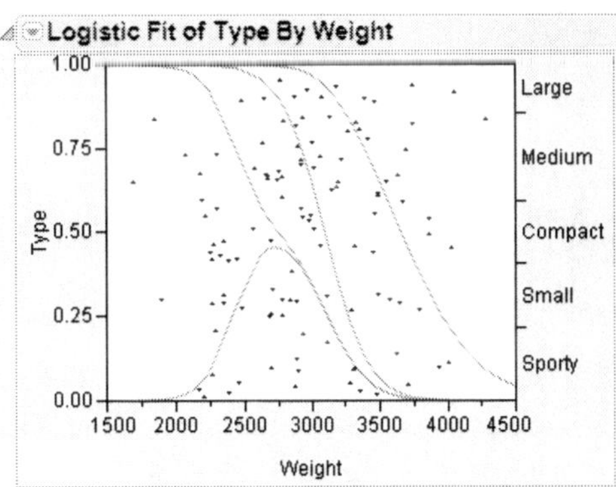

Nominal Logistic Regression

Contents

Overview of Logistic Regression

Logistic regression has a long tradition with widely varying applications such as modeling dose-response data and purchase-choice data. Unfortunately, many introductory statistics courses do not cover this fairly simple method. Many texts in categorical statistics cover it (Agresti 1998), in addition to texts on logistic regression (Hosmer and Lemeshow 1989). Some analysts use the method with a different distribution function, the normal. In that case, it is called *probit analysis*. Some analysts use discriminant analysis instead of logistic regression because they prefer to think of the continuous variables as *Y*s and the categories as *X*s and work backwards. However, discriminant analysis assumes that the continuous data are normally distributed random responses, rather than fixed regressors.

Simple logistic regression is a more graphical and simplified version of the general facility for categorical responses in the Fit Model platform. For examples of more complex logistic regression models, see the *Modeling and Multivariate Methods* book.

Nominal Logistic Regression

Nominal logistic regression estimates the probability of choosing one of the response levels as a smooth function of the *x* factor. The fitted probabilities must be between 0 and 1, and must sum to 1 across the response levels for a given factor value.

In a logistic probability plot, the *y*-axis represents probability. For *k* response levels, *k* - 1 smooth curves partition the total probability (which equals 1) among the response levels. The fitting principle for a logistic regression minimizes the sum of the negative natural logarithms of the probabilities fitted to the response events that occur (that is, maximum likelihood).

Ordinal Logistic Regression

When *Y* is ordinal, a modified version of logistic regression is used for fitting. The cumulative probability of being at or below each response level is modeled by a curve. The curves are the same for each level except that they are shifted to the right or left.

The ordinal logistic model fits a different intercept, but the same slope, for each of *r* - 1 cumulative logistic comparisons, where *r* is the number of response levels. Each parameter estimate can be examined and tested individually, although this is seldom of much interest.

The ordinal model is preferred to the nominal model when it is appropriate because it has fewer parameters to estimate. In fact, it is practical to fit ordinal responses with hundreds of response levels.

Example of Nominal Logistic Regression

This example uses the Penicillin.jmp sample data table. The data in this example comes from an experiment where 5 groups, each containing 12 rabbits, were injected with streptococcus bacteria. Once the rabbits were confirmed to have the bacteria in their system, they were given different doses of penicillin. You want to find out whether the natural log (ln(dose)) of dosage amounts has any effect on whether the rabbits are cured.

1. Open the Penicillin.jmp sample data table.
2. Select **Analyze > Fit Y by X**.
3. Select Response and click **Y, Response**.
4. Select ln(Dose) and click **X, Factor**.

   Notice that JMP automatically fills in Count for **Freq**. Count was previously assigned the role of Freq.
5. Click **OK**.

Figure 7.2 Example of Nominal Logistic Report

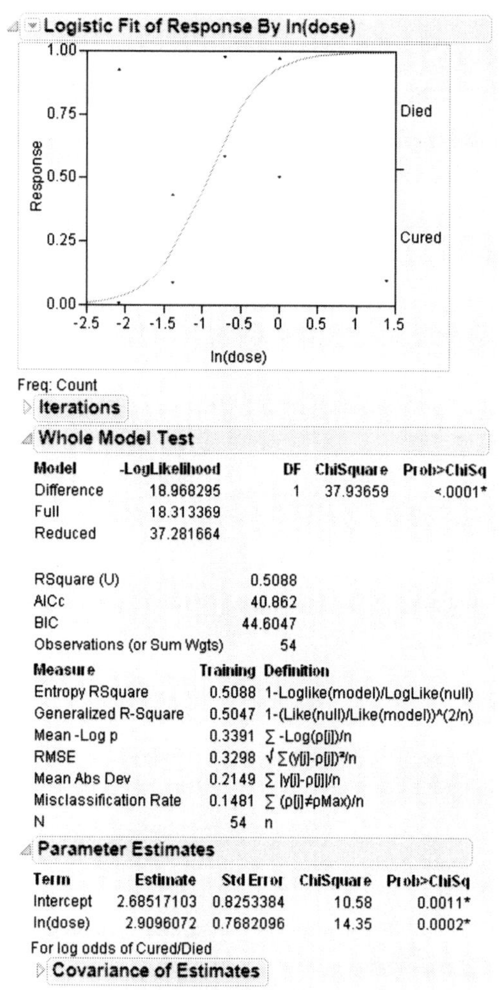

Model	-LogLikelihood	DF	ChiSquare	Prob>ChiSq
Difference	18.968295	1	37.93659	<.0001*
Full	18.313369			
Reduced	37.281664			

RSquare (U)	0.5088
AICc	40.862
BIC	44.6047
Observations (or Sum Wgts)	54

Measure	Training	Definition
Entropy RSquare	0.5088	1-Loglike(model)/LogLike(null)
Generalized R-Square	0.5047	1-(Like(null)/Like(model))^(2/n)
Mean -Log p	0.3391	∑ -Log(p[j])/n
RMSE	0.3298	√ ∑(y[j]-p[j])²/n
Mean Abs Dev	0.2149	∑ |y[j]-p[j]|/n
Misclassification Rate	0.1481	∑ (p[j]≠pMax)/n
N	54	n

Parameter Estimates

Term	Estimate	Std Error	ChiSquare	Prob>ChiSq
Intercept	2.68517103	0.8253384	10.58	0.0011*
ln(dose)	2.9096072	0.7682096	14.35	0.0002*

For log odds of Cured/Died

Covariance of Estimates

The plot shows the fitted model, which is the predicted probability of being cured, as a function of ln (dose). The p-value is significant, indicating that the dosage amounts have a significant effect on whether the rabbits are cured.

Launch the Logistic Platform

You perform a logistic analysis using either the Fit Y by X platform or the Logistic platform. The two approaches are equivalent.

- To launch the Fit Y by X platform, select **Analyze > Fit Y by X**.

  or

- To launch the Logistic platform, from the JMP Starter window, click on the **Basic** category and then click **Logistic**.

Figure 7.3 The Logistic Launch Window

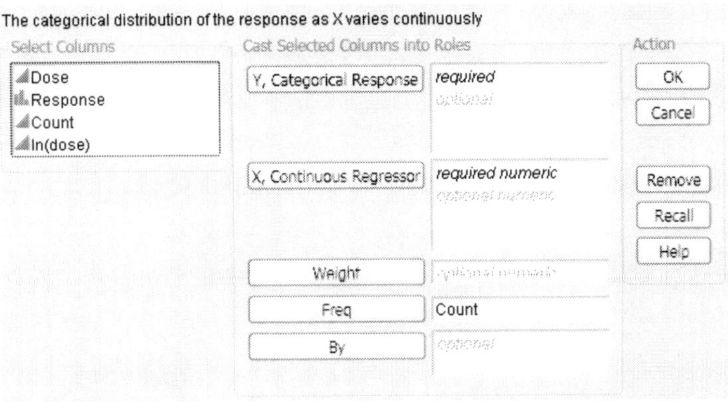

For more information about this launch window, see "Introduction to the Fit Y by X Platform" chapter on page 87.

The Logistic Report

To produce the plot shown in Figure 7.4, follow the instructions in "Example of Nominal Logistic Regression" on page 217.

Figure 7.4 Example of a Logistic Report

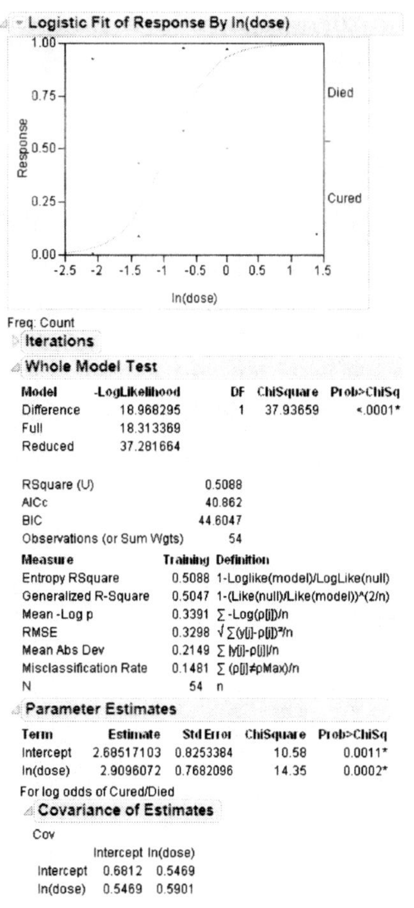

Freq: Count

Iterations

Whole Model Test

Model	-LogLikelihood	DF	ChiSquare	Prob>ChiSq
Difference	18.968295	1	37.93659	<.0001*
Full	18.313369			
Reduced	37.281664			

RSquare (U)	0.5088
AICc	40.862
BIC	44.6047
Observations (or Sum Wgts)	54

Measure	Training	Definition
Entropy RSquare	0.5088	1-Loglike(model)/LogLike(null)
Generalized R-Square	0.5047	1-(Like(null)/Like(model))^(2/n)
Mean -Log p	0.3391	∑ -Log(ρ[j])/n
RMSE	0.3298	√ ∑(y[j]-ρ[j])²/n
Mean Abs Dev	0.2149	∑ |y[j]-ρ[j]|/n
Misclassification Rate	0.1481	∑ (ρ[j]≠ρMax)/n
N	54	n

Parameter Estimates

Term	Estimate	Std Error	ChiSquare	Prob>ChiSq
Intercept	2.68517103	0.8253384	10.58	0.0011*
ln(dose)	2.9096072	0.7682096	14.35	0.0002*

For log odds of Cured/Died

Covariance of Estimates

Cov

	Intercept	ln(dose)
Intercept	0.6812	0.5469
ln(dose)	0.5469	0.5901

The Logistic report window contains the Logistic plot, the Iterations report, the Whole Model Test report, and the Parameter Estimates report.

Note: The red triangle menu provides more options that can add to the initial report window. See "Logistic Platform Options" on page 224.

Logistic Plot

The logistic probability plot gives a complete picture of what the logistic model is fitting. At each *x* value, the probability scale in the *y* direction is divided up (partitioned) into probabilities for each response category. The probabilities are measured as the vertical distance between the curves, with the total across all *Y* category probabilities summing to 1.

Replace variables in the plot in one of two ways: swap existing variables by dragging and dropping a variable from one axis to the other axis; or, click on a variable in the Columns panel of the associated data table and drag it onto an axis.

Related Information

- "Additional Example of a Logistic Plot" on page 228

Iterations

The Iterations report shows each iteration and the evaluated criteria that determine whether the model has converged. Iterations appear only for nominal logistic regression.

Whole Model Test

The Whole Model Test report shows if the model fits better than constant response probabilities. This report is analogous to the Analysis of Variance report for a continuous response model. It is a specific likelihood-ratio Chi-square test that evaluates how well the categorical model fits the data. The negative sum of natural logs of the observed probabilities is called the negative log-likelihood (**–LogLikelihood**). The negative log-likelihood for categorical data plays the same role as sums of squares in continuous data. Twice the difference in the negative log-likelihood from the model fitted by the data and the model with equal probabilities is a Chi-square statistic. This test statistic examines the hypothesis that the x variable has no effect on the responses.

Values of the **Rsquare (U)** (sometimes denoted as R^2) range from 0 to 1. High R^2 values are indicative of a good model fit, and are rare in categorical models.

Table 7.1 Description of the Whole Model Test

Model (sometimes called Source)	• The **Reduced** model only contains an intercept. • The **Full** model contains all of the effects as well as the intercept. • The **Difference** is the difference of the log likelihoods of the full and reduced models.
DF	Records the degrees of freedom associated with the model.

Table 7.1 Description of the Whole Model Test *(Continued)*

<table>
<tr><td>–LogLikelihood</td><td>Measures variation, sometimes called uncertainty, in the sample.

Full (the full model) is the negative log-likelihood (or uncertainty) calculated after fitting the model. The fitting process involves predicting response rates with a linear model and a logistic response function. This value is minimized by the fitting process.

Reduced (the reduced model) is the negative log-likelihood (or uncertainty) for the case when the probabilities are estimated by fixed background rates. This is the background uncertainty when the model has no effects.

The difference of these two negative log-likelihoods is the reduction due to fitting the model. Two times this value is the likelihood-ratio Chi-square test statistic.</td></tr>
<tr><td>Chi-Square</td><td>The likelihood-ratio Chi-square test of the hypothesis that the model fits no better than fixed response rates across the whole sample. It is twice the –LogLikelihood for the Difference Model. It is two times the difference of two negative log-likelihoods, one with whole-population response probabilities and one with each-population response rates.

For more information, see “Statistical Details for the Whole Model Test Report” on page 233.</td></tr>
<tr><td>Prob>ChiSq</td><td>The observed significance probability, often called the p value, for the Chi-square test. It is the probability of getting, by chance alone, a Chi-square value greater than the one computed. Models are often judged significant if this probability is below 0.05.</td></tr>
<tr><td>Rsquare (U)</td><td>The proportion of the total uncertainty that is attributed to the model fit.

To test that the factor variable has no effect on the response, look at the difference between the following:

• the log-likelihood from the fitted model
• the log-likelihood from the model that uses horizontal lines

For more information, see “Statistical Details for the Whole Model Test Report” on page 233.</td></tr>
<tr><td>AICc</td><td>The corrected Akaike Information Criterion.</td></tr>
<tr><td>BIC</td><td>The Bayesian Information Criterion.</td></tr>
<tr><td>Observations

(sometimes called Sum Wgts)</td><td>The total sample size used in computations. If you specified a Weight variable, this is the sum of the weights.</td></tr>
</table>

Table 7.1 Description of the Whole Model Test *(Continued)*

Measure	The available measures of fit are as follows: **Entropy RSquare** compares the log-likelihoods from the fitted model and the constant probability model. **Generalized RSquare** is a generalization of the Rsquare measure that simplifies to the regular Rsquare for continuous normal responses. It is similar to the Entropy RSquare, but instead of using the log-likelihood, it uses the 2/n root of the likelihood. **Mean -Log p** is the average of -log(p), where p is the fitted probability associated with the event that occurred. **RMSE** is the root mean square error, where the differences are between the response and p (the fitted probability for the event that actually occurred). **Mean Abs Dev** is the average of the absolute values of the differences between the response and p (the fitted probability for the event that actually occurred). **Misclassification Rate** is the rate for which the response category with the highest fitted probability is not the observed category. For Entropy RSquare and Generalized RSquare, values closer to 1 indicate a better fit. For Mean -Log p, RMSE, Mean Abs Dev, and Misclassification Rate, smaller values indicate a better fit.
Training	The value of the measure of fit.
Definition	The algebraic definition of the measure of fit.

Parameter Estimates

The nominal logistic model fits a parameter for the intercept and slope for each of $k - 1$ logistic comparisons, where k is the number of response levels. The Parameter Estimates report lists these estimates. Each parameter estimate can be examined and tested individually, although this is seldom of much interest.

Table 7.2 Description of the Parameter Estimates Report

Term	Lists each parameter in the logistic model. There is an intercept and a slope term for the factor at each level of the response variable, except the last level.
Estimate	Lists the parameter estimates given by the logistic model.
Std Error	Lists the standard error of each parameter estimate. They are used to compute the statistical tests that compare each term to zero.

Table 7.2 Description of the Parameter Estimates Report *(Continued)*

Chi-Square	Lists the Wald tests for the hypotheses that each of the parameters is zero. The Wald Chi-square is computed as (Estimate/Std Error)^2.
Prob>ChiSq	Lists the observed significance probabilities for the Chi-square tests.

Covariance of Estimates

Reports the estimated variances of the parameter estimates, and the estimated covariances between the parameter estimates. The square root of the variance estimates is the same as those given in the **Std Error** section.

Logistic Platform Options

The following table describes the options within the Logistic Fit red triangle menu.

Odds Ratios	Adds odds ratios to the Parameter Estimates report. For more details, see the *Modeling and Multivariate Methods* book. This option is available only for a response with two levels.
Inverse Prediction	Prediction of *x* values from given *y* values. For more information, see "Inverse Prediction" on page 226. This option is available only for a response with two levels.
Logistic Plot	Shows or hides the logistic plot.
Plot Options	The Plot Options menu includes the following options: **Show Points** Toggles the points on or off. **Show Rate Curve** Is useful only if you have several points for each *x*-value. In these cases, you get reasonable estimates of the rate at each value, and compare this rate with the fitted logistic curve. To prevent too many degenerate points, usually at zero or one, JMP only shows the rate value if there are at least three points at the *x*-value. **Line Color** Enables you to pick the color of the plot curves.
ROC Curve	A Receiver Operating Characteristic curve is a plot of sensitivity by (1 – specificity) for each value of x. See "ROC Curves" on page 225.

Lift Curve	Produces a lift curve for the model. A lift curve shows the same information as a ROC curve, but in a way to dramatize the richness of the ordering at the beginning. The *Y*-axis shows the ratio of how rich that portion of the population is in the chosen response level compared to the rate of that response level as a whole. See the *Modeling and Multivariate Methods* book for details about lift curves.
Save Probability Formula	Creates new data table columns that contain formulas. See "Save Probability Formula" on page 226.
Script	This menu contains options that are available to all platforms. They enable you to redo the analysis or save the JSL commands for the analysis to a window or a file. For more information, see *Using JMP*.

ROC Curves

Suppose you have an *x* value that is a diagnostic measurement and you want to determine a threshold value of *x* that indicates the following:

- A condition exists if the *x* value is greater than the threshold.
- A condition does not exist if the *x* value is less than the threshold.

For example, you could measure a blood component level as a diagnostic test to predict a type of cancer. Now consider the diagnostic test as you vary the threshold and, thus, cause more or fewer false positives and false negatives. You then plot those rates. The ideal is to have a very narrow range of *x* criterion values that best divides true negatives and true positives. The Receiver Operating Characteristic (ROC) curve shows how rapidly this transition happens, with the goal being to have diagnostics that maximize the area under the curve.

Two standard definitions used in medicine are as follows:

- *Sensitivity*, the probability that a given *x* value (a test or measure) correctly predicts an existing condition. For a given *x*, the probability of incorrectly predicting the existence of a condition is 1 – sensitivity.
- *Specificity*, the probability that a test correctly predicts that a condition does not exist.

A ROC curve is a plot of sensitivity by (1 – specificity) for each value of *x*. The area under the ROC curve is a common index used to summarize the information contained in the curve.

When you do a simple logistic regression with a binary outcome, there is a platform option to request a ROC curve for that analysis. After selecting the **ROC Curve** option, a window asks you to specify which level to use as *positive*.

If a test predicted perfectly, it would have a value above which the entire abnormal population would fall and below which all normal values would fall. It would be perfectly sensitive and then pass through the point (0,1) on the grid. The closer the ROC curve comes to this ideal point, the better its discriminating ability. A test with no predictive ability produces a curve that follows the diagonal of the grid (DeLong, et al. 1988).

The ROC curve is a graphical representation of the relationship between false-positive and true-positive rates. A standard way to evaluate the relationship is with the area under the curve, shown below the plot in the report. In the plot, a yellow line is drawn at a 45 degree angle tangent to the ROC Curve. This marks a good cutoff point under the assumption that false negatives and false positives have similar costs.

Related Information

- "Example of ROC Curves" on page 230

Save Probability Formula

The **Save Probability Formula** option creates new data table columns. These data table columns save the following:

- formulas for linear combinations (typically called logits) of the x factor
- prediction formulas for the response level probabilities
- a prediction formula that gives the most likely response

Inverse Prediction

Inverse prediction is the opposite of prediction. It is the prediction of x values from given y values. But in logistic regression, instead of a y value, you have the probability attributed to one of the Y levels. This feature only works when there are two response categories (a binary response).

The Fit Model platform also has an option that gives an inverse prediction with confidence limits. The *Modeling and Multivariate Methods* book gives more information about inverse prediction.

Related Information

- "Example of Inverse Prediction Using the Crosshair Tool" on page 231
- "Example of Inverse Prediction Using the Inverse Prediction Option" on page 232

Additional Examples of Logistic Regression

This section contains additional examples using logistic regression.

Example of Ordinal Logistic Regression

This example uses the AdverseR.jmp sample data table to illustrate an ordinal logistic regression. Suppose you want to model the severity of an adverse event as a function of treatment duration value.

1. Open the AdverseR.jmp sample data table.
2. Right-click on the icon to the left of ADR SEVERITY and change the modeling type to ordinal.
3. Select **Analyze > Fit Y by X**.

4. Select ADR SEVERITY and click **Y, Response**.
5. Select ADR DURATION and click **X, Factor**.
6. Click **OK**.

Figure 7.5 Example of Ordinal Logistic Report

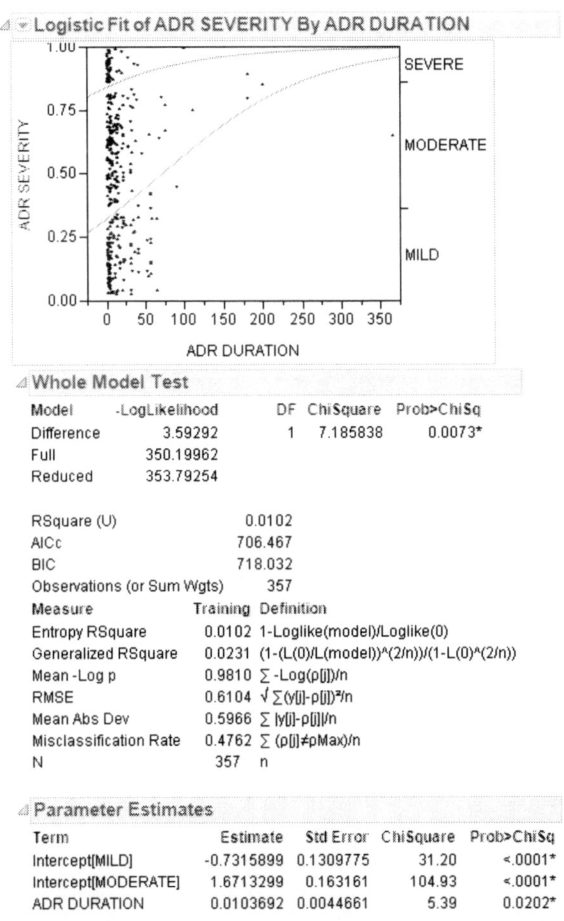

Whole Model Test

Model	-LogLikelihood	DF	ChiSquare	Prob>ChiSq
Difference	3.59292	1	7.185838	0.0073*
Full	350.19962			
Reduced	353.79254			

RSquare (U)	0.0102
AICc	706.467
BIC	718.032
Observations (or Sum Wgts)	357

Measure	Training	Definition
Entropy RSquare	0.0102	1-Loglike(model)/Loglike(0)
Generalized RSquare	0.0231	(1-(L(0)/L(model))^(2/n))/(1-L(0)^(2/n))
Mean -Log p	0.9810	∑ -Log(p[j])/n
RMSE	0.6104	√ ∑(y[j]-p[j])²/n
Mean Abs Dev	0.5966	∑ |y[j]-p[j]|/n
Misclassification Rate	0.4762	∑ (p[j]≠pMax)/n
N	357	n

Parameter Estimates

Term	Estimate	Std Error	ChiSquare	Prob>ChiSq
Intercept[MILD]	-0.7315899	0.1309775	31.20	<.0001*
Intercept[MODERATE]	1.6713299	0.163161	104.93	<.0001*
ADR DURATION	0.0103692	0.0044661	5.39	0.0202*

You interpret this report the same way as the nominal report. See "The Logistic Report" on page 219.

In the plot, markers for the data are drawn at their *x*-coordinate. When several data points appear at the same *y* position, the points are jittered. That is, small spaces appear between the data points so you can see each point more clearly.

Where there are many points, the curves are pushed apart. Where there are few to no points, the curves are close together. The data pushes the curves in that way because the criterion that is maximized is the product of the probabilities fitted by the model. The fit tries to avoid points attributed to have a small probability, which are points crowded by the curves of fit. See the *Modeling and Multivariate Methods* book for more information about computational details.

For details about the Whole Model Test report and the Parameter Estimates report, see "The Logistic Report" on page 219. In the Parameter Estimates report, an intercept parameter is estimated for every response level except the last, but there is only one slope parameter. The intercept parameters show the spacing of the response levels. They always increase monotonically.

Additional Example of a Logistic Plot

This example uses the Car Physical Data.jmp sample data table to show an additional example of a logistic plot. Suppose you want to use weight to predict car size (Type) for 116 cars. Car size can be one of the following, from smallest to largest: Sporty, Small, Compact, Medium, or Large.

1. Open the Car Physical Data.jmp sample data table.
2. In the Columns panel, right-click on the icon to the left of Type, and select **Ordinal**.
3. Right-click on Type, and select **Column Info**.
4. From the Column Properties menu, select **Value Ordering**.
5. Move the data in the following top-down order: Sporty, Small, Compact, Medium, Large.
6. Click **OK**.
7. Select **Analyze > Fit Y by X**.
8. Select Type and click **Y, Response**.
9. Select Weight and click **X, Factor**.
10. Click **OK**.

    The report window appears.

Figure 7.6 Example of Type by Weight Logistic Plot

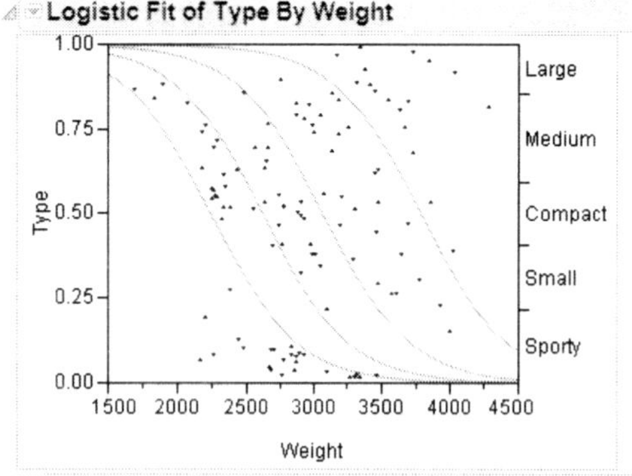

In Figure 7.6, note the following observations:

- The first (bottom) curve represents the probability that a car at a given weight is Sporty.
- The second curve represents the probability that a car is Small or Sporty. Looking only at the distance between the first and second curves corresponds to the probability of being Small.
- As you might expect, heavier cars are more likely to be Large.
- Markers for the data are drawn at their *x*-coordinate, with the *y* position jittered randomly within the range corresponding to the response category for that row.

If the *x* -variable has no effect on the response, then the fitted lines are horizontal and the probabilities are constant for each response across the continuous factor range. Figure 7.7 shows a logistic plot where **Weight** is not useful for predicting **Type**.

Figure 7.7 Examples of Sample Data Table and Logistic Plot Showing No *y* by *x* Relationship

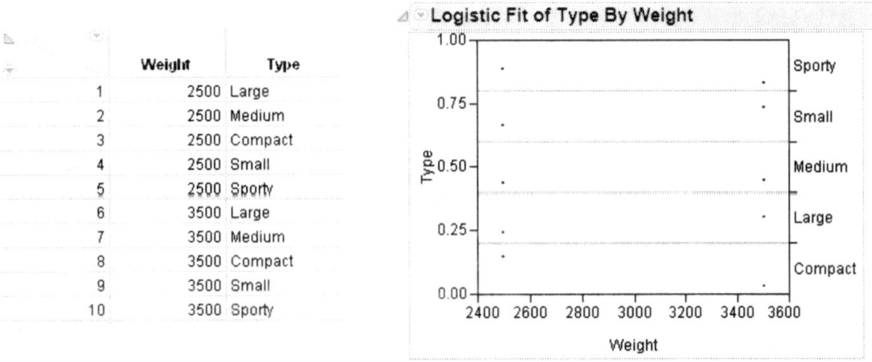

	Weight	Type
1	2500	Large
2	2500	Medium
3	2500	Compact
4	2500	Small
5	2500	Sporty
6	3500	Large
7	3500	Medium
8	3500	Compact
9	3500	Small
10	3500	Sporty

Note: To re-create the plots in Figure 7.7 and Figure 7.8, you must first create the data tables shown here, and then perform steps 7-10 at the beginning of this section.

If the response is completely predicted by the value of the factor, then the logistic curves are effectively vertical. The prediction of a response is near certain (the probability is almost 1) at each of the factor levels. Figure 7.8 shows a logistic plot where **Weight** almost perfectly predicts **Type**.

Note: In this case, the parameter estimates become very large and are marked ***unstable*** in the regression report.

Figure 7.8 Examples of Sample Data Table and Logistic Plot Showing an Almost Perfect y by x Relationship

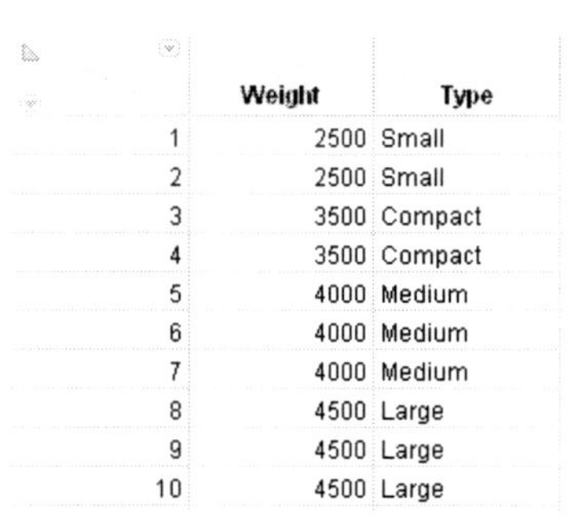

	Weight	Type
1	2500	Small
2	2500	Small
3	3500	Compact
4	3500	Compact
5	4000	Medium
6	4000	Medium
7	4000	Medium
8	4500	Large
9	4500	Large
10	4500	Large

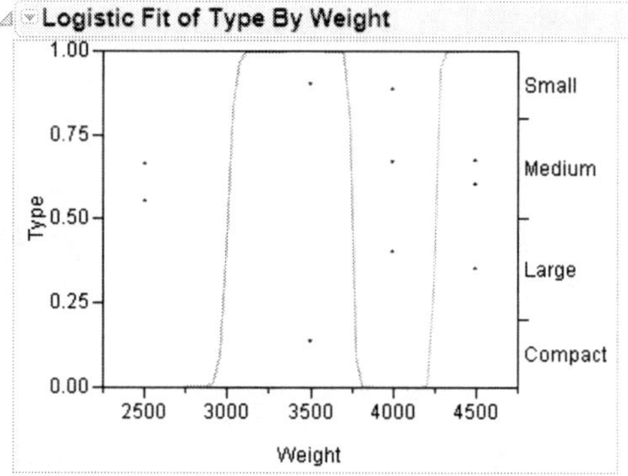

Example of ROC Curves

To demonstrate ROC curves, proceed as follows:

1. Open the Penicillin.jmp sample data table.
2. Select **Analyze > Fit Y by X**.
3. Select Response and click **Y, Response**.
4. Select ln(Dose) and click **X, Factor**.

   Notice that JMP automatically fills in Count for **Freq**. Count was previously assigned the role of Freq.
5. Click **OK**.
6. From the red triangle menu, select **ROC Curve**.
7. Select Cured as the positive.
8. Click **OK**.

Note: This example shows a ROC Curve for a nominal response. For details about ordinal ROC curves, see the Recursive Partitioning chapter in the *Modeling and Multivariate Methods* book.

The results for the response by ln(Dose) example are shown here. The ROC curve plots the probabilities described above, for predicting response. Note that in the ROC Table, the row with the highest Sens-(1-Spec) is marked with an asterisk.

Figure 7.9 Examples of ROC Curve and Table

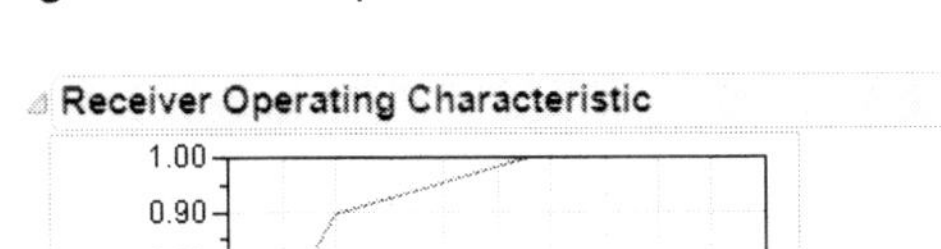

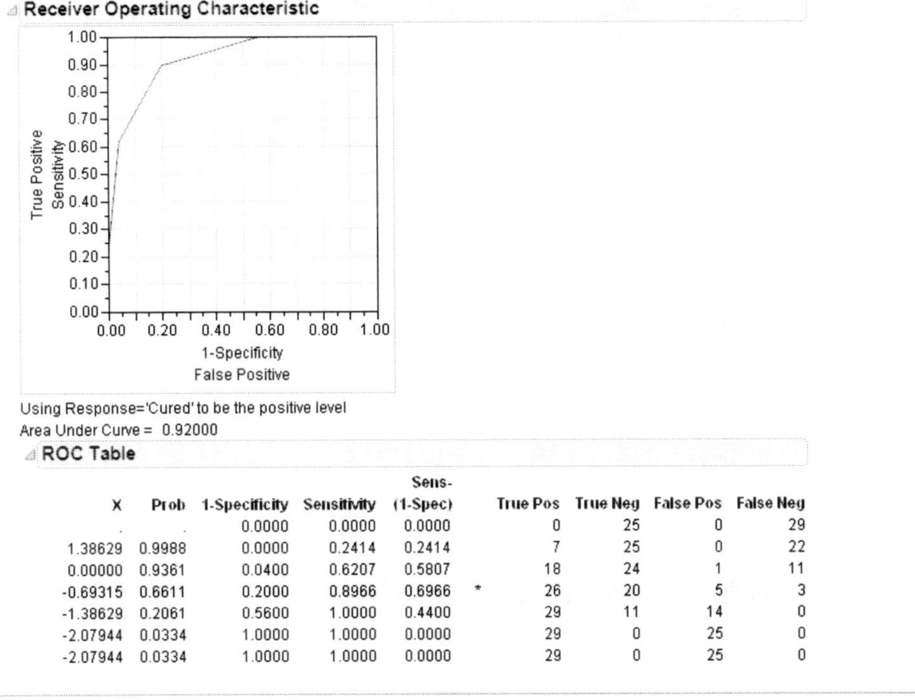

X	Prob	1-Specificity	Sensitivity	Sens-(1-Spec)		True Pos	True Neg	False Pos	False Neg
.	.	0.0000	0.0000	0.0000		0	25	0	29
1.38629	0.9988	0.0000	0.2414	0.2414		7	25	0	22
0.00000	0.9361	0.0400	0.6207	0.5807		18	24	1	11
-0.69315	0.6611	0.2000	0.8966	0.6966	*	26	20	5	3
-1.38629	0.2061	0.5600	1.0000	0.4400		29	11	14	0
-2.07944	0.0334	1.0000	1.0000	0.0000		29	0	25	0
-2.07944	0.0334	1.0000	1.0000	0.0000		29	0	25	0

Since the ROC curve is well above a diagonal line, you conclude that the model has good predictive ability.

Example of Inverse Prediction Using the Crosshair Tool

In a study of rabbits who were given penicillin, you want to know what dose of penicillin results in a 0.5 probability of curing a rabbit. In this case, the inverse prediction for 0.5 is called the *ED50,* the effective dose corresponding to a 50% survival rate. Use the crosshair tool to visually approximate an inverse prediction.

To see which value of ln(dose) is equally likely either to cure or to be lethal, proceed as follows:

1. Open the Penicillin.jmp sample data table.
2. Select **Analyze > Fit Y by X**.
3. Select Response and click **Y, Response**.
4. Select ln(Dose) and click **X, Factor**.

   Notice that JMP automatically fills in Count for **Freq**. Count was previously assigned the role of Freq.
5. Click **OK**.
6. Click on the crosshairs tool.
7. Place the horizontal crosshair line at about 0.5 on the vertical (Response) probability axis.

8. Move the cross-hair intersection to the prediction line, and read the ln(dose) value that shows on the horizontal axis.

In this example, a rabbit with a ln(dose) of approximately -0.9 is equally likely to be cured as it is to die.

Figure 7.10 Example of Crosshair Tool on Logistic Plot

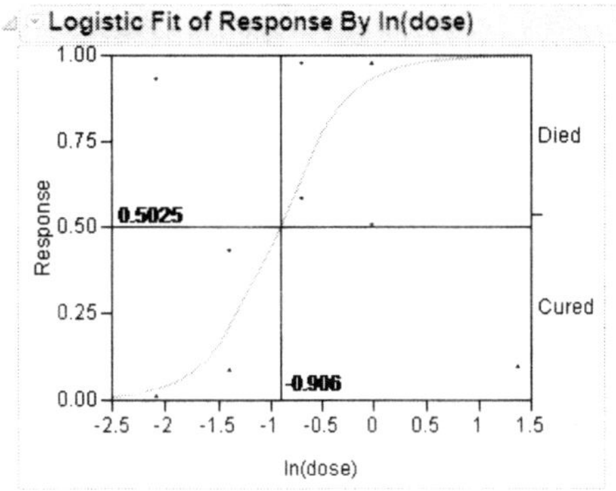

Example of Inverse Prediction Using the Inverse Prediction Option

If your response has exactly two levels, the **Inverse Prediction** option enables you to request an exact inverse prediction. You are given the *x* value corresponding to a given probability of the lower response category, as well as a confidence interval for that *x* value.

To use the **Inverse Prediction** option, proceed as follows:

1. Open the Penicillin.jmp sample data table.
2. Select **Analyze > Fit Y by X**.
3. Select Response and click **Y, Response**.
4. Select ln(Dose) and click **X, Factor**.

   Notice that JMP automatically fills in Count for **Freq**. Count was previously assigned the role of Freq.
5. Click **OK**.
6. From the red triangle menu, select **Inverse Prediction**. See Figure 7.11.
7. Type 0.95 for the **Confidence Level.**
8. Select **Two sided** for the confidence interval.
9. Request the response probability of interest. Type 0.5 and 0.9 for this example, which indicates you are requesting the values for ln(Dose) that correspond to a 0.5 and 0.9 probability of being cured.
10. Click **OK**.

    The Inverse Prediction plot appears.

Figure 7.11 Inverse Prediction Window

Specify one or more probability values you want to inverse-predict for.

ln(dose) (to predict) Confidence Level

0.95

Two sided

Probability(Response=Cured)

OK Cancel Help

Figure 7.12 Example of Inverse Prediction Plot

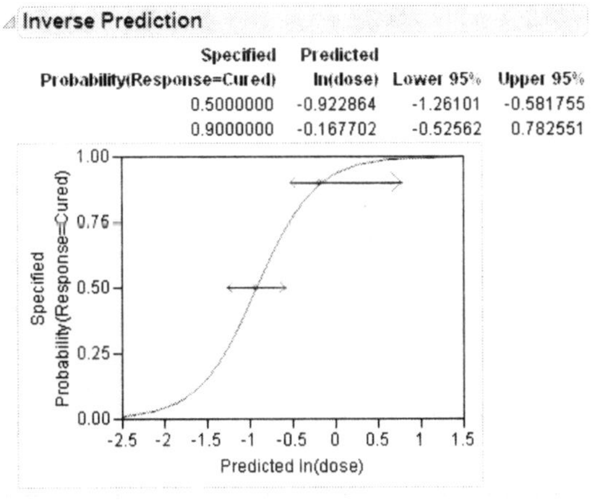

Inverse Prediction

Specified Probability(Response=Cured)	Predicted ln(dose)	Lower 95%	Upper 95%
0.5000000	-0.922864	-1.26101	-0.581755
0.9000000	-0.167702	-0.52562	0.782551

The estimates of the x values and the confidence intervals are shown in the report as well as in the probability plot. For example, the value of ln(Dose) that results in a 90% probability of being cured is estimated to be between -0.526 and 0.783.

Statistical Details for the Logistic Platform

Statistical Details for the Whole Model Test Report

Chi-Square

The Chi-Square statistic is sometimes denoted G^2 and is written as follows:

$$G^2 = 2(\sum -\ln p(\text{background}) - \sum -\ln p(\text{model}))$$

where the summations are over all observations instead of all cells.

Rsquare (U)

The ratio of this test statistic to the background log-likelihood is subtracted from 1 to calculate R^2. More simply, Rsquare (U) is computed as follows:

$$\frac{-\text{loglikelihood for Difference}}{-\text{loglikelihood for Reduced}}$$

using quantities from the Whole Model Test report.

Chapter 8

Comparing Paired Data

Using the Matched Pairs Platform

The Matched Pairs platform compares the means between two or more correlated variables and assesses the differences. For example, you might compare a blood pressure measurement taken on the same subject before a treatment and again after the treatment. A statistical method called the *paired t-test* takes the correlated responses into account.

The platform produces a graph of the paired differences by the paired means, and the paired *t*-test results for all three alternative hypotheses. Additional features provide for more than two matched responses and for a grouping column to test across samples, in a simple version of repeated measures analysis.

Figure 8.1 Example of Matched Pairs Analysis

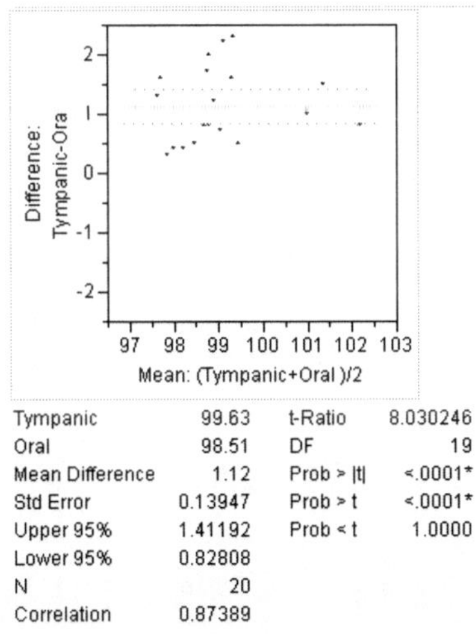

Tympanic	99.63	t-Ratio	8.030246
Oral	98.51	DF	19
Mean Difference	1.12	Prob > |t|	<.0001*
Std Error	0.13947	Prob > t	<.0001*
Upper 95%	1.41192	Prob < t	1.0000
Lower 95%	0.82808		
N	20		
Correlation	0.87389		

Contents

Overview of the Matched Pairs Platform

The Matched Pairs platform compares row-by-row differences between two response columns using a paired *t*-test. Often, the two columns represent measurements on the same subject before and after some treatment. Alternatively, the measurements could represent data taken on the same subject with two different instruments.

If you have paired data arranged in two data table columns, then you are ready to use the Matched Pairs platform. However, if all of your measurements are in a single column, then perform one of the following tasks:

- Use the **Split** option in the **Tables** menu to split the column of measurements into two columns. Then you can use the Matched Pairs platform.
- For two response columns, create a third column that calculates the difference between the two responses. Then test that the mean of the difference column is zero with the Distribution platform.
- For the two responses stored in a single column, you can do a two-way analysis of variance. One factor (the ID variable) identifies the two responses and the other factor identifies the subject. Use the Fit Y by X Oneway platform with a blocking variable (the subject column), or use the Fit Model platform to do a two-way ANOVA. The test on the ID factor is equivalent to the paired *t*-test.

Note: If the data are paired, do not do a regular independent *t*-test. Do not stack the data into one column and use the Fit Y by X One-way ANOVA on the ID without specifying a block variable. To do this has the effect of ignoring the correlation between the responses. This causes the test to overestimate the effect if responses are negatively correlated, or to underestimate the effect if responses are positively correlated.

Example of Comparing Matched Pairs

This example uses the Therm.jmp sample data table. The data contains temperature measurements on 20 people. Temperature is measured using two types of thermometers: oral and tympanic (ear). You want to determine whether the two types of thermometers produce equal temperature readings. Note that the differences in temperature between the different people are not important. The matched pairs analysis is testing the differences between the thermometers.

1. Open the Therm.jmp sample data table.
2. Select **Analyze > Matched Pairs.**
3. Select Oral and Tympanic and click **Y, Paired Response.**
4. Click **OK.**

   The report window appears.

Figure 8.2 The Matched Pairs Report Window

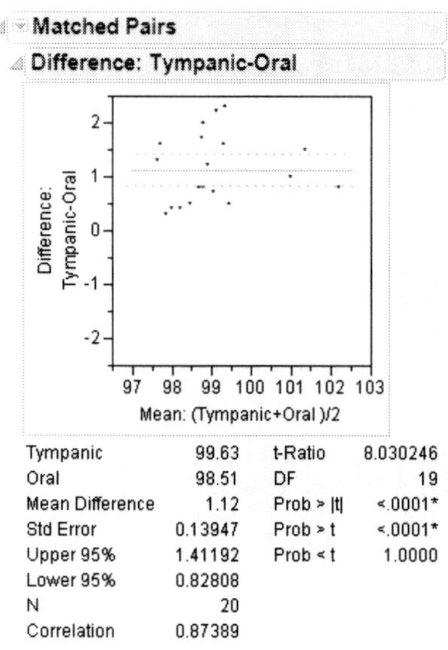

Tympanic	99.63	t-Ratio	8.030246
Oral	98.51	DF	19
Mean Difference	1.12	Prob > |t|	<.0001*
Std Error	0.13947	Prob > t	<.0001*
Upper 95%	1.41192	Prob < t	1.0000
Lower 95%	0.82808		
N	20		
Correlation	0.87389		

The results show that, on average, the tympanic thermometer measures 1.12 degrees higher than the oral thermometer. The small p-value (Prob > |t|) indicates that this difference is statistically significant, and not due to chance.

Note that this matched pairs analysis does not indicate which thermometer is correct (if either), but only indicates that there is a difference between the thermometers.

Launch the Matched Pairs Platform

Launch the Matched Pairs platform by selecting **Analyze > Matched Pairs**.

Figure 8.3 The Matched Pairs Launch Window

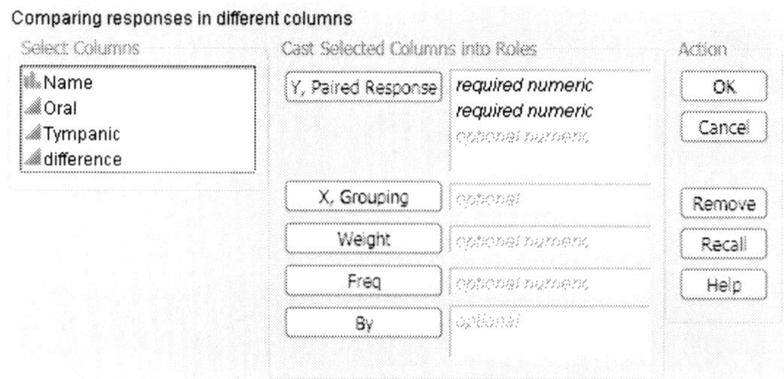

Table 8.1 Description of the Matched Pairs Launch Window

Y, Paired Response	Provide the two response columns. For information about analyzing more than two responses, see "Multiple Y Columns" on page 239.
X, Grouping	(Optional) Provide a grouping variable to compare the differences across groups. For more information, see "Across Groups" on page 240.
Weight	(Optional) Identifies one column whose numeric values assign a weight to each row in the analysis.
Freq	(Optional) Identifies one column whose numeric values assign a frequency to each row in the analysis.
By	(Optional) Performs a separate matched pairs analysis for each level of the By variable.

For more information about the launch window, see *Using JMP*.

Multiple Y Columns

You can have more than two responses. If the number of responses is odd, all possible pairs are analyzed. The following table shows an example for three responses.

Y1 by Y2	Y1 by Y3
	Y2 by Y3

If the number of responses is even, the Matched Pairs platform asks whether you want to do all possible pairs. If you do not do all possible pairs, adjacent responses are analyzed as a pair. The following table shows the arrangement of analyses for four responses.

Y1 by Y2	Y3 by Y4

The Matched Pairs Report

Follow the instructions in "Example of Comparing Matched Pairs" on page 237 to produce the report window shown in Figure 8.4.

The Matched Pairs report shows a Tukey mean-difference plot, summary statistics, and the results of the paired t-test. See "Difference Plot and Report" on page 240. If you specified an **X, Grouping** variable, the report also includes the Across Groups report. See "Across Groups" on page 240.

Figure 8.4 Example of Matched Pairs Report

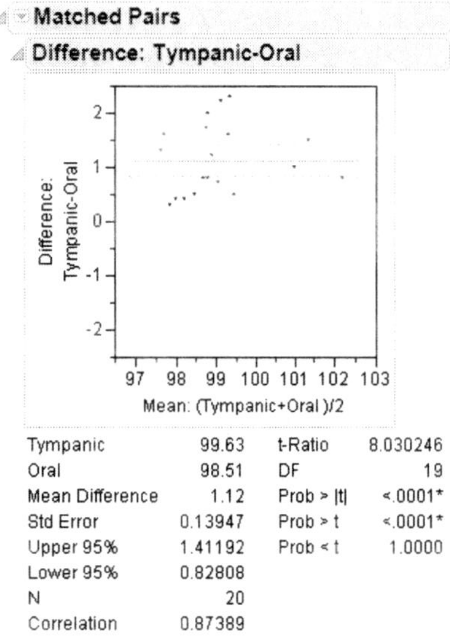

Tympanic	99.63	t-Ratio	8.030246
Oral	98.51	DF	19
Mean Difference	1.12	Prob > |t|	<.0001*
Std Error	0.13947	Prob > t	<.0001*
Upper 95%	1.41192	Prob < t	1.0000
Lower 95%	0.82808		
N	20		
Correlation	0.87389		

Note: The red triangle menu provides additional options that can add reports to the initial report window. See "Matched Pairs Platform Options" on page 241.

Difference Plot and Report

The Difference plot shows differences by means. In the Difference plot, note the following:

- The mean difference is shown as the horizontal line, with the 95% confidence interval above and below shown as dotted lines. If the confidence region includes zero, then the means are not significantly different at the 0.05 level. In this example the difference is significant.
- If you add a reference frame, the mean of pairs is shown by the vertical line. For details about a reference frame, see "Matched Pairs Platform Options" on page 241.

The Difference report shows the mean of each response, the difference of the means, and a confidence interval for the difference. The Difference report also shows the results of the paired t-test.

Across Groups

Note: The Across Groups report appears only if you have specified an **X, Grouping** variable.

The Across Groups analysis corresponds to a simple repeated measures analysis. (You can get the same test results using the **Manova** personality of the Fit Model platform.)

Table 8.2 Description of the Across Groups Report

Mean Difference	Shows the mean of the difference across rows in each group between the two paired columns. In other words, this is the within-subject by across-subject interaction, or split-plot by whole-plot interaction.
Mean Mean	Shows the mean of the mean across rows in each group across the two paired columns. In other words, this is the across-subject or whole-plot effect.
Test Across Groups	Two *F*-tests determine whether the across-groups values are different: • **Mean Difference** tests that the change across the pair of responses is different in different groups. • **Mean Mean** tests that the average response for a subject is different in different groups

Related Information

- "Example Comparing Matched Pairs across Groups" on page 242

Matched Pairs Platform Options

The following table describes the options in the Matched Pairs red triangle menu.

Plot Dif by Mean	Shows or hides the plot of the paired differences by paired means. For a detailed description of this plot, see "Difference Plot and Report" on page 240.
Plot Dif by Row	Shows or hides the plot of paired differences by row number.
Reference Frame	Shows or hides the reference frame on the Plot Dif by Mean plot. A rectangle showing where a plot of Y2 by Y1 would be located inside the plot, tilted and possibly squished. A vertical red line is shown representing the mean of means. The reference frame is shown initially when the range of the differences is greater than half the range of the data.
Wilcoxon Signed Rank	Shows or hides the Wilcoxon Signed Rank test. This is a nonparametric version of the paired t-test that compares the sizes of the positive differences to the sizes of the negative differences. This test does assume that the distribution of the differences is symmetric. (See Conover 1999, page 350.)

Sign Test	Shows or hides the Sign Test. This is a nonparametric version of the paired t-test that uses only the sign (positive or negative) of the difference for the test.
Set α Level	Changes the alpha level used in the analyses. Affects the confidence intervals in the report and on the plot.
Script	This menu contains options that are available to all platforms. They enable you to redo the analysis or save the JSL commands for the analysis to a window or a file. For more information, see *Using JMP*.

Example Comparing Matched Pairs across Groups

This example uses the Dogs.jmp sample data table. This example shows you how to produce both a Matched Pairs Across Groups report and the corresponding MANOVA report using Fit Model.

1. Open the Dogs.jmp sample data table.
2. Select **Analyze > Matched Pairs**.
3. Select LogHist0 and LogHist1 and click **Y, Paired Response**.
4. Select drug and click **X, Grouping**.
5. Click **OK**.

The report on the left in Figure 8.5 appears.

Now produce the Fit Model report using the same data table.

1. Select **Analyze > Fit Model**.
2. Select LogHist0 and LogHist1 and click **Y**.
3. Select drug and click **Add**.
4. Select the **Manova** personality.
5. Click **Run Model**.
6. In the Response Specification report, select **Repeated Measures** from the **Choose Response** menu.
7. Click **OK**.

Figure 8.5 Examples of Matched Pairs Across Groups and Fit Model MANOVA with Repeated Measures

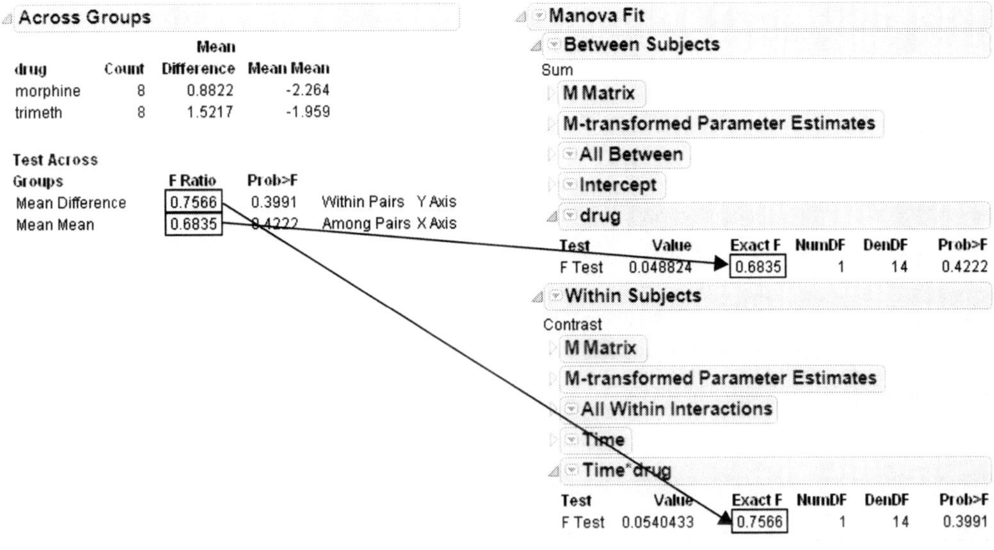

The F Ratio for the Mean Difference in the Across Groups report corresponds to the F Ratio for Time*drug under the Within Subjects report. The F Ratio for the Mean Mean in the Across Groups report corresponds to the F Ratio for drug under Between Subjects in the Manova Fit report.

Statistical Details for the Matched Pairs Platform

This section contains statistical details for Matched Pairs analyses.

Graphics for Matched Pairs

The primary graph in the platform is a Tukey mean-difference (Cleveland 1994, p. 130), which plots the difference of the two responses on the *y*-axis against the mean of the two responses on the *x*-axis. This graph is the same as a scatterplot of the two original variables, but turned 45 degrees. A 45 degree rotation and rescaling turns the original coordinates into a difference and a mean.

Figure 8.6 Example of Transforming to Difference by Mean, Rotated by 45 Degrees

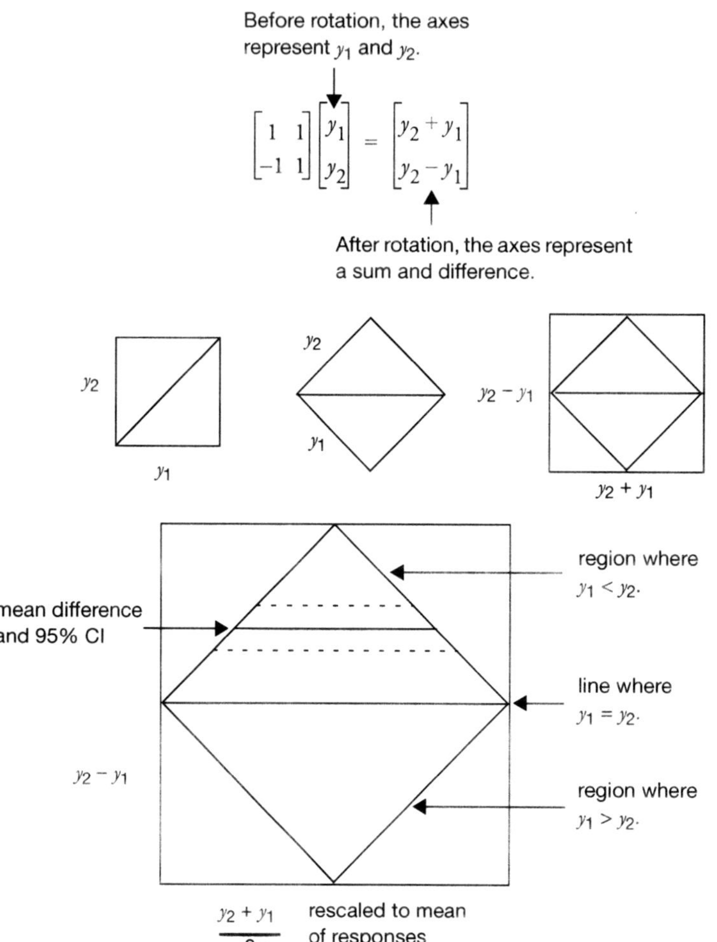

Correlation of Responses

In most cases where the pair of measurements is taken from the same individual at different times, they are positively correlated. However, if they represent competing responses, the correlation can be negative.

Figure 8.7 shows how the positive correlation of the two responses becomes the small variance on the difference (the y-axis). If the correlation is negative, the ellipse is oriented in the other direction and the variance of the rotated graph is large on the y-axis.

Figure 8.7 Examples of Positive Correlation Before and After Rotation

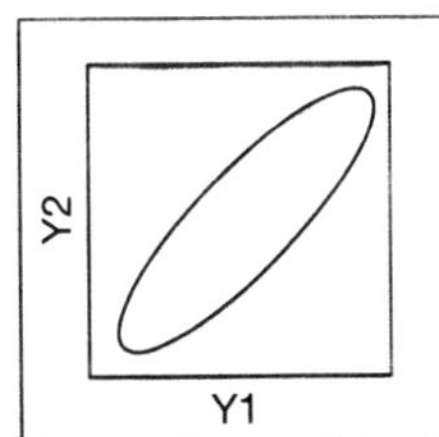

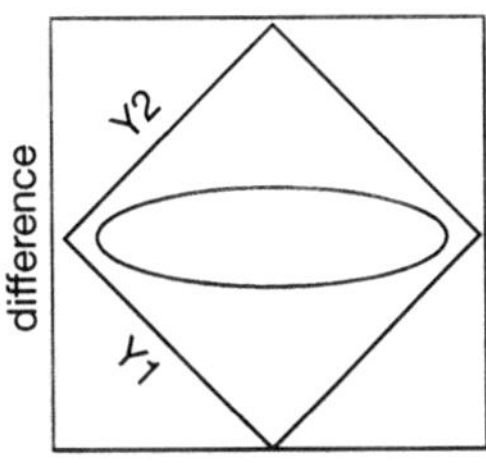

Chapter 9

Bootstrapping

Approximate the Distribution of Statistics

JMP PRO

Bootstrapping is a re-sampling method for approximating the sampling distribution of a statistic. The data is re-sampled with replacement and the statistic is computed. This process is repeated to produce a distribution of values for the statistic.

Bootstrapping is useful when estimating properties of a statistic (mean, standard error, and so on) and performing inference, in the following situations:

- The theoretical distribution of the statistic is complicated or unknown.
- The sample size is too small for regular statistical inference.
- Inference using parametric methods is not possible due to violations of assumptions.

JMP provides bootstrapping for statistical platforms that support Frequency columns in which the rows are assumed to be independent.

The Bootstrap option is on the right-click menu, separate from standard platform commands.

Figure 9.1 Example of the Distribution for Bootstrapping Results

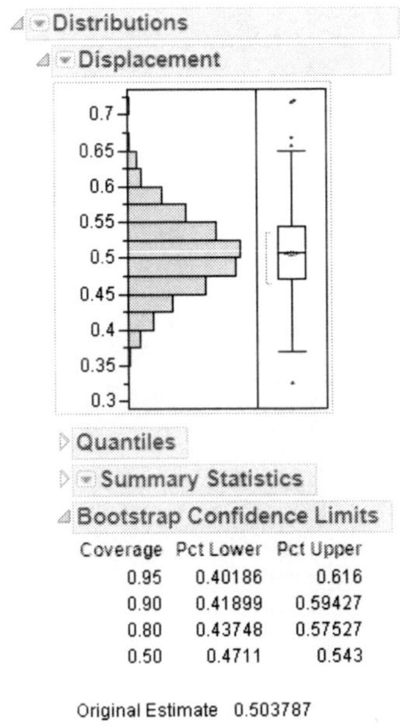

Coverage	Pct Lower	Pct Upper
0.95	0.40186	0.616
0.90	0.41899	0.59427
0.80	0.43748	0.57527
0.50	0.4711	0.543

Original Estimate 0.503787

Contents

Example of Bootstrapping

A tire manufacturer wants to predict an engine's horsepower from the engine's displacement (in^3). The company is most interested in estimating the slope of the relationship between the variables. The slope values help the company predict the corresponding change in horsepower when the displacement changes.

1. Open the Car Physical Data.jmp sample data table.
2. Select **Analyze > Fit Y by X**.
3. Assign Horsepower to the **Y, Response** role.
4. Assign Displacement to the **X, Factor** role.
5. Click **OK**.
6. Select **Fit Line** from the Bivariate Fit red triangle menu.
7. In the **Parameter Estimates** report, right-click the **Estimate** column and select **Bootstrap** (Figure 9.2).

Figure 9.2 Select the Bootstrap Option

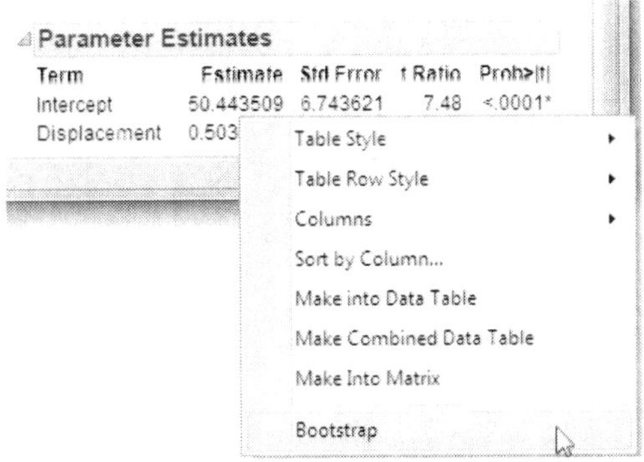

The column that you right-click is relevant when the **Split Selected Column** option is selected. For more information, see "Bootstrap Window Options" on page 251.

8. Type 1000 for the **Number of Bootstrap Samples**.
9. Select **Fractional Weights**. Leave the **Split Selected Column** option selected.
10. Click **OK**.

    The bootstrap process runs and produces two tables.

    Next, analyze the bootstrapped slope.

11. Click on the table with the unstacked results for the slope and intercept, the table with 1,001 rows. Figure 9.3 shows part of the table.

Note: The values in the results tables differ due to the random sampling of observations and assignment of frequencies. Row 1 is the estimate from the original data, and is marked with an X and has the excluded row state.

Figure 9.3 Bootstrap Results Table

	X	Y	BootID•	Displacement	Intercept
1	Displacement	Horsepower	0	0.5037874592	50.443509471
2	Displacement	Horsepower	1	0.4517015789	56.584586818
3	Displacement	Horsepower	2	0.4475978885	57.347393677
4	Displacement	Horsepower	3	0.5664813841	44.234841147
5	Displacement	Horsepower	4	0.5030703865	52.669912731
6	Displacement	Horsepower	5	0.4266247813	61.783063823
7	Displacement	Horsepower	6	0.5102777592	47.216419996
8	Displacement	Horsepower	7	0.4707442785	55.618710427
9	Displacement	Horsepower	8	0.5375189707	47.262722025

12. Select **Analyze > Distribution**.
13. Assign Displacement to the **Y, Columns** role.
14. Click **OK**.

 The Distribution report includes the Bootstrap Confidence Limits report (Figure 9.4).

Figure 9.4 Bootstrap Report

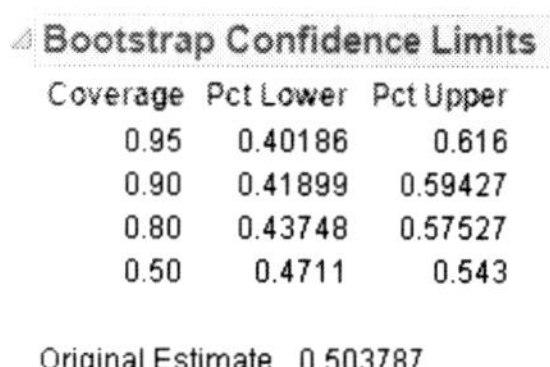
Bootstrap Confidence Limits

Coverage	Pct Lower	Pct Upper
0.95	0.40186	0.616
0.90	0.41899	0.59427
0.80	0.43748	0.57527
0.50	0.4711	0.543

Original Estimate 0.503787

In this example, the estimate of the slope from the original data is 0.504. For 95% coverage of the bootstrap results, the company can estimate the slope to be between 0.398 and 0.605. When the displacement is changed by one unit, the horsepower changes by some amount between 0.398 and 0.605.

Perform a Bootstrap Analysis

To perform a bootstrap analysis, right-click on a report in a platform report window and select **Bootstrap.** Specify the number of bootstrap samples and choose whether to use fractional weights or to split the selected column. For more information, see “Bootstrap Window Options” on page 251. Every statistic in the report is bootstrapped.

The bootstrapping process samples from the original data. An invisible data table is used to store the sampled data and perform the analysis.

Bootstrap Window Options

After you right-click a column in a report and select **Bootstrap**, the Bootstrapping window appears with the following options:

Number of Bootstrap Samples The number of times that you want to re-sample the data and compute the statistics. A higher number results in more precise estimates of the statistics' properties.

Fractional Weights Uses fractional weights in the re-sampling process.

If **Fractional Weights** is not selected, the following observations are true:

- The sampling weights assigned to the original observations are integers.
- The influence of an observation is determined by the relative weights. For example, if an observation is assigned a weight of 2, then that observation appears twice in the sample for the current bootstrap iteration.
- The sampling weights can be zero. Observations with a weight of zero are not included in the estimation process for the current bootstrap iteration.

If **Fractional Weights** is selected, the following observations are true:

- The sampling weights can be non-integers. The influence of an observation is determined by the relative weights.
- The sampling weights cannot be zero. All of the observations are included in the bootstrap process.

In both cases, the sum of the weights is n (the number of observations in the original data).

Split Selected Column Splits a stacked column of bootstrap results into separate columns, one for each statistic. For example, say that the column that you right-clicked before choosing **Bootstrap** has three statistics in it. In addition to the main results table with all the statistics stacked, another results table is created with the results for that column, unstacked into separate columns. For an example, see “Stacked Results Table” on page 251.

Stacked Results Table

The initial results of a bootstrap analysis appear in a stacked results table (Figure 9.5).

Figure 9.5 Stacked Results Table

	X	Y	Term	~Bias	Estimate	Std Error	t Ratio	Prob>|t|	BootID•
1	Displacement	Horsepower	Intercept		50.443509471	6.7436209999	7.48	<.0001	
2	Displacement	Horsepower	Displacement		0.5037874592	0.0398202611	12.65	<.0001	
3	Displacement	Horsepower	Intercept		56.584586818	6.8160870497	8.30	<.0001	
4	Displacement	Horsepower	Displacement		0.4517015789	0.0388523839	11.63	<.0001	
5	Displacement	Horsepower	Intercept		57.347393677	6.4946394861	8.83	<.0001	
6	Displacement	Horsepower	Displacement		0.4475978885	0.0377574205	11.85	<.0001	
7	Displacement	Horsepower	Intercept		44.234841147	7.9668982119	5.55	<.0001	
8	Displacement	Horsepower	Displacement		0.5664813841	0.0450359362	12.58	<.0001	
9	Displacement	Horsepower	Intercept		52.669912731	7.2407550365	7.27	<.0001	

Note the following about the stacked results table:

- The data table columns used in the analysis appear in the table.
- The results table includes every element in the platform report that you are bootstrapping. In this example, the report is the Parameter Estimates report from a Bivariate model in Fit Y by X. The elements are **Term**, **Bias**, **Estimate**, **Std Error**, **t Ratio**, **Prob>|t|**.
- The **BootID•** column identifies the iteration of the bootstrap process. The rows where **BootID•** = 0 correspond to the original estimates. Those rows are marked with an X and have the excluded row state.

If you selected the **Split Selected Column** option, an unstacked results table also appears. See "Unstacked Results Table" on page 252.

Unstacked Results Table

When you right-click a report column and select **Bootstrap**, select **Split Selected Column** to create a data table of unstacked results. The report item that you right-click on is the one that gets unstacked into a separate table. For example, in Figure 9.6, the **Estimate** column from Figure 9.5 is split into two columns (**Displacement** and **Intercept**), one for each level of **Term**.

Figure 9.6 Unstacked Results Table

	X	Y	BootID•	Displacement	Intercept
1	Displacement	Horsepower	0	0.5037874592	50.443509471
2	Displacement	Horsepower	1	0.4517015789	56.584586818
3	Displacement	Horsepower	2	0.4475978885	57.347393677
4	Displacement	Horsepower	3	0.5664813841	44.234841147
5	Displacement	Horsepower	4	0.5030703865	52.669912731
6	Displacement	Horsepower	5	0.4266247813	61.783063823
7	Displacement	Horsepower	6	0.5102777592	47.216419996
8	Displacement	Horsepower	7	0.4707442785	55.618710427
9	Displacement	Horsepower	8	0.5375189707	47.262722025

The **BootID•** column identifies the iteration of the bootstrap process. The row where **BootID•** = 0 correspond to the original estimates. That row is marked with an X and has the excluded row state.

Analysis of Bootstrap Results

Once you have the bootstrap results, you can analyze them using the Distribution platform. Select **Analyze > Distribution** and then assign the columns of interest to the appropriate roles. The Distribution platform searches for a **BootID•** column to produce the Bootstrap Confidence Limits report (Figure 9.7).

Figure 9.7 Bootstrap Confidence Limits Report

Bootstrap Confidence Limits

Coverage	Pct Lower	Pct Upper
0.95	0.40186	0.616
0.90	0.41899	0.59427
0.80	0.43748	0.57527
0.50	0.4711	0.543

Original Estimate 0.503787

The Bootstrap Confidence Limits report shows the following quantiles of the bootstrap results:

Coverage percentage associated with the quantile

Pct Lower the 100((1-p)/2)th quantile of the bootstrap results, where p is the Coverage value

Pct Upper the 100(p + (1-p)/2)th quantile of the bootstrap results, where p is the Coverage value

Original Estimate estimate of the statistic using the original data

For more information about interpreting the Bootstrap Confidence Limits report, see “Example of Bootstrapping” on page 249.

Chapter 10

Interactive Data Visualization

Using Graph Builder

Graph Builder provides a workspace where you can interactively create and modify graphs. Graph types include points, lines, bars, histograms, and more. You can use techniques such as trellising and overlay to examine relationships between several variables on the same graph.

Figure 10.1 Examples Using Graph Builder

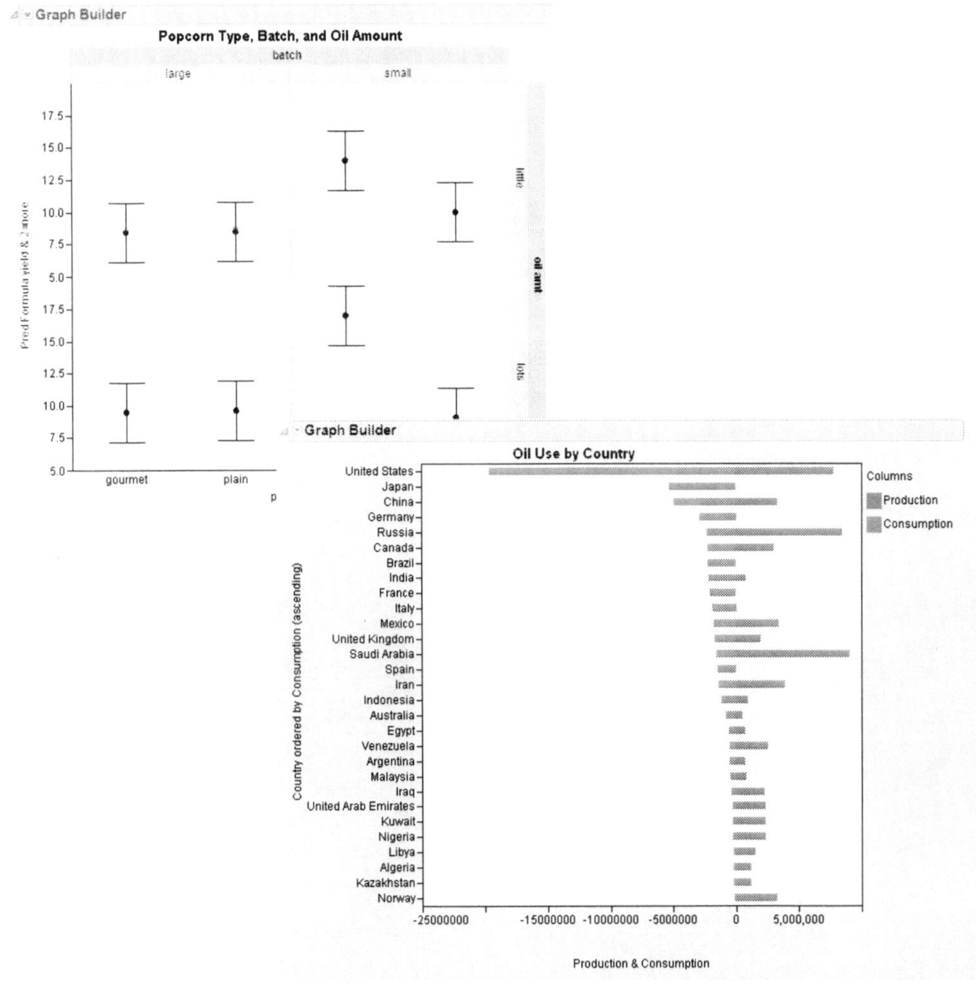

Contents

Overview of Graph Builder

You can interact with Graph Builder to create visualizations of your data. You start with a blank slate and drag and drop variables to place them where you want them. Instant feedback encourages exploration and discovery. Change your mind and move variables to new positions, or right-click to change your settings.

Graph Builder helps you see multi-dimensional relationships in your data with independent grouping variables for side-by-side or overlaid views. With many combinations to compare, you can create a trellis display of small graphs. Graph elements supported by Graph Builder include points, lines, bars, histograms, box plots, and contours.

The underlying philosophy of Graph Builder is to *see your data*. To that end, the default visualization elements impose no assumptions, such as normality. If there are not too many observations, you see all of them as marks on the graph. A smooth trend curve follows the data instead of an equation.

Once you see the data, you can draw conclusions directly, or decide where further analysis is needed to quantify relationships.

Example Using Graph Builder

You have data about nutrition information for candy bars. You want to find out which factors can best predict calorie levels. Working from a basic knowledge of food science, you believe that the fat content is a good place to start.

1. Open the Candy Bars.jmp sample data table.
2. Select **Graph > Graph Builder**.
3. Drag and drop Total fat g into the **X** zone.
4. Drag and drop Calories into the **Y** zone.

Figure 10.2 Example of Calories versus Total fat g

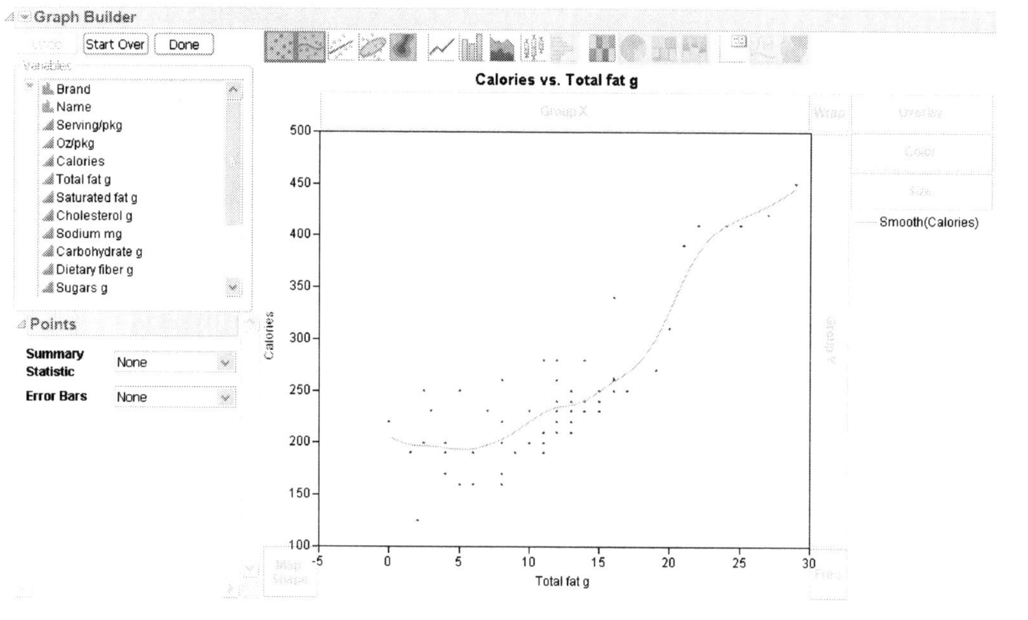

As you suspected, the candy bars with higher fat grams also have higher calories. But the relationship is not perfect. You can add other factors to try to increase the correlation. Next, determine whether cholesterol has an effect.

5. Drag and drop **Cholesterol g** into the **Group X** zone.

Figure 10.3 Example of Calories versus Total fat g by Cholesterol g in the Group X Zone

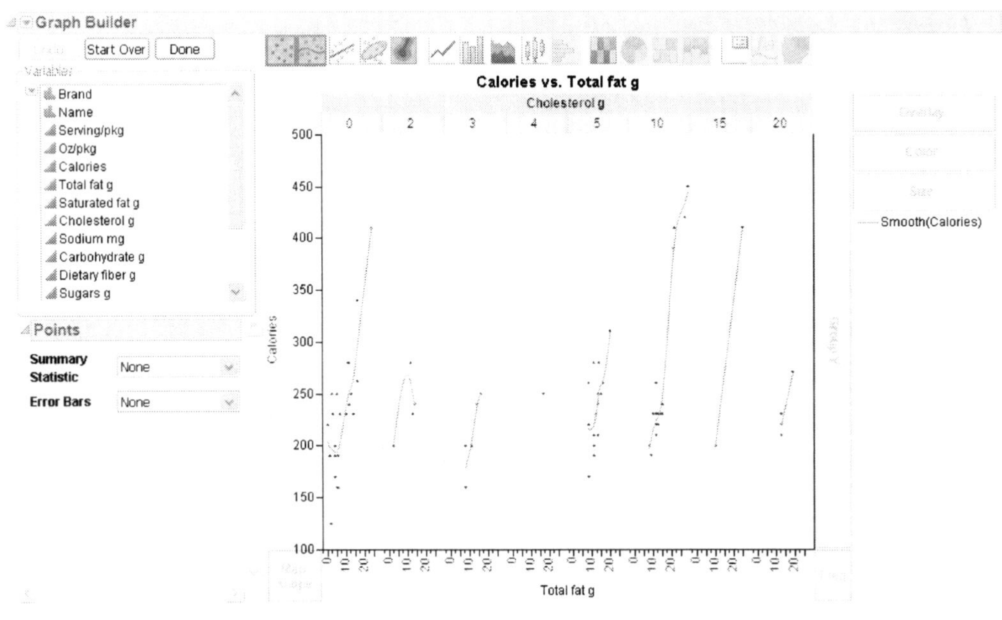

Eight levels of the variable make the graph difficult to read. Try putting Cholesterol g into the Wrap zone instead.

6. Click on Cholesterol g in the **Group X** zone and drag and drop it into the **Wrap** zone.

Figure 10.4 Example of Calories Versus Total fat g by Cholesterol g in the Wrap Zone

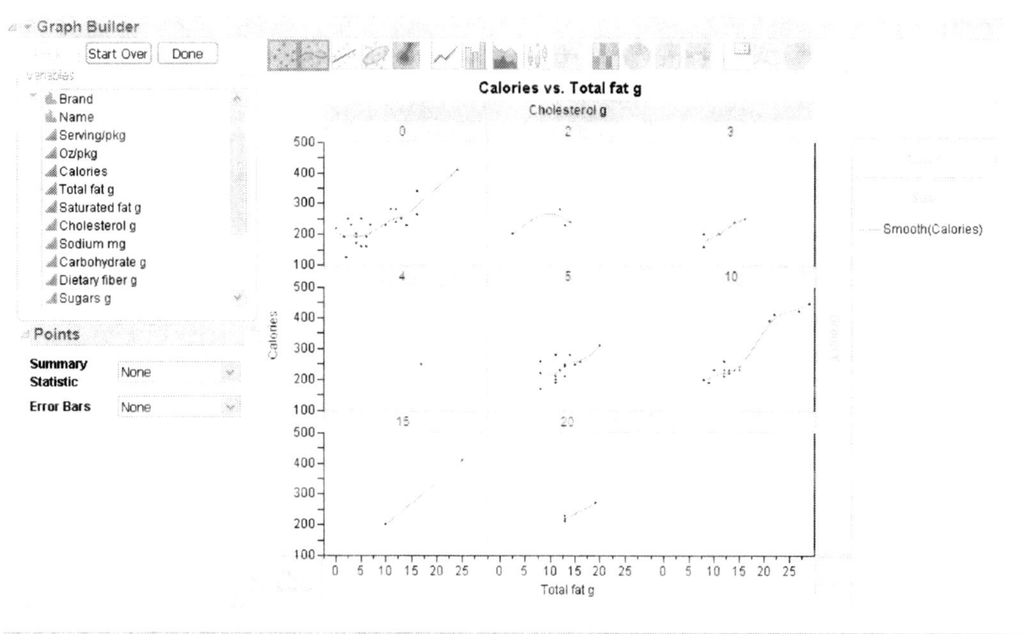

A scatterplot of Calories versus Total fat g is created for every level of Cholesterol g.

You can see that some of the cells have very little data; other cells have a lot of data. Among the cells that have a lot of data (cholesterol equals 0, 5, 10), there is still considerable variation in calories. So you decide to remove Cholesterol g.

7. Remove Cholesterol g by right-clicking on the Cholesterol g label in the **Group X** zone and selecting **Remove**.

Determine whether carbohydrates has any effect.

8. Drag and drop Carbohydrate g into the **Wrap** zone.

Figure 10.5 Example of Calories Versus Total fat g by Carbohydrate g

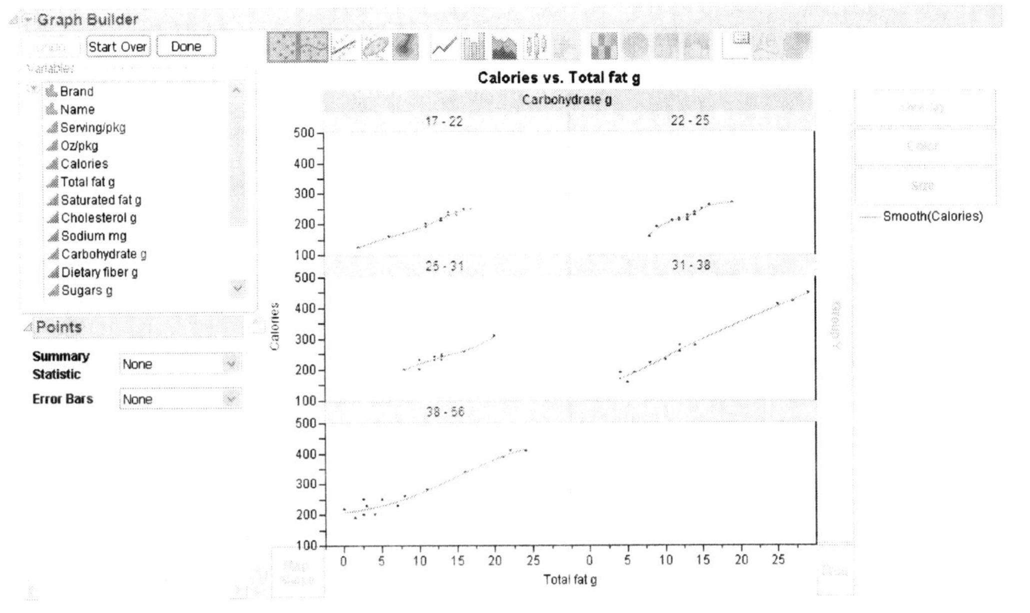

Carbohydrate g is a continuous variable with many values, so Graph Builder uses the percentiles to create five ranges of carbohydrate g levels. About the same number of points are displayed in each group. You can see that the relationship between calories and fat is relatively strong for each level of carbohydrate. It appears that carbohydrates adds additional predictive ability.

Now that you have determined that carbohydrates have a significant impact on calories, combine the five scatterplots into one scatterplot to directly compare the lines. You still want to identify the carbohydrate levels.

9. Drag and drop the **Carbohydrate g** label from the **Group X** zone to the **Overlay** zone.

Figure 10.6 Example of Carbohydrates Overlay

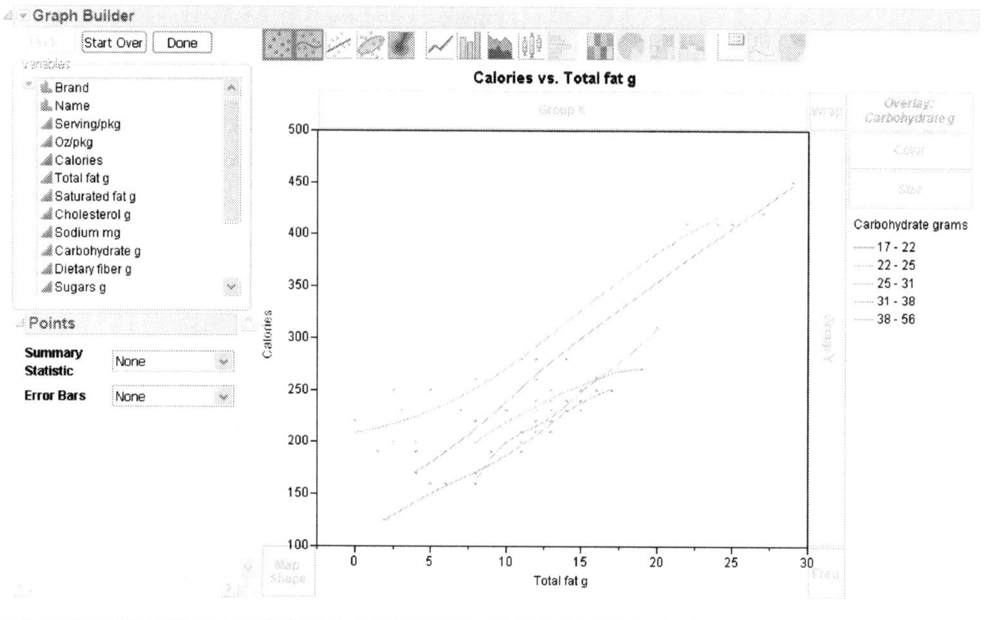

The scatterplots combine into one, and the carbohydrate levels are individually colored.

Modify the legend title.

10. Right-click on the legend title (Carbohydrate g) and select **Legend Settings**.
11. Rename the Title to Carbohydrate grams.
12. Click **OK**.

Note: For details about making changes to the legend, see “Modify the Legend” on page 274.

13. Now that you are satisfied with this graph, click **Done**.

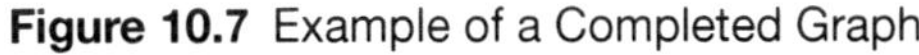

Figure 10.7 Example of a Completed Graph

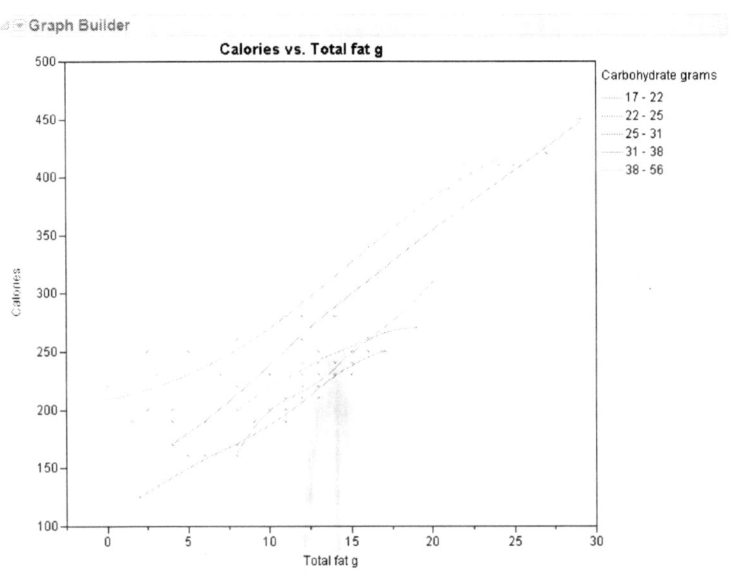

You now have a presentation-friendly graph that you can copy and paste outside of JMP.

To copy the entire graph:

14. Click the Selection Tool.
15. Click anywhere on the Graph Builder title bar.

The entire area is highlighted and ready to copy.

Launch Graph Builder

Launch Graph Builder by selecting **Graph > Graph Builder**.

Figure 10.8 The Graph Builder Window

The Graph Area and Zones

The primary element in the Graph Builder window is the graph area. The graph area contains drop zones, and you can drag and drop variables from the Select Columns box into the zones. The following table describes the Graph Builder drop zones.

X, Y	Drop variables here to assign them the X or Y role.
Group X	Subsets or partitions the data based on the variable or variables that you select. Displays the variable horizontally. Once a variable is placed here, no variable can be placed in **Wrap**.
Group Y	Subsets or partitions the data based on the variable or variables that you select. Displays the variable vertically.
Map Shape	Drop variables here to create map shapes. See "Create Map Shapes" on page 275. If you have a variable in the **Map Shape** zone, the **X** and **Y** zones disappear.
Wrap	Subsets or partitions the data based on the variable or variables that you select. Wraps the data horizontally and vertically. Once a variable is placed here, no variable can be placed in **Group X**.
Freq	Drop a variable here to use it as a frequency or weight for graph elements that use statistics, such as mean or counts.

Overlay	Groups the Y variables by the selected variable, overlays the responses, and marks the levels with different colors.
Color	Drop variables here to color the graph: • If you are using a map, the map shapes are colored. See "Change Colors and Transparency" on page 277. • If you are using a contour plot, colored contours appear. • If your graph contains points, they are colored.
Size	(Use with Map Shapes) Scales map shapes according to the size variable, minimizing distortion.
Legend	Shows descriptions of graph elements. If you attempt to drop a variable here, the variable defaults to Overlay. See "Modify the Legend" on page 274.

If you drop variables into the center area, JMP guesses the drop zone to put them into, based on whether the variables are continuous, ordinal or nominal.

- The X, Y, and Map Shape zones are positional, and influence the types of graph elements that are available.
- The Group X, Group Y, Wrap, and Overlay zones partition the data into subsets and lay out multiple graphs by either dividing the graph space or by overlaying the graphs.
- The Color and Freq zones modify certain graph elements.

Related Information

For the X, Y, Group X, and Group Y zones, see also:

- "Add Variables" on page 270
- "Change Variable Roles" on page 271
- "Remove Variables" on page 271
- "Add Multiple Variables to Grouping Zones" on page 274

Element Type Icons

You can change the element type by clicking on an element type icon. Use the SHIFT key to apply multiple elements at once. Once you select an element, only compatible elements are active.

Choose from the following element types:

The **Points** element shows data values.

The **Smoother** element shows a smooth curve through the data. The smoother is a cubic spline with a lambda of 0.05 and standardized *X* values. See "Using the Smoother" on page 277.

The **Line of Fit** element shows a linear regression with confidence intervals.

The **Ellipse** element shows a bivariate normal density ellipse.

The **Contour** element shows regions of density or value contours. If you specify only one continuous variable for *X* or *Y*, a violin plot appears instead of a contour plot.

The **Line** element shows a response summarized by categories.

The **Bar** element shows a response summarized by categories.

The **Area** element shows a response summarized by categories.

The **Box Plot** element shows a compact view of a variable's distribution, with quartiles and outliers.

The **Histogram** element shows a variable's distribution using binning. If you specify the same variable for *X* and *Y*, then the *Y* role is ignored and a single histogram appears.

The **Heatmap** element shows counts using color for *X* and *Y* categories.

The **Pie** element shows portions of a whole.

The **Treemap** element shows a response summarized by categories.

The **Mosaic** element shows counts using size for *X* and *Y* categories.

The **Caption Box** element shows a summary statistic value for the data.

The **Formula** element shows a function defined by a column formula.

The **Map Shapes** element creates a map on the graph.

When applicable, properties for each element appear and can be adjusted in the Graph Builder window. For more information, see "Right-Click Menu for a Graph" on page 267.

Buttons

There are three buttons on the Graph Builder window:

- **Undo** reverses the last change made to the window.
- **Start Over** returns the window to the default condition, removing all data and graph elements from the window, and all variables from the drop zones.

- **Done** hides the buttons and Select Columns box and removes all drop zone outlines. In this presentation-friendly format, you can copy the graph to other programs. To copy the graph, select **Edit > Copy**. To restore the window to the interactive mode, click **Show Control Panel** on the Graph Builder red triangle menu.

Graph Builder Options

The red triangle menu for Graph Builder contains these options:

- **Show Control Panel** shows or hides the platform buttons, the Select Columns box, and the drop zone borders.
- **Show Legend** shows or hides the legend.
- **Show Footer** shows or hides the footer, which contains informative messages such as missing map shapes, error bar notes, freq notes, and WHERE clauses.
- **Lock Scales** prevents axis scales and gradient legend scales from automatically adjusting in response to data or filtering changes.
- **Auto Stretching** enables multiple graphs in the same window to be sized properly.
- **Sampling** uses a random sample of the data to speed up graph drawing. If the sample size is zero, or greater than or equal to the number of rows in the data table, then sampling is turned off.
- **Include Missing Categories** enables a graph to collect and display missing values for categorical variables.
- **Launch Analysis** launches the Fit Model platform with the variables on the graph placed into roles. It launches the Distribution platform when only one variable is placed.
- **Script** contains options that are available to all platforms. They enable you to redo the analysis or save the JSL commands for the analysis to a window or a file. For more information, see *Using JMP*.

Graph Builder Right-Click Menus

Graph Builder contains various right-click menus, depending on the area you right-click on. Any changes that you make to a graph element apply to all graphs for that variable, across all grouping variables.

Right-Click Menu for a Graph

Right-clicking on a graph shows a menu of the available graph elements and other options. The first menus that appear reflect the elements that you have selected. For example, if you have selected the Points and Line of Fit icons, the first menus are Points and Line of Fit. Each of these elements have specific submenus. The following table describes the right-click menu options and shows which graph elements each option is applicable to.

Option	Graph Element	Description
Add	All graph elements	Add an element to the graph.

Option	Graph Element	Description
Area Style	Area	Changes the area style.
Bar Style	Bar	Changes the bar style.
Box Style	Box Plot	Changes the box plot style.
Change to	All graph elements	Change the existing element to another element.
Confidence of Fit	Line of Fit	Shows confidence lines for the fit.
Confidence of Prediction	Line of Fit	Shows confidence lines for the prediction.
Coverage	Ellipse	Change the percentage of the ellipse coverage.
Degree	Line of Fit	Change the degree of the line.
Error Bars	Line, Bar, and Points	Adds error bars to the graph.
Horizontal	Mosaic, Histogram	Changes the primary direction of the mosaic plot to horizontal.
Jitter	Box Plot and Points	Turns jitter on or off.
Label	Pie, Treemap	Changes the labels on the treemap.
Move Backward	All graph elements	If you have multiple graph elements, you can move them backward (or move them to the back).
Move Forward	All graph elements	If you have multiple graph elements, you can move them forward (or move them to the front).
Number of Levels	Contour	Specify the number of contour levels to display.
Outliers	Box Plot	Shows or hides outliers.
Pie Style	Pie	Changes the style of the pie chart.
Remove	All graph elements	Removes the graph element.
Root Mean Square Error	Line of Fit	Adds the root mean square error to the graph.
Row order	Line	Connects line points in the order of their row numbers instead of the order of their *X* values.
Show Missing Shapes	Map Shapes	Shows or hides missing data from the map. This option is turned off by default.

Option	Graph Element	Description
Summary Statistic	Line, Bar, Points, Area, Pie, Treemap, Caption Box, Map Shapes	Provides options for changing the statistic being plotted.
Vertical	Mosaic, Histogram	Changes the primary direction of the mosaic plot to vertical.
X	All graph elements	This option only appears if you have multiple X variables. You can hide or show each one.
X Position	Caption box	Changes the horizontal position of the caption.
Y	All graph elements	This option only appears if you have multiple Y variables. You can hide or show each one.
Y Position	Caption box	Changes the vertical position of the caption.

Note: For a description of the **Rows**, **Graph**, **Customize**, and **Edit** menus, see the *Using JMP book*.

Additional Right-Click Menus

Depending on the area that you click on in Graph Builder, you see different right-click options.

Right-Click Area	Description
The Y or X axis	The **Remove** and **Swap** commands appear. See "Remove Variables" on page 271 or "Change Variable Roles" on page 271. For descriptions of the options below the line, see the *Using JMP* book.
A variable label	The **Remove** and **Swap** commands appear. See "Remove Variables" on page 271 or "Change Variable Roles" on page 271. For descriptions of the options below the line, see the *Using JMP* book.

Right-Click Area	Description
A zone	These options might appear: **Number of Levels** changes the number of levels. See "Separate Variables into Groups" on page 271. **Levels per Row** (use with a **Wrap** variable) changes the number of columns included in the graph. **Color** changes the background color of the grouping zone. **Remove** removes a variable. See "Remove Variables" on page 271. **X or Y Group Edge** moves the grouping variable labels. See "Move Grouping Variable Labels" on page 270. **Swap** swaps the position of two variables. See "Change Variable Roles" on page 271.
The legend title	See "Modify the Legend" on page 274.
An item in the legend	For descriptions of these options, see the *Using JMP* book.
The empty space below Variables	See "Graph Builder Options" on page 267.
The empty space below the legend or above or below the graph	For descriptions of these options, see the *Using JMP* book.

Add Variables

To add a variable to a drop zone, click and drag the column name from the **Variables** box and drop it into the desired drop zone. Alternatively, you can highlight the column name in the **Variables** box and click in the desired drop zone. Both numeric and categorical variables can be added to all of the drop zones.

Related Information

- "Example of Adding Variables" on page 278

Move Grouping Variable Labels

Grouping variable labels can be relocated to another position on the graph. The Group X labels can be either on the top or the bottom of the graph. The Group Y labels can be either on the right or the left of the graph.

- To relocate a Group X label, right-click on the variable in the Group X zone and select **X Group Edge > Top** or **Bottom**.
- To relocate a Group Y label, right-click on the variable in the Group Y zone and select **Y Group Edge > Left** or **Right**.

Separate Variables into Groups

When you add a categorical variable to the Group X or Group Y zone, a partition is created for each level of the variable.

When you add a continuous variable to a grouping zone, Graph Builder uses quantiles of the data to divide the variable into five groups. Once the variable is added to the display, you can change the number of groups as follows:

1. Right-click on the grouping variable label and select **Number of Levels**.
2. Type in the number of levels that you want to display.
3. Click **OK**.

Change Variable Roles

To change variable roles, perform one of the following methods:

- Drag and drop a variable from one zone to another.
- Right-click on a variable in a zone and select **Swap**. Select the variable that you want to switch with.

Remove Variables

To remove a variable, perform one of the following methods:

- Drag and drop a variable into blank space.
- Right-click on a variable in a zone and select **Remove**.

Add Multiple Variables to the X or Y Zone

You can assign more than one variable to the same drop zone. By visualizing multiple Y variables across multiple X variables, you can discover multivariate relationships and interactions.

To add a variable to the left or right of the existing X variable, drag and drop the variable into the X zone. Use the same method for adding a variable to the top or bottom of the existing Y variable.

When you click and drag a variable, a blue shape appears, symbolizing where the variable is added. The following table illustrates what each shape looks like for each task.

Task	Shape
Add an X variable to the left of another X variable	

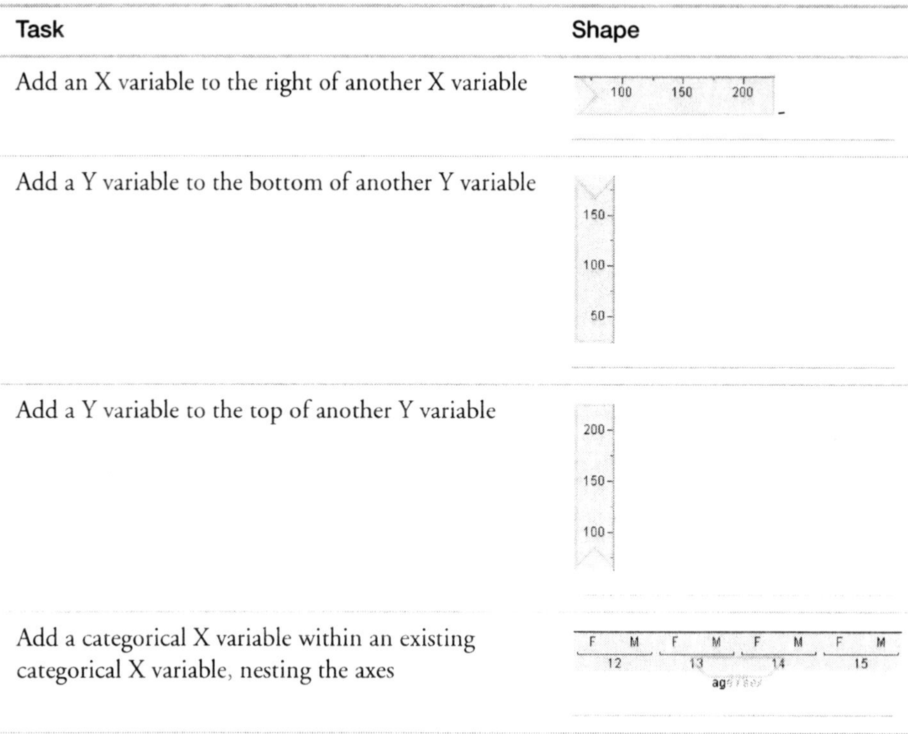

Task	Shape
Add an X variable to the right of another X variable	100 150 200
Add a Y variable to the bottom of another Y variable	150 100 50
Add a Y variable to the top of another Y variable	200 150 100
Add a categorical X variable within an existing categorical X variable, nesting the axes	F M F M F M F M / 12 13 14 15

Related Information

- "Example of Adding Multiple Variables to the X or Y Zone" on page 280

Merge Variables

Merging variables places both variables on the same axis and creates a graph for both variables. Merging is similar to adding variables, in that a graph element is added for both variables. But adding variables maintain separate axes and scales, and merging variables use the same axis and the same scale for the variables.

If you have no variables in the **X** or **Y** zone, you can drag multiple variables from the Variables list into the **X** or **Y** zone. This merges the selected variables.

If you have existing variables in the **X** or **Y** zone where you want to add variables, note the following:

- If you merge a categorical variable with an existing continuous variable, then the categorical variable is transformed into integer values. For example, using the Students.jmp data, if sex is merged with height, the values of sex (F and M) are transformed into 0 and 1. The transformation allows the two variables to use the same axis and scale.
- If you merge a numeric variable with an existing categorical variable, the result is ordered. See "Order Variables" on page 273.

Related Information

- "Example of Merging Variables" on page 281

Order Variables

The levels of a nominal or ordinal variable on a graph are ordered alphabetically by default. There are two ways that you can modify this order:

- Use the **Value Ordering** column property. See the *Using JMP* book.
- Use a second variable. See "Use a Second Variable" on page 273.

Use a Second Variable

A nominal or ordinal variable can be ordered only by a numeric variable. The variable that does the ordering has to be numeric so that an average can be computed for each level of the categorical variable.

Note: If you try to order a numeric variable with another numeric variable, JMP defaults to a merge. See "Merge Variables" on page 272.

- To order by a second variable, drag the second (numeric) variable into the graph area next to the existing (nominal or ordinal) variable.
- To change the order from ascending to descending, right-click on the variable and deselect **Ascending**.
- The default ordering statistic is the mean. To use another statistic, right-click on the variable and select **Order Statistic**. The available statistics are N, Mean, Median, Min, Max, Range, Sum, and % of Total.
- To remove the ordering, right-click and select **Remove Order**.

Related Information

- "Example of Ordering Variables Using a Second Variable" on page 282

Replace Variables

You can replace an existing variable with an incoming variable, maintaining a single graph element. If grouping variables exist, a single graph element is maintained for the incoming variable for each combination of grouping variables.

To replace an existing variable, drag and drop the variable from the Select Columns box atop the existing variable. Before you drop the variable, a hexagon appears.

Create a Second Y Axis

If you have two Y variables, you can move one of the variables to create a second Y axis. Right-click on the Y variables and select **Move Right**. Select the variable that you want to move.

Add Multiple Variables to Grouping Zones

You can add more than one variable to the **Group X** or **Group Y** zones. You place the incoming variable below or above the existing variable, depending on how you want your data to be ordered.

To add an additional variable to a grouping zone, simply drag and drop the variable into the drop zone.

When you drag a variable, a blue shape appears, symbolizing where the variable is added. The following table illustrates what each shape looks like for each task.

Task	Shape
Add the incoming variable below the existing variable	M
Add the incoming variable above the existing variable	11 12
Replace the existing variable with the incoming variable	age 12 13 14 15

Related Information

- "Example of Adding Multiple Variables to Grouping Zones" on page 284
- "Example of Replacing Variables" on page 286

Order Grouping Variables

Ordering the levels of grouping variables works the same as ordering the levels of variables in the **X** zone or **Y** zone. See "Order Variables" on page 273.

Modify the Legend

To modify a legend, right-click or double-click on the legend title. The following commands appear:

- **Legend Settings** opens a window where you can modify legend settings, such as the title and title position. See Figure 10.9.
- **Revert Legend** returns the legend to the default condition (if you have changed it).

Figure 10.9 The Legend Settings Window

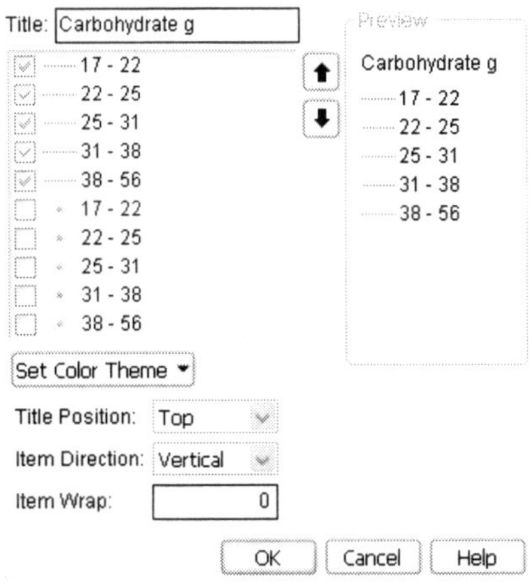

Table 10.1 Description of the Legend Settings Window

Title	The name of the legend.
Check boxes	Shows or hides items in the legend.
Up and down arrows	Changes the order of items in the legend.
Set Color Theme	Select a different color theme.
Title Position	Places the legend title on top or to the left of the items in the legend.
Item Direction	Displays the legend horizontally or vertically.
Item Wrap	Sets the legend to be *n* items tall (if vertical) or *n* items wide (if horizontal).
Preview	Shows your changes to the legend.
OK	Commits your changes to the legend.
Cancel	Cancels your changes to the legend.
Help	Opens the online Help.

Create Map Shapes

Use Graph Builder to create maps using the **Map Shape** zone. When a column contains the names of geographical regions (such as countries, states, provinces, counties), you can assign the column to the **Map Shape** zone. This creates a map of the regions.

Related Information

- "Creating Maps" chapter on page 403
- "Example of Creating Map Shapes" on page 287

Built-in Map Files

When you use the **Map Shape** zone, JMP searches through its built-in map files for matching names. Built-in map files include the following:

- world countries
- states and counties in the United States
- first-level divisions for Canada, China, the United Kingdom, France, Italy, Japan, and Germany

By default, JMP installs built-in map files in the following directory:

- Windows: C:\Program Files\SAS\JMP\10\Maps
- Macintosh: /Library/Application Support/JMP/10/Maps

Each map consists of two JMP data files with a common prefix:

- -Name file that contains the unique names for the different regions
- -XY file that contains the latitude and longitude coordinates of the boundaries

The two files are implicitly linked by a Shape ID column. For example:

- The World-Country-Name.jmp file contains the exact names and abbreviations for different countries throughout the world.
- The World-Country-XY.jmp file contains the latitude and longitude numbers for each country, by Shape ID.

Note: If JMP does not recognize the names in your data, check the built-in map files for the spelling that JMP recognizes. For example, United Kingdom is a country name, but not Great Britain.

Create Custom Map Files

You can create your own map files by following the same pattern as the built-in files. In order for JMP to automatically find your files, place them in the following directory:

- Windows XP: *<username>*\Local Settings\Application Data\SAS\JMP\Maps
- Windows 7 and Vista: *<username>*\AppData\Local\SAS\JMP\Maps
- Macintosh: *<username>*/Library/Application Support/JMP/Maps

Or, you can link the map files to your data files explicitly with the **Map Role** column property.

Note the following when creating map files:

- Each set of map files that you create must contain a -Name file and a -XY file.

- The first column in both files must be the ascending, numeric **Shape ID** variable. The **-Name** file can contain any other columns.
- For the **Map Role** column property, columns that are marked with the **Shape Name Definition** are searched for shape identification and must contain unique values.
- For the **Map Role** column property, mark the column with **Shape Name Use** for explicitly linked shape data. Provide the location and column of the map file data to use.
- If you import an ESRI SHP file, it is opened in this format. **-Name** files commonly have a .dbf extension.
- The *Using JMP* book includes more information about how to structure the map data tables and your analysis data table. Steps for adding the **Map Role** column property are also included.

Related Information

- "Creating Maps" chapter on page 403
- "Examples of Creating Custom Maps" on page 301.

Change Colors and Transparency

To change colors and transparency for a map, right-click on the color bar in the legend. The right-click options vary, depending on whether the Color variable is continuous or categorical (nominal or ordinal). However, for both types of variables, you can change the transparency.

Continuous variables use a color gradient. To change the color theme, proceed as follows:

1. Right-click on the color bar and select **Gradient**.
2. In the Gradient Settings window, select a different **Color Theme**.

Categorical (nominal or ordinal) variables use a singular coloring system, where each level of the variable is colored differently.

To change the color of one of the variable levels, proceed as follows:

1. Right-click on the color of the variable level that you want to change and select **Fill Color**.
2. Select the new color.

Note: For more details about color options, see the *Using JMP* book.

Additional Examples Using Graph Builder

This section contains examples that further illustrate how to use Graph Builder.

Using the Smoother

The smoother is a cubic spline with a lambda of 0.05 and standardized *X* values. You can obtain the same spline in the Bivariate platform, by performing the following steps:

1. Select **Analyze** > **Fit Y by X**.
2. Select continuous variables for **Y, Response** and **X, Factor**.
3. Click **OK**.
4. From the red triangle menu for Bivariate Fit, select **Fit Spline** > **Other**.
5. Type in the smoothness value of 0.05.
6. Check the **Standardize X** box.
7. Click **OK**.

For details about the Bivariate platform, see the "Performing Bivariate Analysis" chapter on page 91.

Example of Adding Variables

To show an example of adding variables to Graph Builder, make a scatterplot of weight versus height for males and females.

1. Open the Students.jmp sample data table.
2. Select **Graph > Graph Builder.**
3. Click height and drag and drop it into the **X** zone.

Figure 10.10 Example of a height Plot

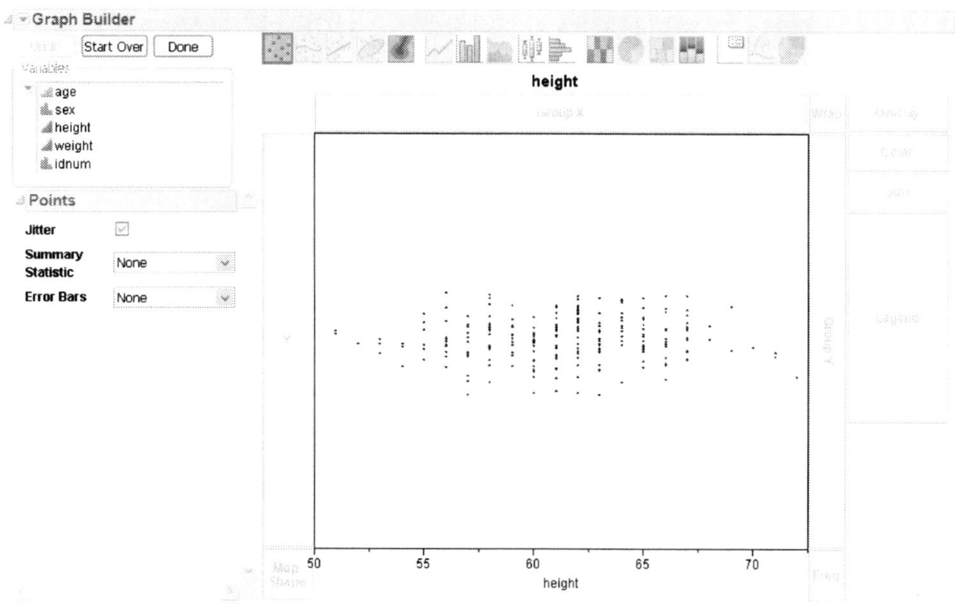

A plot of the raw data appears. Random jitter is added in the Y direction.

4. Click weight and drag and drop it into the **Y** zone.

Figure 10.11 Example of a Scatterplot of weight Versus height

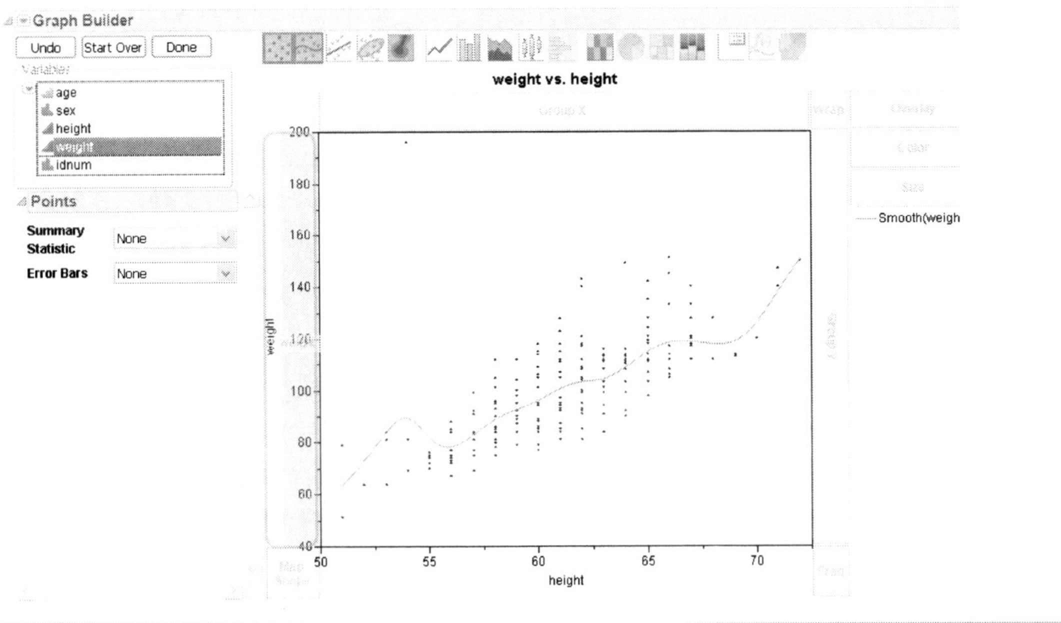

Notice the following additional elements:

- A smoother (line) shows the general pattern of weight as a function of height.
- A legend describes the elements on the graph (in this case, the smoother).

5. Click sex and drag and drop it into the **Group X** zone.

Figure 10.12 Example of a Scatterplot for Each Level of the sex Variable

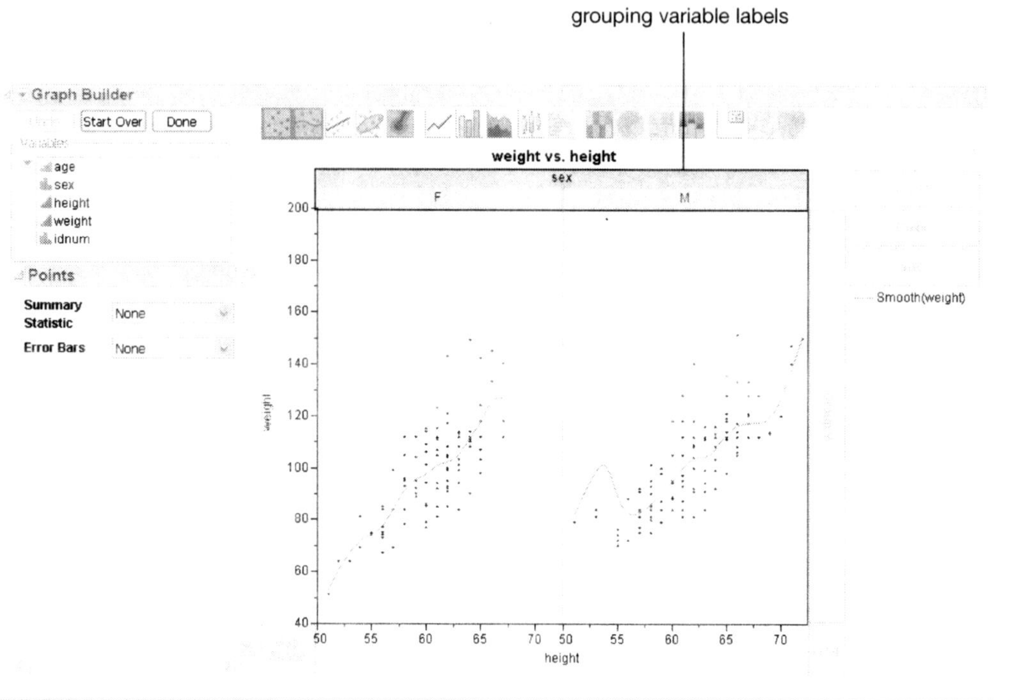

Side-by-side scatterplots (one for each level of sex) replace the initial scatterplot. You now see weight versus height for males and females.

Example of Adding Multiple Variables to the X or Y Zone

Start from the graph in Figure 10.10. Add the weight variable to the left of the height variable. Click weight and drag and drop it into the X axis, to the left of height.

Figure 10.13 Dragging and Dropping the weight Variable

A graph is created for both variables, with separate axes and scales. If grouping variables exist, a graph is created for both the existing variable and incoming variable for each combination of grouping variables.

Example of Merging Variables

To demonstrate combining two continuous variables, start from the graph in Figure 10.10. Merge the weight variable with the height variable. Drag and drop weight to the center of the X zone, slightly above the axis. Before you drop the variable, a blue pentagon appears.

Figure 10.14 Merging height and weight

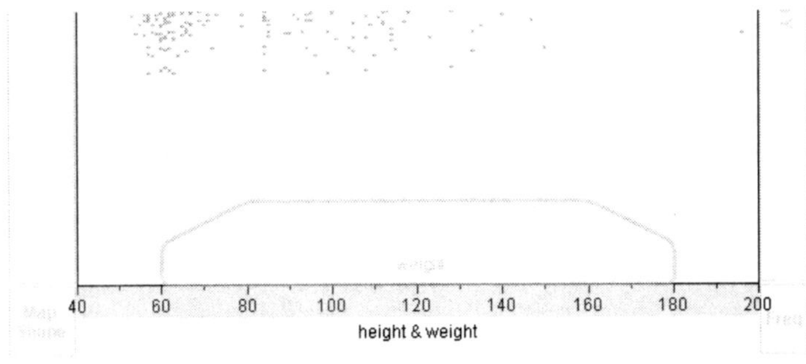

A graph element is added for weight that uses the same scale as height.

Figure 10.15 Example of weight Combined with height

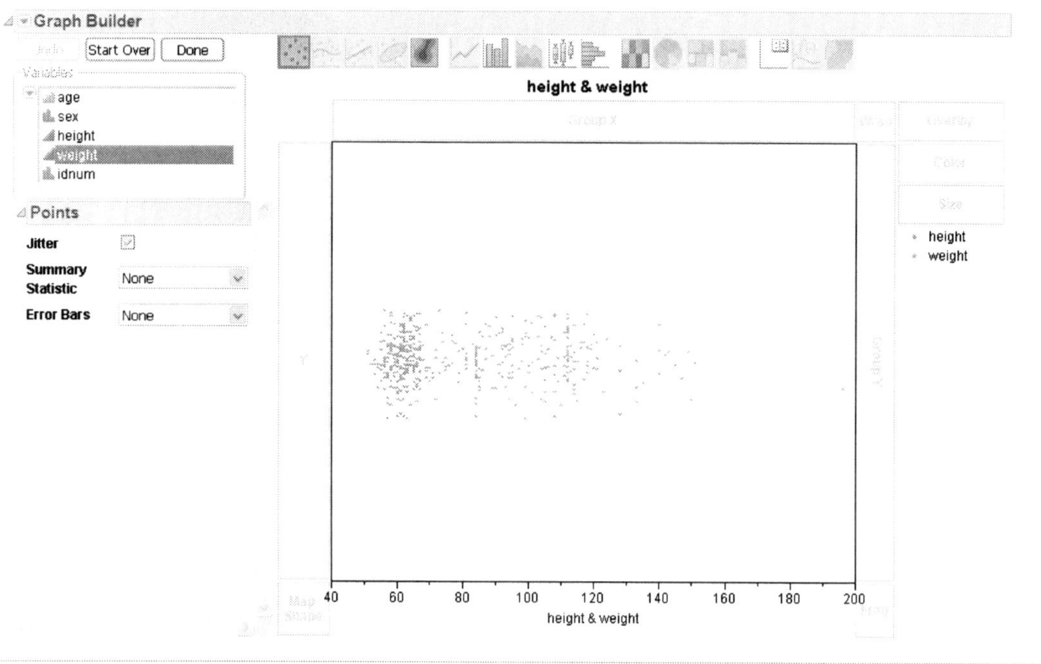

Example of Ordering Variables Using a Second Variable

To order a nominal or ordinal variable by a numeric variable, consider data about the sizes of cars. You want to see the different car sizes in ascending order.

1. Open the Cars.jmp sample data table.
2. Select **Graph > Graph Builder**.
3. Click Size and drag it into the **X** zone.

This variable represents the size of the vehicle. Eight levels are listed alphabetically on the X axis: compact (comp), heavy (hev), lightweight (lt), medium (med), mini, multi-purpose (mpv), pick-up truck (pu), and van. Since the levels are listed alphabetically, they are not ordered by size. Heavy comes before mini and lightweight. Another variable, Wt (weight) can act as a good substitute for size.

4. Click Wt and drag and drop it just above the **X** axis. Before you drop the variable, a blue pentagon appears.

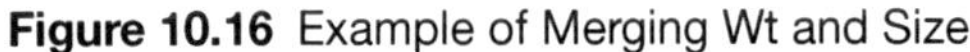

Figure 10.16 Example of Merging Wt and Size

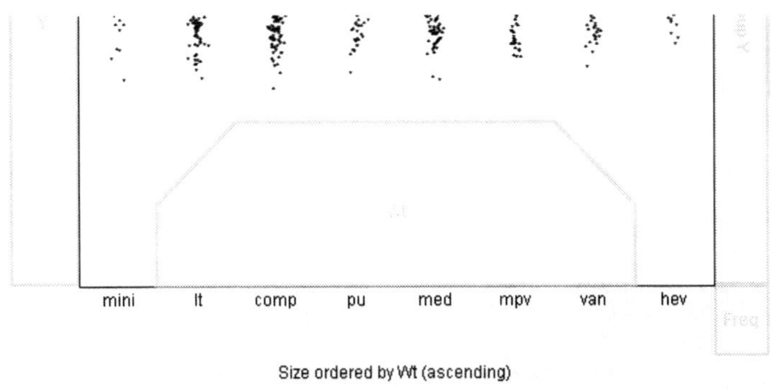

The levels of Size are now ordered according to the average Wt of all vehicles in that level, in ascending order. Notice that mini and lightweight are now ordered before heavy. The axis label is updated, signifying that an ordering variable is in use.

To verify that Size is actually ordered by Wt, click on Wt under **Variables** and drag and drop it into the **Y** zone. Figure 10.17 shows that the average Wt increases from the left to right.

Figure 10.17 Example of Size Ordered by Wt

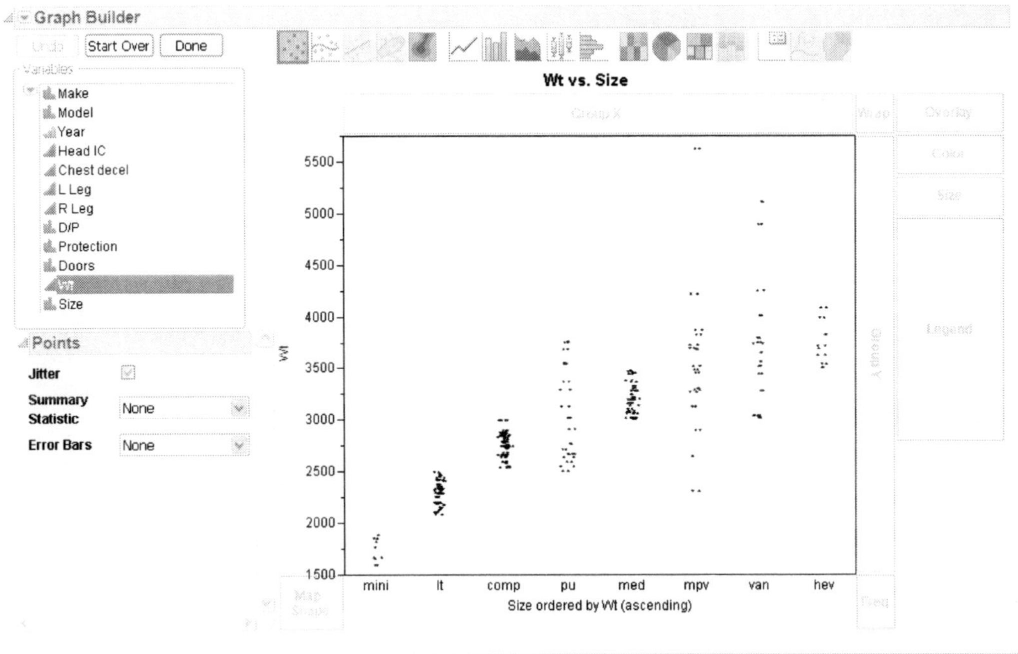

To change the order from ascending to descending, proceed as follows:

1. Right-click in the **X** zone.
2. Deselect **Ascending.** See Figure 10.18.

The default ordering statistic is the mean. To use another statistic, select it from the **Order Statistic** menu shown in Figure 10.18. The available statistics are N, Mean, Min, Max, Sum, and % of Total.

To remove the ordering, right-click and select **Remove Order**.

Figure 10.18 The Right-Click Menu for the X Zone

Note: For details about the Axis and Edit menu options, see the *Using JMP* book.

Example of Adding Multiple Variables to Grouping Zones

You have data on popcorn. You want to see the popcorn yields, grouped by popcorn type (gourmet or plain) and by batch size (small or large).

1. Open the Popcorn.jmp sample data table.
2. Select **Graph > Graph Builder**.
3. Click popcorn and drag and drop it into **Group X**.

   This is the popcorn type.
4. Click yield and drag and drop it into the **Y** zone.

   You can see the yield by popcorn type.

Figure 10.19 Example of yield by popcorn

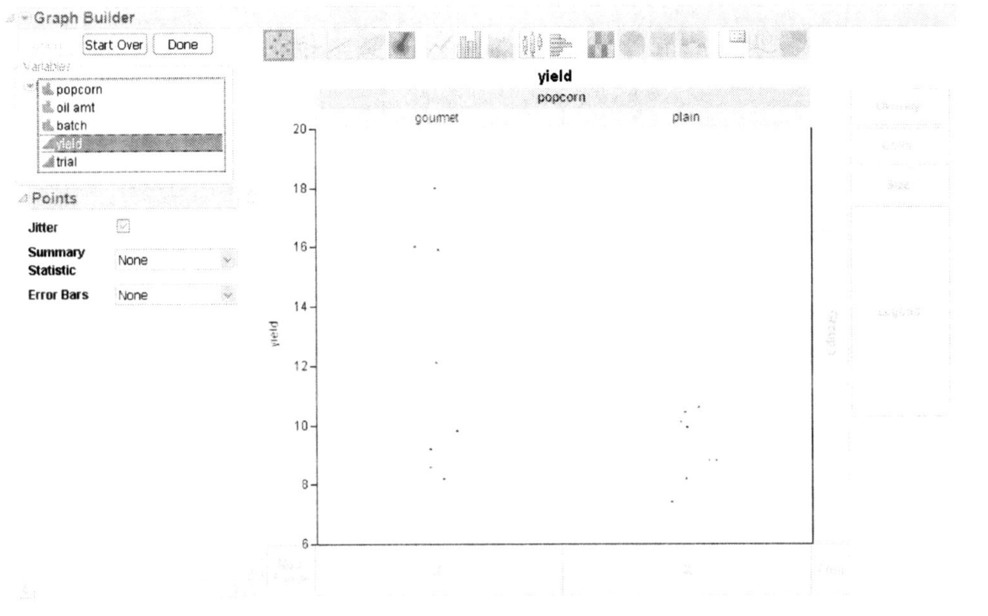

To add the second variable, perform either step 5 or step 6, depending on where you want to place the incoming variable (above or below the existing variable).

5. To place the batch variable **above** the popcorn variable, drag and drop batch into the left side of the **Group X** zone. Before you drop the variable, a left-justified blue polygon appears.

Figure 10.20 Example of Adding batch Above popcorn

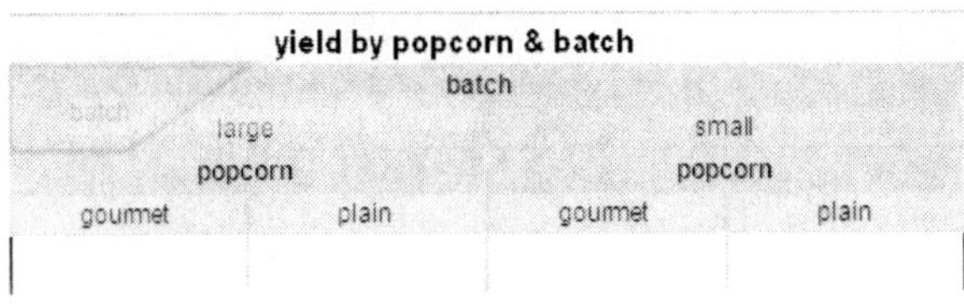

6. To place the batch variable **below** the popcorn variable, drag and drop batch into the right side of the **Group X** zone. Before you drop the variable, a right-justified blue polygon appears.

Figure 10.21 Example of Adding batch Below popcorn

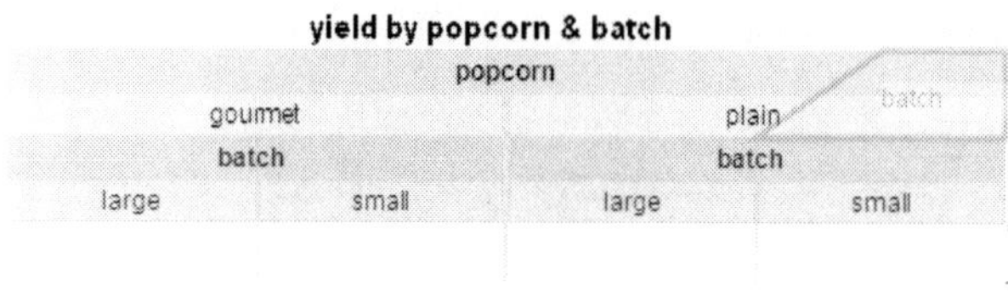

7. To see the data shape more clearly, right-click in the graph and select **Add > Box Plot**.

Figure 10.22 Examples of Popcorn yield Grouped by popcorn and batch

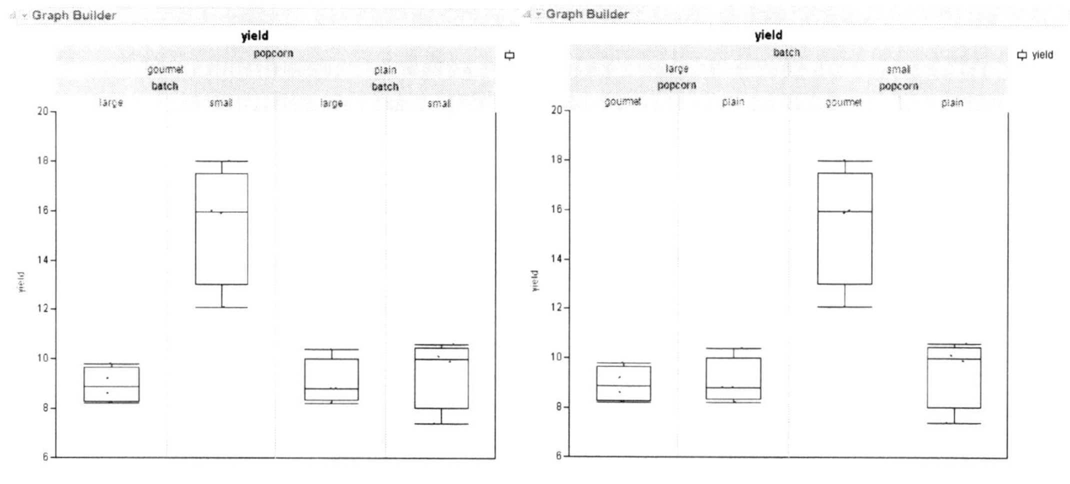

Depending on where you place the incoming variable, you can look at the data in different ways. But you could come to the same conclusion: that batch size varies for gourmet popcorn, but batch size does not vary as much for plain popcorn.

Example of Replacing Variables

You can replace an existing variable with an incoming variable. To demonstrate replacing variables, start from the graph in Figure 10.19. Replace popcorn with batch in the **Group X** zone. Drag and drop batch into the center of the **Group X** zone. Before you drop the variable, a blue quadrilateral appears.

Figure 10.23 Example of Replacing popcorn with batch

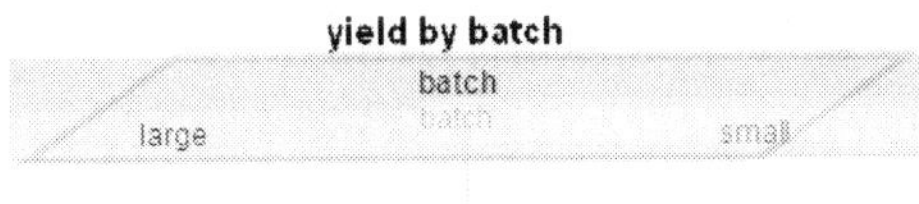

Once you drop the variable, groups are created for batch instead of popcorn.

Example of Creating Map Shapes

This example uses the Crime.jmp sample data table, which contains data on crime rates for each US state.

1. Open the Crime.jmp sample data table.
2. Select **Graph > Graph Builder**.
3. Drag and drop State into the **Map Shape** zone.
4. Drag and drop murder into the **Color** zone.

Figure 10.24 Example of murder by state

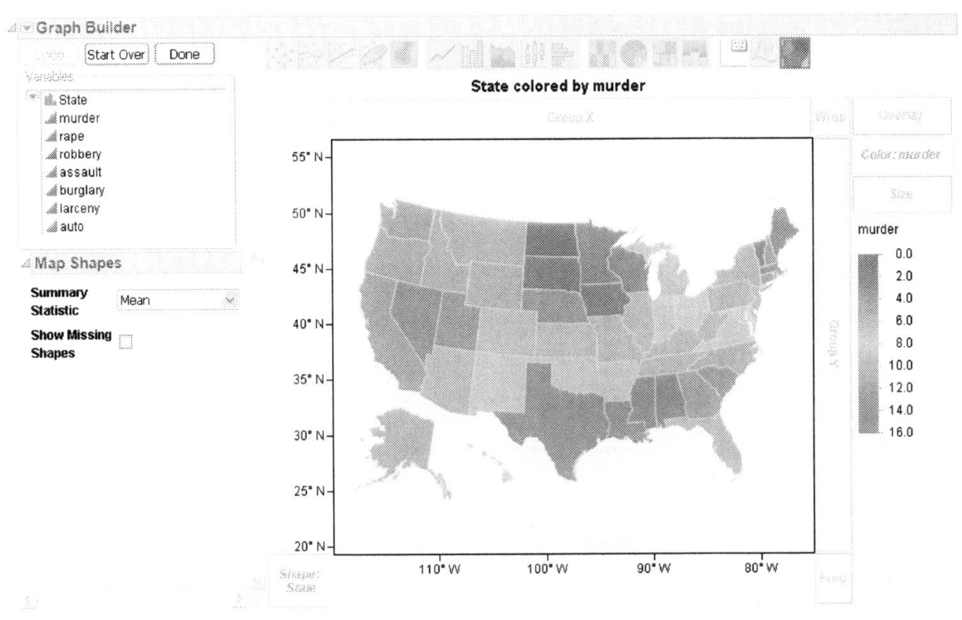

Note the following:

- The latitude and longitude appear on the Y and X axes.
- The legend shows the colors that correspond to the murder rates. Since murder is a continuous variable, the colors are on a gradient.

Measure Global Oil Consumption and Production

You have data about oil consumption and production, measured in barrels per day, for selected countries. You want to find out which countries consume the most oil, and which countries produce the most oil.

1. Open the Oil Use.jmp sample data table.
2. Select **Graph > Graph Builder**.
3. Drag and drop Country into the **Y** zone. You can resize the Graph Builder window and the graph if necessary.

Note: Notice that the default ordering for Country is ascending alphabetical (starting point is at the bottom). You can change the sorting order within a data table by using either the Value Ordering or Row Order Levels commands. For details, see the *Using JMP* book.

Figure 10.25 Example of Country Assigned to the Y Zone

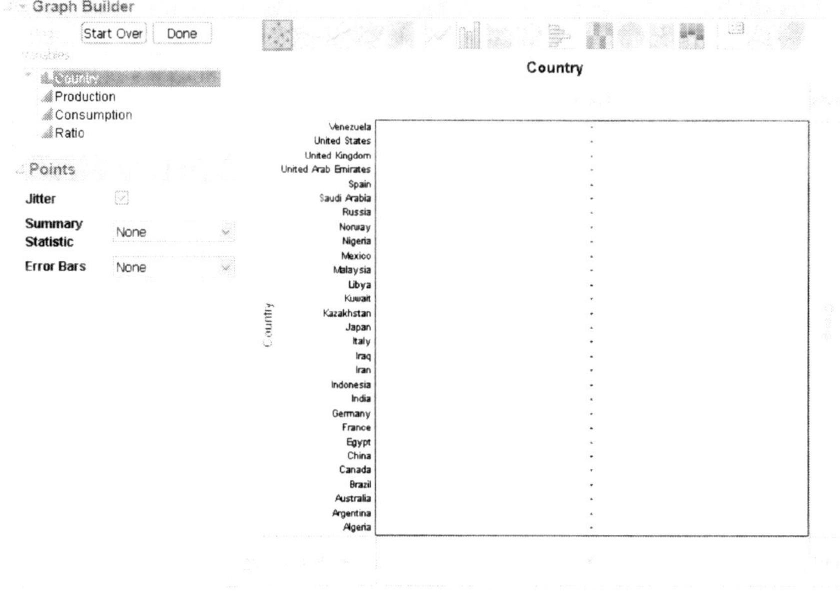

4. Drag and drop Production and Consumption to the **X** zone.

   Markers appear on the graph for both variables, along with a legend that identifies the different colors.

Figure 10.26 Example of Country versus Production and Consumption

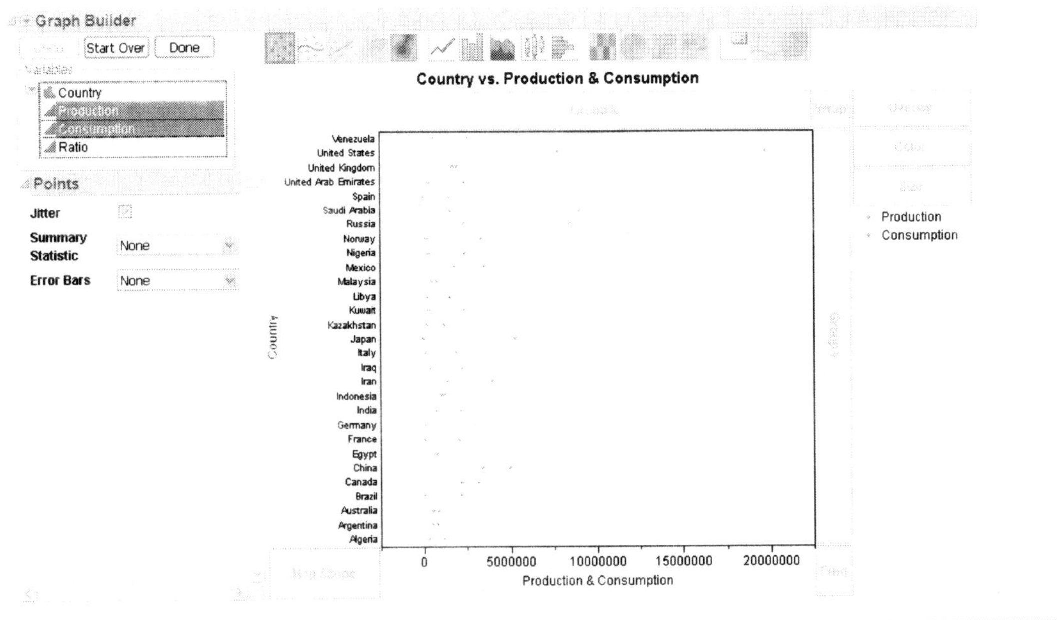

Because the default graph element is set to points, it is difficult to read and learn from the graph. Change the points to bars to make the graph easier to interpret.

5. To change the points to bars, right-click on the graph and select **Points** > **Change to** > **Bar.**

Figure 10.27 Example of Side-by-Side Bars for Production and Consumption

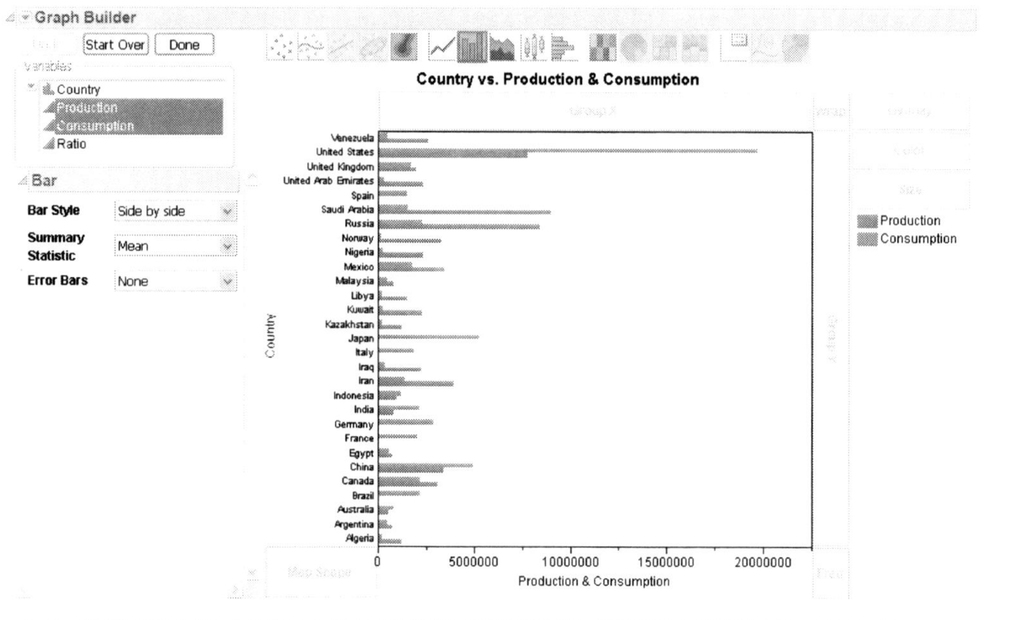

Experiment with the presentation of the bar chart.

6. Change the default side-by-side bars to stacked bars by right-clicking on the graph and selecting **Bar > Bar Style > Stacked**.

Figure 10.28 Example of Stacked Bars for Production and Consumption

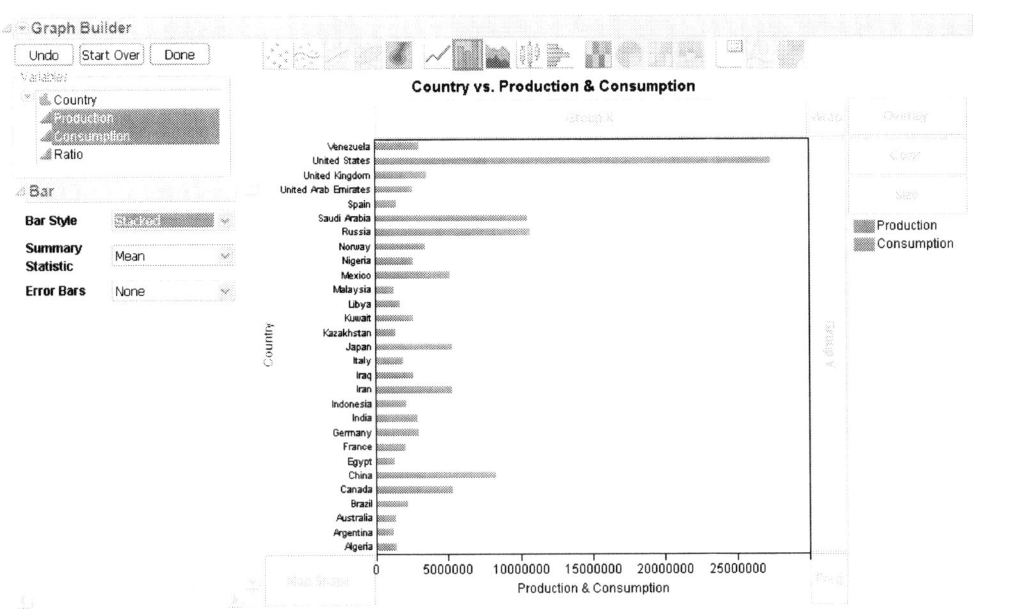

This still is not exactly what you want. It would be clearer to see the production and consumption bars going in different directions.

7. Go back to the **Oil Use.jmp** sample data table.
8. Double-click on an empty column heading to create a new column.
9. Double-click on the new column title, and rename it Negative Consumption. Click **OK**.
10. With the new column highlighted, select **Cols > Formula**.
11. Click **Consumption**.
12. Click **+/-**.
13. Click **OK**.

The new Negative Consumption column contains the negative consumption values. Use this column to see the consumption bar going in the opposite direction from the production bar. Launch Graph Builder and create the graph again.

14. Select **Graph > Graph Builder**.
15. Drag and drop **Country** into the **Y** zone. You can resize the Graph Builder window and the graph if necessary.
16. Drag and drop **Production** and **Negative Consumption** to the **X** zone.
17. Right-click on the graph and select **Points > Change to > Bar**.
18. Right-click on the graph and select **Bar > Bar Style > Stacked**.

Figure 10.29 Example of Stacked Bars for Production and Negative Consumption

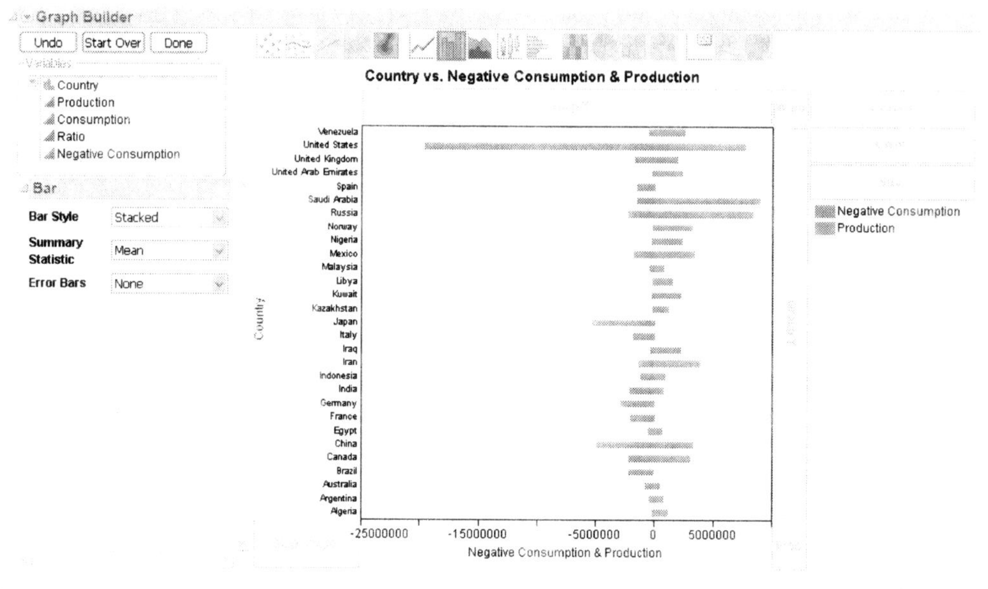

You can clearly see consumption on the left in red, and production on the right in blue. Sort the countries by their oil consumption. You can use the Consumption column as an ordering variable.

19. Drag and drop Consumption into the graph, to the right of the Y zone.

    Before you drop the variable, a blue polygon appears.

Figure 10.30 Example of Dragging Consumption into the Y Axis

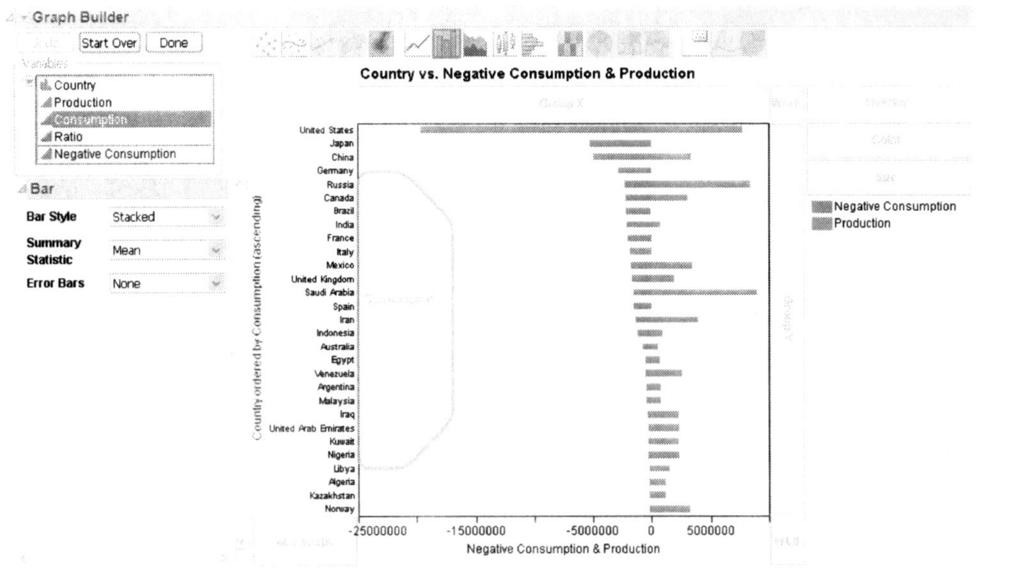

The Country variables are now ordered by Consumption.

Figure 10.31 Example of Country Organized by Consumption

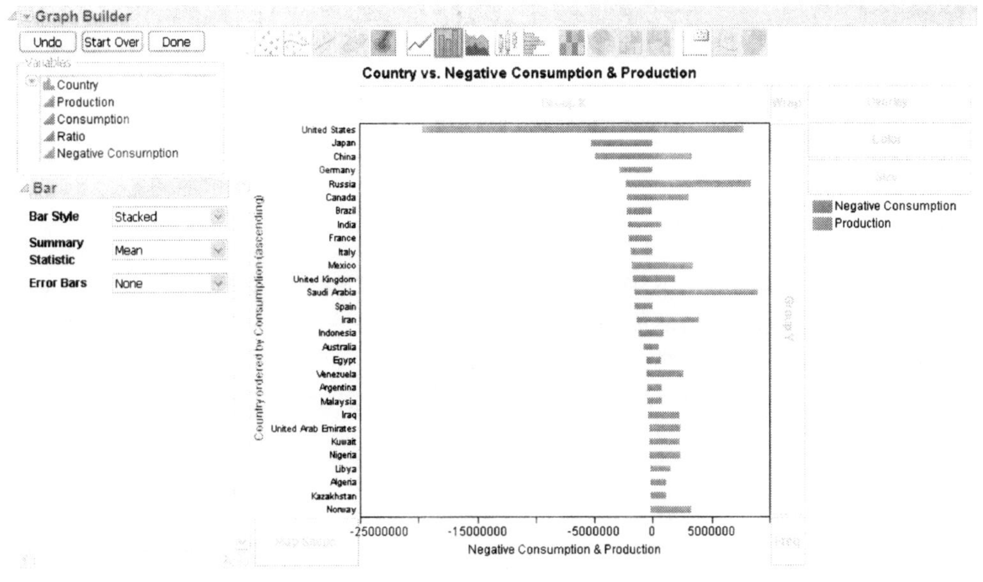

You can clearly see the countries that consume the most oil at the top of the graph. You can also see each country's oil production.

Analyze Popcorn Yield

You have data about popcorn. The data includes popcorn type (plain or gourmet), how much oil was used, the batch size (small or large), the yield, and the trial number. You want to determine how these factors affect the popcorn yield: popcorn type, oil, and batch size.

Note: These data are artificial, but inspired from an experiment reported in Box, Hunter, and Hunter (1978).

1. Open the Popcorn.jmp sample data table.
2. Run the attached Fit Model script attached to the data table by clicking on the red triangle next to **Fit Model** and select **Run Script**.

   A three-way ANOVA model is fit to the data.
3. Click on the disclosure icon next to **Effect Tests** to open the report.

The popcorn*batch interaction has a small p-value (0.0026). From this, you conclude that there is a significant interaction between popcorn and batch.

Figure 10.32 Example of Effect Tests Output

Effect Tests

Source	Nparm	DF	Sum of Squares	F Ratio	Prob > F
popcorn	1	1	34.810000	17.3184	0.0032*
oil amt	1	1	4.410000	2.1940	0.1768
popcorn*oil amt	1	1	4.000000	1.9900	0.1960
batch	1	1	49.000000	24.3781	0.0011*
popcorn*batch	1	1	37.210000	18.5124	0.0026*
oil amt*batch	1	1	0.010000	0.0050	0.9455
popcorn*oil amt*batch	1	1	4.000000	1.9900	0.1960

4. Save the model predictions to the data table. From the red triangle menu next to Response yield, select **Save Columns > Prediction Formula**. See Figure 10.33.

   Notice that a new column is added to the data table, **Pred Formula yield**.
5. Save the prediction intervals to the data table. From the red triangle menu next to Response yield, select **Save Columns > Mean Confidence Interval**. See Figure 10.33.

   Notice that two new columns are added to the data table: **Lower 95% Mean yield** and **Upper 95% Mean yield**.

Figure 10.33 The Save Columns Menu

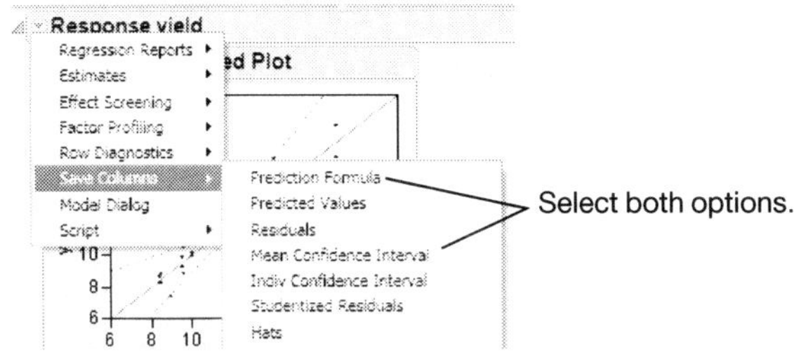

Now you can use Graph Builder to visualize the interaction between popcorn and batch.

6. Select **Graph > Graph Builder**.
7. Drag and drop these columns into the **Y** zone:
   - Pred Formula yield
   - Lower 95% Mean yield
   - Upper 95% Mean yield
8. Drag and drop popcorn into the **X** zone.
9. Drag and drop batch into the **Group X** zone.
10. Drag and drop oil amt into the **Group Y** zone.
11. Because the Jitter option is turned on, there are two data points per combination of factors, and the prediction at a given combination is the same. Turn off the jitter by right-clicking in the graph and deselecting **Points** > **Jitter**.

Figure 10.34 Example of Three Responses versus Three Factors

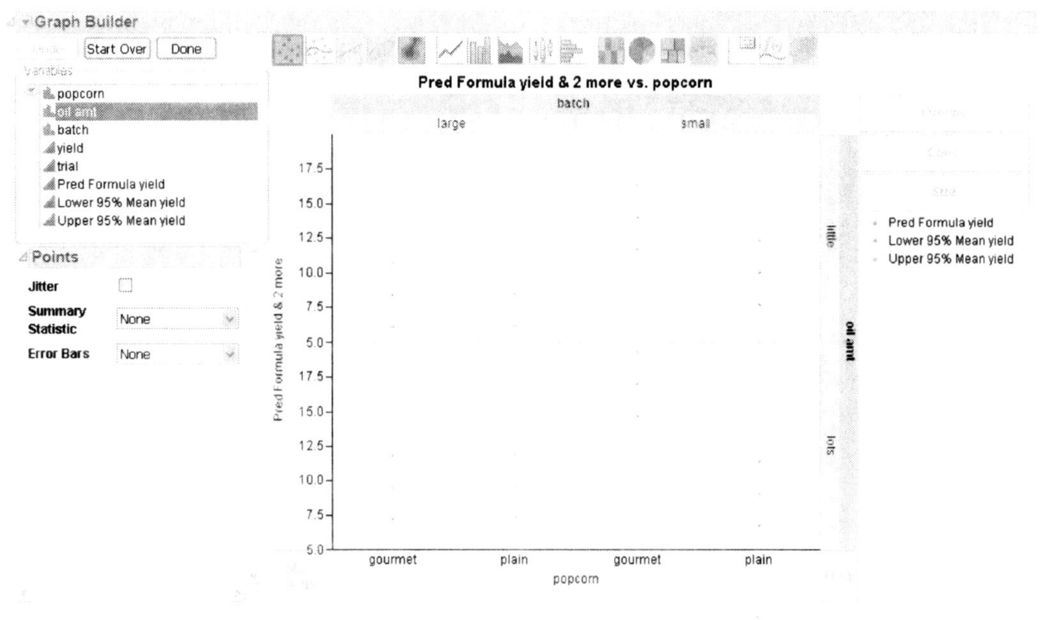

Format the graph to see interval bars for Lower 95% Mean yield and Upper 95% Mean yield, and to see points for Pred Formula yield.

12. Right-click on the graph and select **Add** > **Bar**.

    This adds bars for all three responses.

13. Change the bar style to interval by right-clicking on the graph and selecting **Bar** > **Bar Style** > **Interval**.

Figure 10.35 The Interval Bar Style

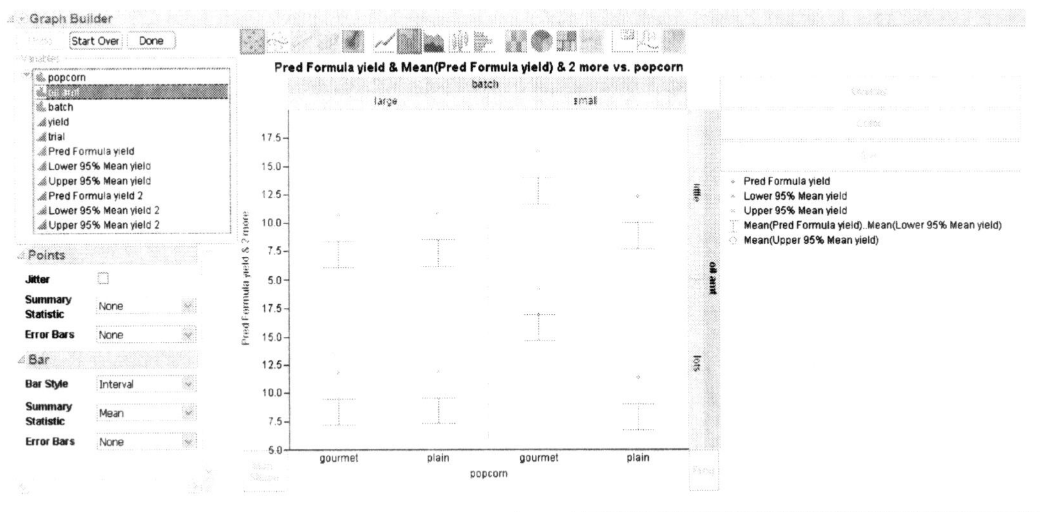

The interval bar style currently spans from Lower 95% Mean yield to Pred Formula yield, but you want it to span up to Upper 95% Mean yield.

14. Remove the bar element for Pred Formula yield by right-clicking on the graph and deselecting **Bar** > **Y** > **Pred Formula yield**.

    Now the confidence interval spans from the lower to upper value.

Figure 10.36 Example of Correct Interval Span

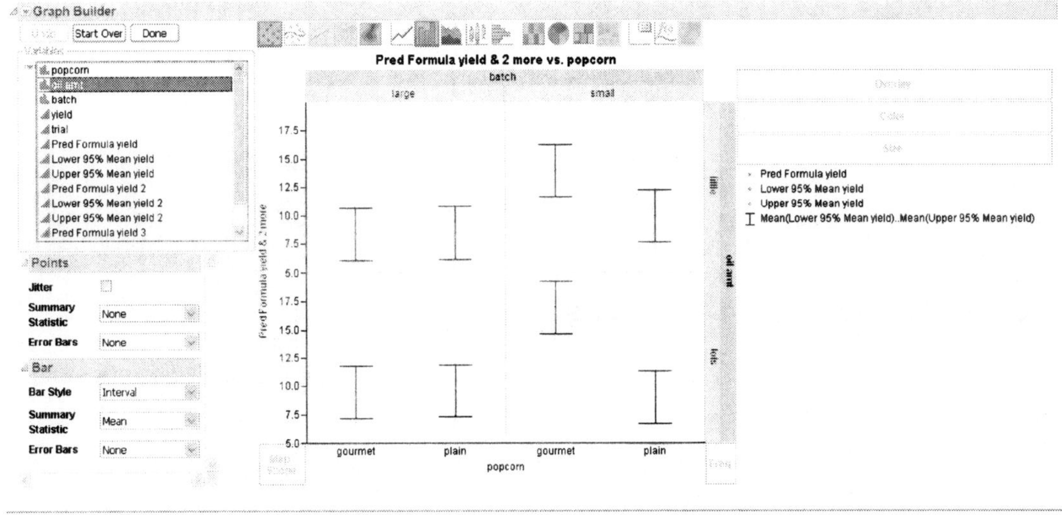

Remove the point graph element for Lower 95% Mean yield and Upper 95% Mean yield.

15. Right-click on the graph and select **Points** > **Y**, and deselect the Lower 95% Mean yield and Upper 95% Mean yield individually.

To make the points easier to see, increase the size of the points.

16. Right-click on the graph, and select the XXL option under **Graph** > **Marker Size**. Do this for each quadrant of the graph.

Note: You can also change the graph title and labels for the X and Y zones. Click on the label and type in the new text.

Figure 10.37 Example of Predicted Values and Confidence Intervals

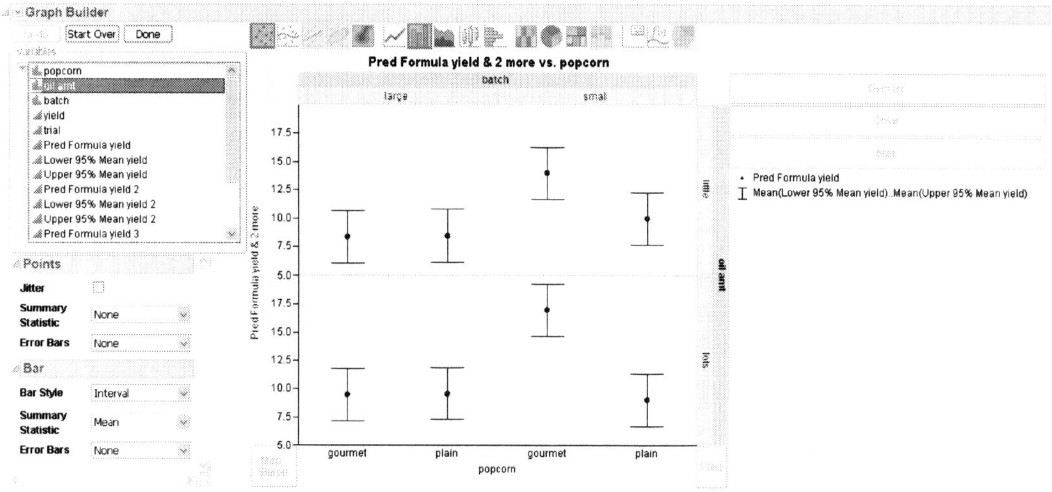

From Figure 10.37, you can see the following relationships:

- For large batches, there is no difference between plain and gourmet popcorn.
- For small batches, the gourmet popcorn has a higher yield than the plain popcorn.
- For the oil amount, the relationship is the same whether the oil amount is little or lots, so there is no three-way interaction.

Examine Diamond Characteristics

You have data about diamonds, including their carat weight and price. Examine the relationship between carat weight and price.

1. Open the Diamonds Data.jmp sample data table.
2. Select **Graph > Graph Builder.**
3. Drag and drop Price into the **Y** zone.

4. Drag and drop **Carat Weight** into the **X** zone.
5. Right-click on the graph and select **Add > Contour**.

Figure 10.38 Example of Contour Plot of Diamond Characteristics

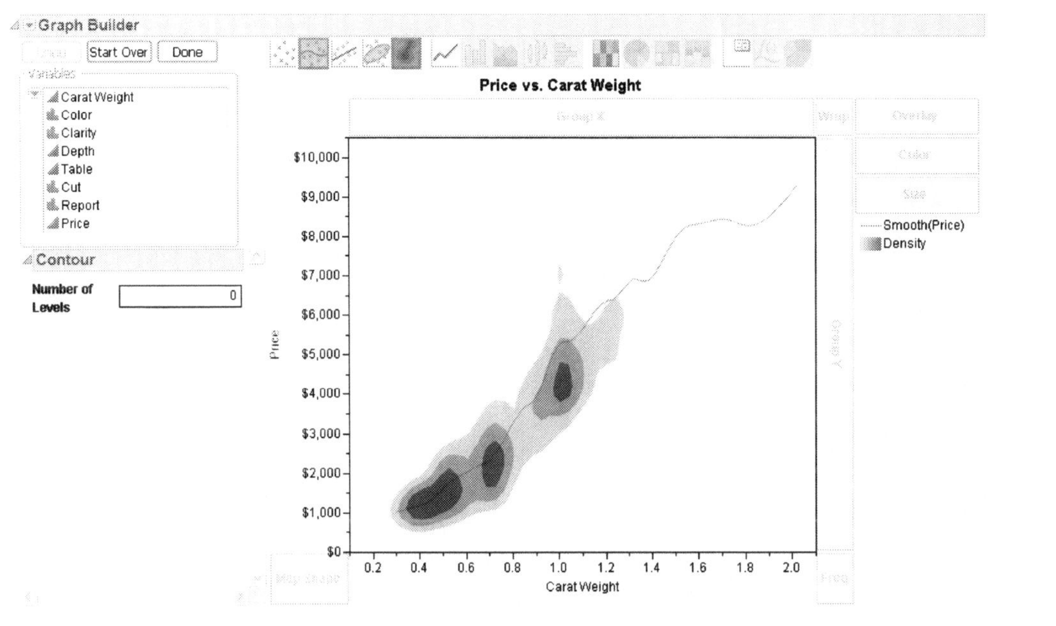

To see how many points there are, temporarily change the contour plot to points.

6. Right-click on the graph and select **Contour > Change to > Points**.

Figure 10.39 Example of Points Showing Diamond Characteristics

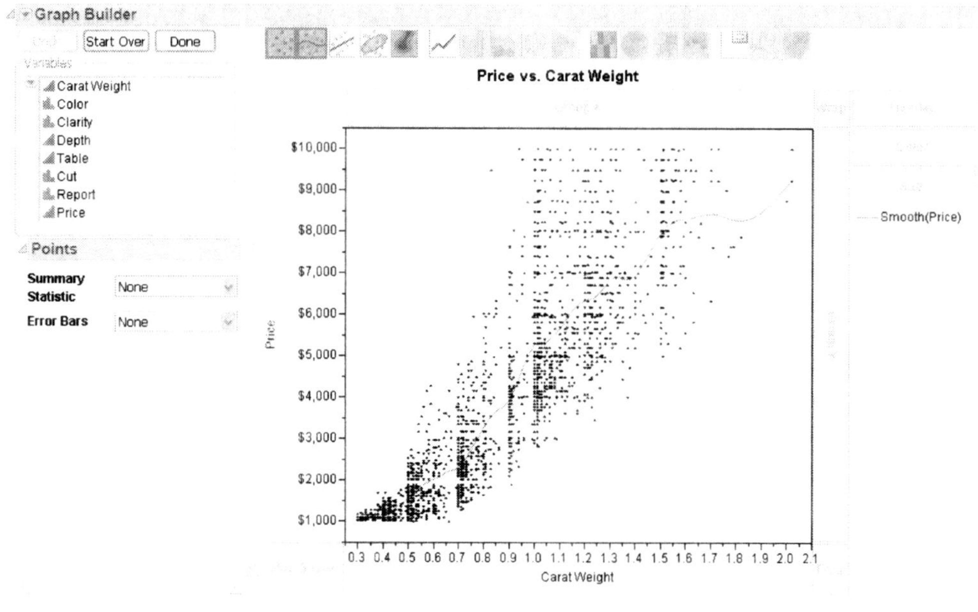

You can see that the points are difficult to interpret. Some points are likely overlapping, making the density unclear. Change the points back to a contour plot.

7. Right-click on the plot and select **Points > Change to > Contour**.

Figure 10.40 Example of Contour Plot of Diamond Characteristics

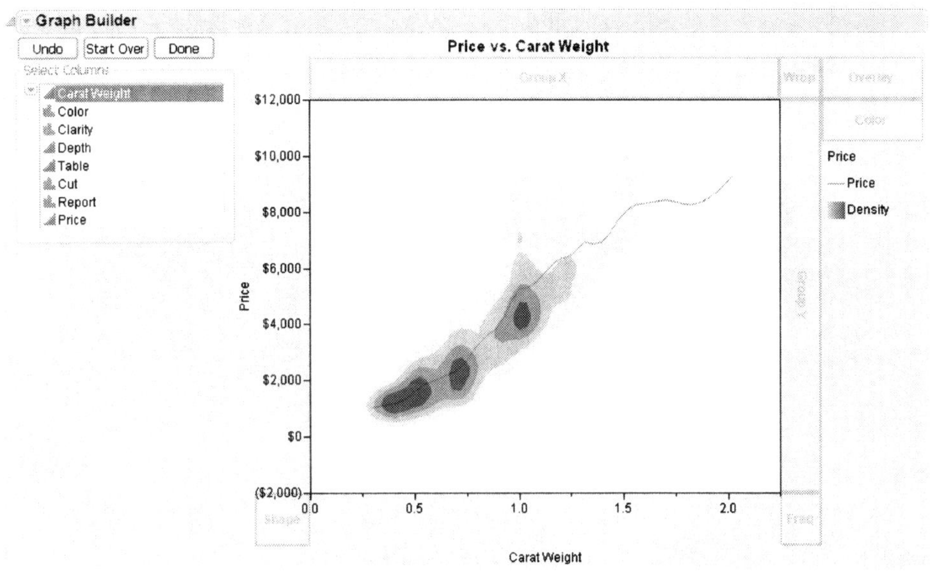

The darker the density, the more data. In Figure 10.40, notice the following:

- Most people in the sample purchased diamonds with a carat weight of about 0.5, 0.75, and 1.0.
- Most people who purchased diamonds with a carat weight of 0.5 paid about $850-$2,000 dollars.
- Most people who purchased diamonds with a carat weight of 0.75 paid about $1,600-$2,700 dollars.
- Most people who purchased diamonds with a carat weight of 1.0 paid about $3,800-$4,800 dollars.

Examples of Creating Custom Maps

Once you have your custom map files, you can create custom maps in one of two ways:

- Set up your custom map files initially and save them in the predetermined location. JMP finds and uses them in the future with any appropriate data.
- Point to specific predefined map files directly from your data. This step might be required each time you want to specify custom maps.

Set Up Automatic Custom Maps

As an example of setting up automatic custom maps, say that you have downloaded ESRI shapefiles from the Internet and you want to use them as your map files in JMP. The shapefiles are named **Parishes.shp** and **Parishes.dbf**. These files contain coordinates and information about the parishes (or counties) of Louisiana.

Save the .shp file with the appropriate name and in the correct directory, as follows:

1. In JMP, open the **Parishes.shp** file from the following default location:
   - On Windows: C:\Program Files\SAS\JMP\10\Samples\Import Data
   - On Mac: /Library/Application Support/JMP/10/Samples/Import Data

Note: If you cannot see the file, you might need to change the file type to **All Files**.

   JMP opens the file as Parishes. The .shp file contains the *x* and *y* coordinates.

2. Save the Parishes file with the following name and extension: Parishes-XY.jmp. Save the file here:
   - On Windows 7 and Vista: C:\Users\user name\AppData\Local\SAS\JMP\Maps
   - On Windows XP: C:\Documents and Settings\user name\Local Settings\Application Data\SAS\JMP\Maps
   - On Mac: /Users/user name/Library/Application Support/JMP/Maps

Note: If the Maps folder does not already exist, you need to create it.

3. Close the Parishes-XY.jmp file.

Perform the initial setup and save the .dbf file, as follows:

1. Open the Parishes.dbf file from the following default location:
   - On Windows: C:\Program Files\SAS\JMP\10\Samples\Import Data
   - On Mac: /Library/Application Support/JMP/10/Samples/Import Data

Note: If you cannot see the file, you might need to change the file type to **All Files**.

   JMP opens the file as Parishes. The .dbf file contains identifying information.

2. In the Parishes file, add a new column. Name it Shape ID. Drag and drop it to be the first column.
3. In the first three rows of the Shape ID column, type 1, 2, and 3.
4. Select all three cells, right-click, and select **Fill > Continue sequence to end of table**.

Figure 10.41 Shape ID Column in Parishes File

	Shape ID	PARISH	FIPS	PARISH_FIP	Parish_Num
1	1	Webster	22119.00000000000	119	60
2	2	East Carroll	22035.00000000000	035	18
3	3	West Carroll	22123.00000000000	123	62
4	4	Morehouse	22067.00000000000	067	34
5	5	Union	22111.00000000000	111	56
6	6	Lincoln	22061.00000000000	061	31
7	7	Ouachita	22073.00000000000	073	37
8	8	Richland	22083.00000000000	083	42
9	9	Bienville	22013.00000000000	013	07
10	10	Franklin	22041.00000000000	041	21

5. Right-click on the PARISH column and select **Column Info**.
6. Select **Column Properties > Map Role**.
7. Select **Shape Name Definition**.
8. Click **OK**.
9. Save the Parishes file with the following name and extension: Parishes-Name.jmp. Save the file here:
   - On Windows 7 and Vista: C:\Users\user name\AppData\Local\SAS\JMP\Maps
   - On Windows XP: C:\Documents and Settings\user name\Local Settings\Application Data\SAS\JMP\Maps
   - On Mac: /Users/user name/Library/Application Support/JMP/Maps
10. Close the Parishes-Name.jmp file.

Once the map files have been set up, you can use them. The Katrina.jmp data table contains data on Hurricane Katrina's impact by parish. You want to visually see how the population of the parishes changed after Hurricane Katrina. Proceed as follows:

11. Open the Katrina.jmp sample data table.
12. Select **Graph > Graph Builder**.
13. Drag and drop Parish into the **Map Shape** zone.

    The map appears automatically, since you defined the Parish column using the custom map files.
14. Drag and drop Population into the **Color** zone.
15. Drag and drop Date into the **Group X** zone.

Figure 10.42 Population of Parishes Before and After Katrina

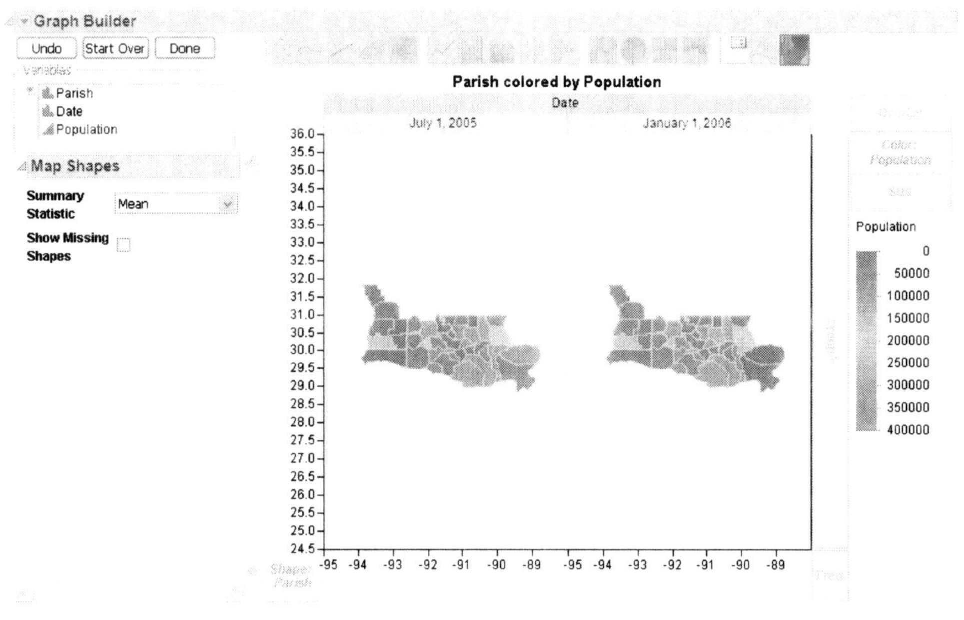

16. Select the **Magnifier** tool to zoom in on the Orleans parish.

Figure 10.43 Orleans Parish

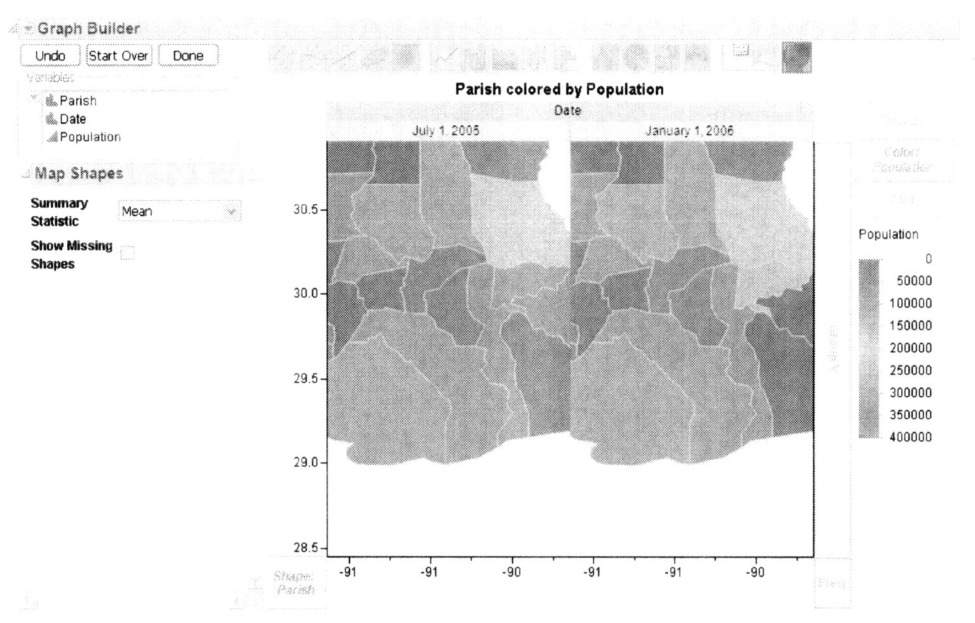

You can clearly see the drop in population as a result of Hurricane Katrina. The population of the Orleans parish went from 437,186 in July 2005 to 158,353 in January 2006.

Point to Specific Map Files Directly from Your Data

As an example of pointing to specific map files directly from your data, say that you already have your custom map files and they are named appropriately. Your map files are US-MSA-Name.jmp and US-MSA-XY.jmp. They are saved in the Sample Data folder.

The PopulationByMSA.jmp data table contains population data from the years 2000 and 2010 for the metropolitan statistical areas (MSAs) of the United States. Set up the **Map Role** column property in the data table, as follows:

1. Open the PopulationByMSA.jmp sample data table.
2. Right-click on the **Metropolitan Statistical Area** column and select **Column Info**.
3. Select **Column Properties > Map Role**.
4. Select **Shape Name Use**.
5. Next to the **Map name data table**, enter the following path:
   - On Windows: C:\Program Files\SAS\JMP\10\Samples\Data\US-MSA-Name.jmp
   - On Mac: /Library/Application Support/JMP/10/Samples/Data/US-MSA-Name.jmp

   This tells JMP where the data tables containing the map information reside.
6. Next to the **Shape definition column**, type MSA_Name.

   MSA_Name is the specific column within the US-MSA-Name.jmp data table that contains the unique names for each metropolitan statistical area. Notice that the MSA_Name column has the **Shape Name Definition Map Role** property assigned, as part of correctly defining the map files.

Note: Remember, the Shape ID column in the -Name data table maps to the Shape column in the -XY data table. This means that indicating where the -Name data table resides links it to the -XY data table, so that JMP has everything that it needs to create the map.

Figure 10.44 Map Role Column Property

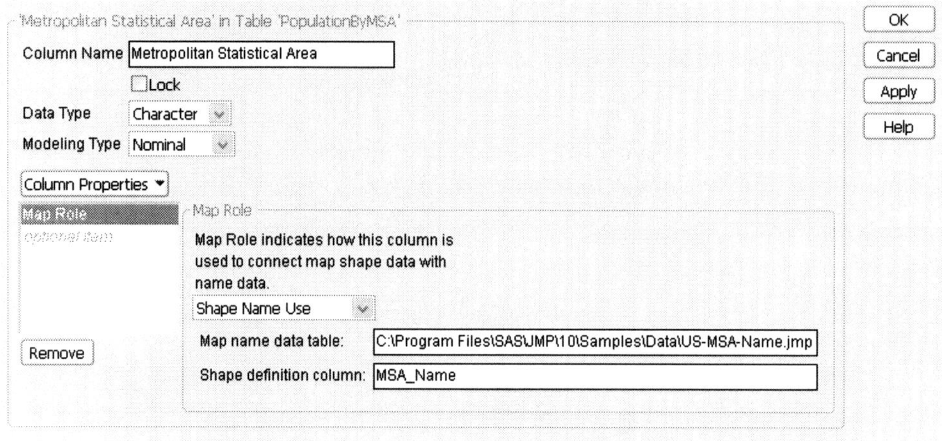

7. Click **OK.**

Once the **Map Role** column property has been set up, you can perform your analysis. You want to visually see how the population has changed in the metropolitan statistical areas of the United States between the years 2000 and 2010.

1. Select **Graph > Graph Builder.**
2. Drag and drop **Metropolitan Statistical Area** into the **Map Shape** zone.

   Since you have defined the **Map Role** column property on this column, the map appears.
3. Drag and drop **Change in Population** to the **Color** zone.

Figure 10.45 Change in Population for Metropolitan Statistical Areas

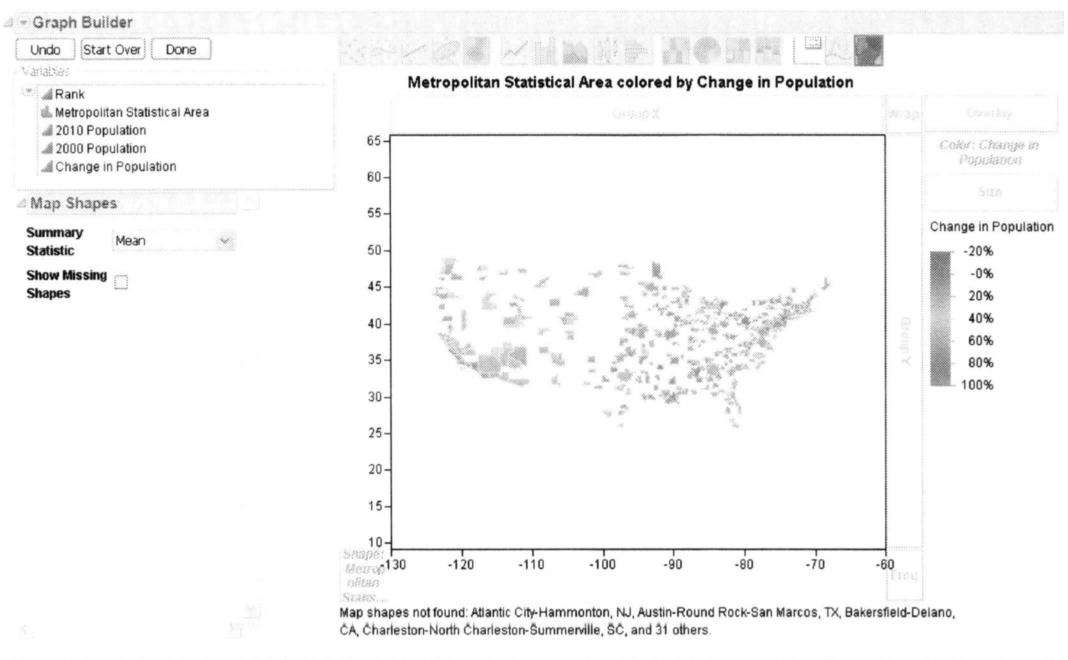

4. Select the **Magnifier** tool to zoom in on the state of Florida.
5. Select the **Arrow** tool and click on the red area.

Figure 10.46 Population Change of Palm Coast, Florida

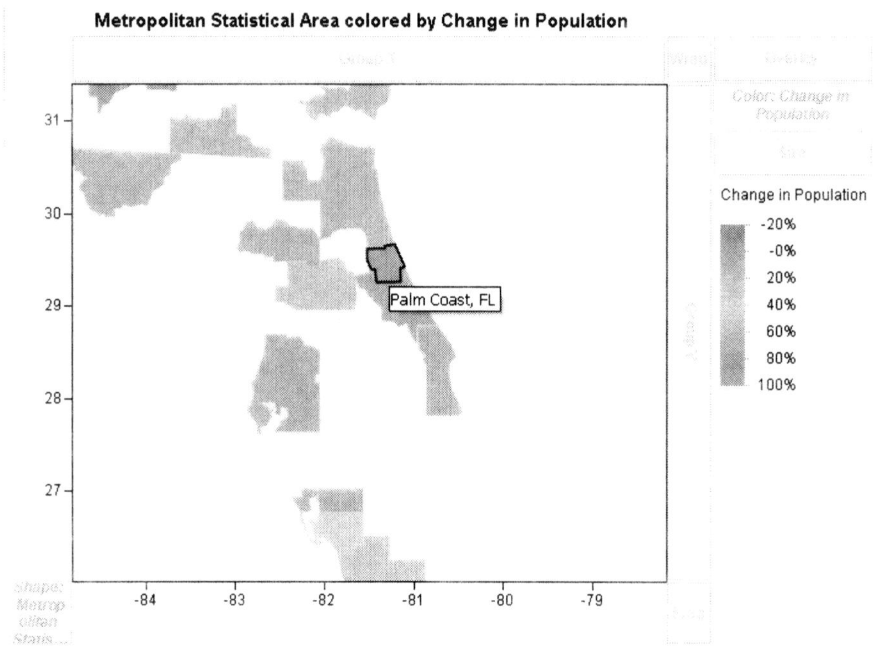

6. Select the **Magnifier** tool and hold down the ALT key while clicking on the map to zoom out.
7. Select the **Magnifier** tool and zoom in on the state of Utah.
8. Select the **Arrow** tool and click on the area that is slightly red.

Figure 10.47 Population Change of St. George, Utah

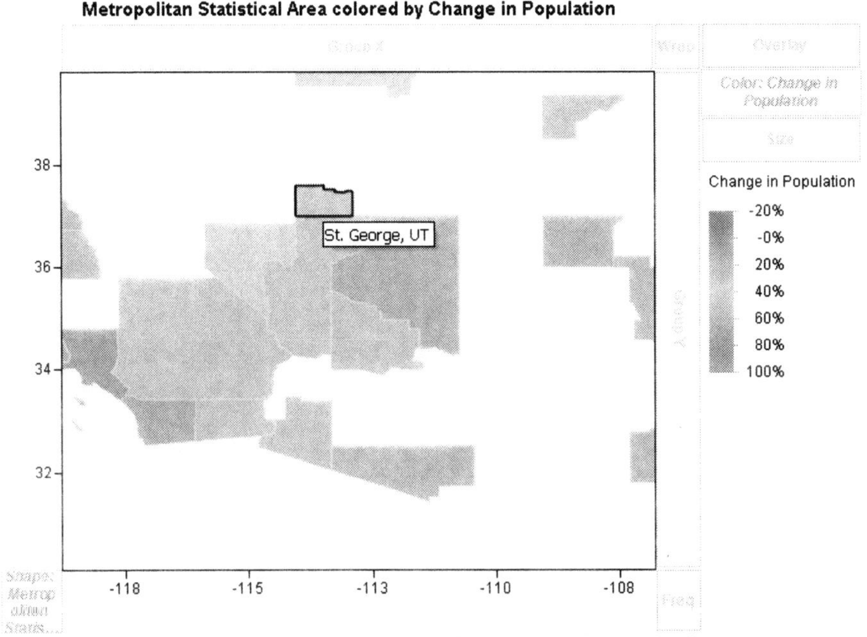

You can see that the areas of Palm Coast, Florida and St. George, Utah had the most population change between 2000 and 2010. The Palm Coast area saw a population change of 92%, and the St. George area saw a population change of about 53%.

Chapter 11

Creating Summary Charts

Using the Chart Platform

The Chart platform on the **Graph** menu charts continuous variables versus categorical variables. The continuous variables are summarized for each categorical level. Chart supports several chart types, such as bar charts, pie charts, and line charts. Chart is similar to the **Tables > Summary** command and is useful for making graphical representations of summary statistics.

If you want to make a plot of individual data points (rather than summaries of data points), we recommend using the Overlay Plot platform. See the "Creating Overlay Plots" chapter on page 337 for details about overlay plots.

Figure 11.1 Examples of Charts

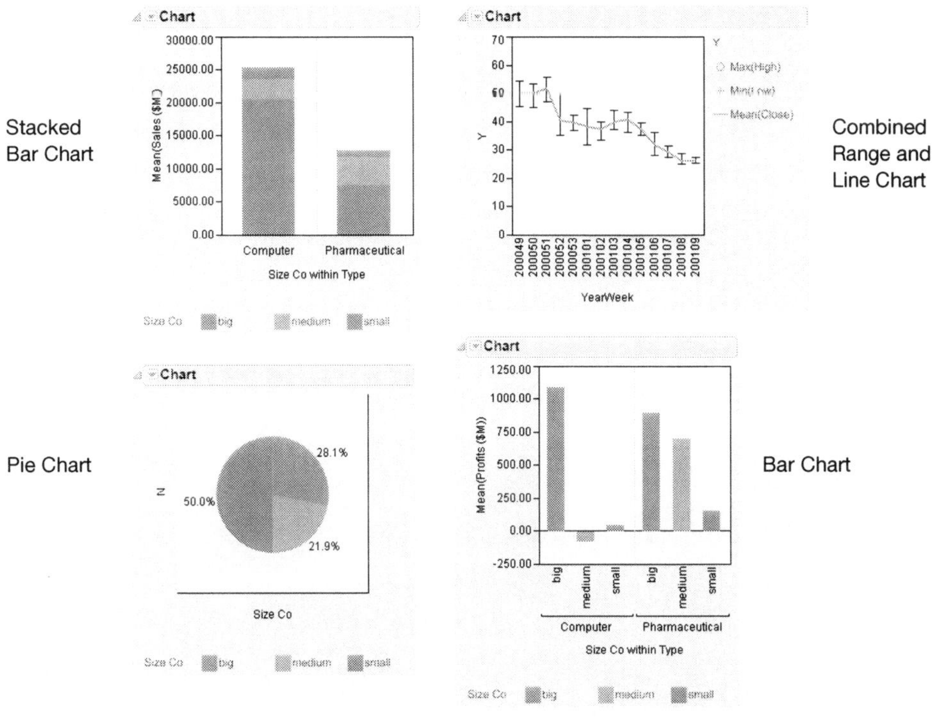

Contents

Example of the Chart Platform

Here is a simple example that shows how to plot the mean height of students based on their age group.

1. Open the Students.jmp sample data table.
2. Select **Graph > Chart**.
3. Select height and click **Statistics**.
4. Select **Mean** from the menu of statistics.
5. Select age and click **Categories, X, Levels**.

Figure 11.2 The Completed Chart Launch Window

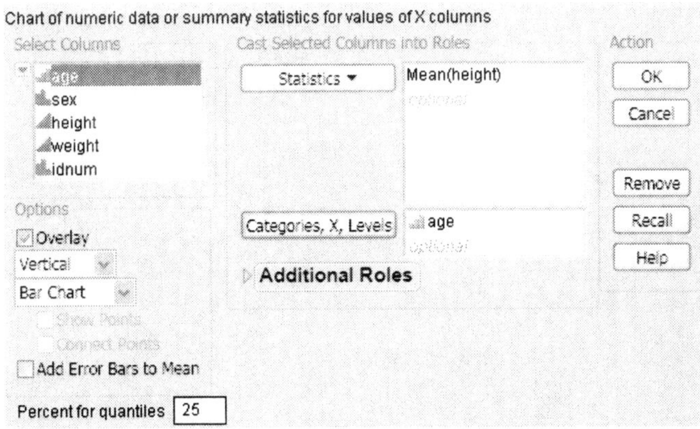

6. Click **OK**.

Figure 11.3 Mean of height by age

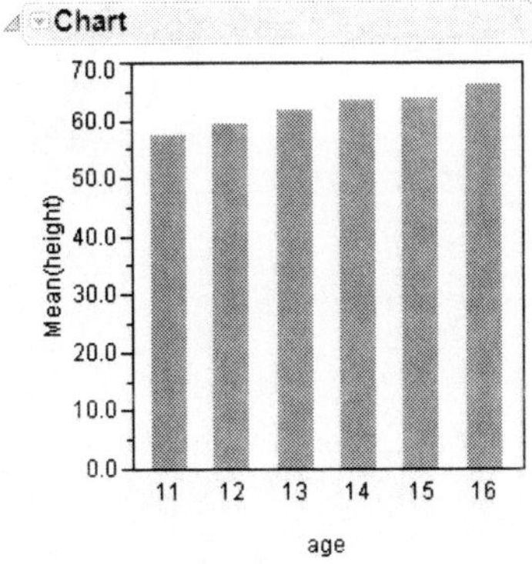

This bar chart shows the following:

- The mean height of the students in this class increases with age.
- The largest increase occurs at the earliest ages.
- The mean changes very little for the older students.

You might expect the change of the mean height to be different for males and females.

7. From the red triangle menu for Chart, select **Script > Relaunch Analysis.**

   The Chart launch window appears, already filled out for you. If you clicked **OK** now, you would see a duplicate of the chart you already have.

8. Open the Additional Roles outline.
9. Select sex and click **Grouping**.
10. Click **OK**.

Figure 11.4 Mean of height by age and Grouped by sex

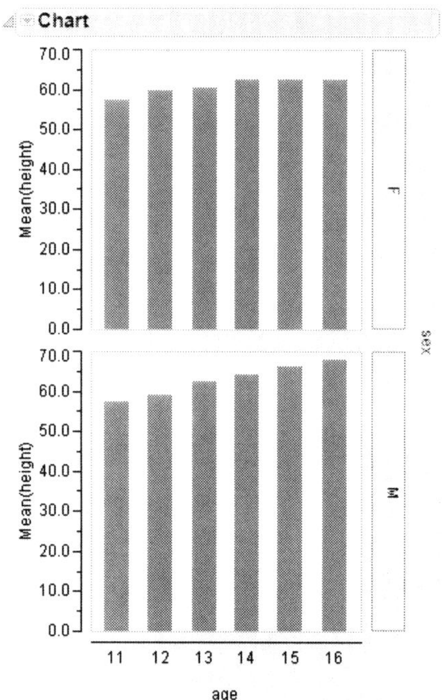

These two bar charts confirm your assumption. The mean of height for girls in the class rises early and then remains stable. The mean of height for boys rises more dramatically overall, and also continues to increase at later ages.

Launch the Chart Platform

Launch the Chart platform by selecting **Graph > Chart**.

Figure 11.5 The Chart Launch Window

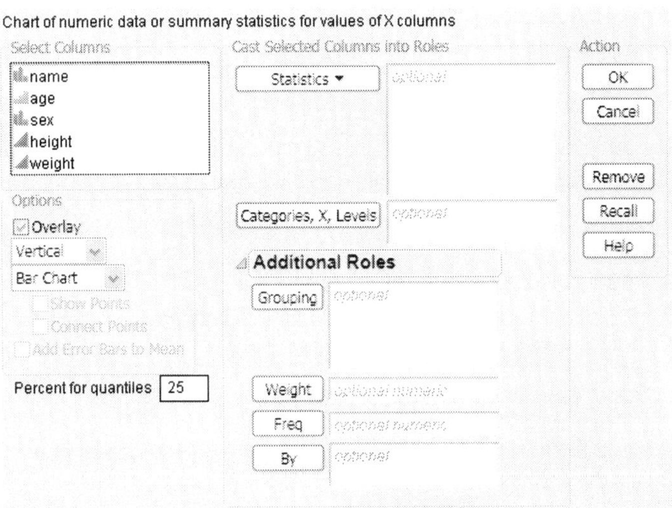

In the Chart launch window, you can assign the following:

- Up to two X variables, which appear on the *x*-axis in the same order that you assign them in the launch window.
- As many Y variables (statistics) as you want. If the data is already summarized, select **Data** as the statistics option.

Table 11.1 Description of the Chart Launch Window

Cast Selected Columns Into Roles:	
Statistics	Use this menu to select the statistic to chart for each Y variable. See "Plot Statistics for Y Variables" on page 317.
Categories, X, Levels	Select up to two X variables whose levels are categories on the *x*-axis. The Chart platform produces a bar for each level or combination of levels of the X variables. If you do not specify an X variable, the chart has a bar for each row in the data table. See "Use Categorical Variables" on page 318.

Table 11.1 Description of the Chart Launch Window *(Continued)*

Additional Roles:	
Grouping	For one or more **Grouping** variables, independent results are produced for each level or combination of levels of the grouping variables. These results appear in the same report window, but in separate plots. See "Use Grouping Variables" on page 319.
Weight	Assigns a variable to give the observations different weights.
Freq	Assigns a frequency variable. This is useful if you have summarized data.
By	By variables cause plots to be created in separate outline nodes.
Options:	
Overlay	If this option is selected, all Y variables are plotted in one graph. If this option is not selected, each Y variable is plotted in its own graph. This option is selected by default where available. This option is available for all chart types except **Pie Chart**.
Chart Orientation	Select Vertical for a vertical chart or Horizontal for a horizontal chart. Vertical is the default option. This option is available for all chart types except **Pie Chart**.
Chart Type	Select the type of chart that you want. Available chart types are **Bar Chart**, **Line Chart**, **Pie Chart**, **Needle Chart**, and **Point Chart**. Selecting a chart controls which of the other options are available. You can always change these options after the chart appears. See "Y Options" on page 323.
Show Points	Shows the points in the plot. This option is selected by default where available. This option is available for all chart types except **Bar Chart** and **Pie Chart**.
Connect Points	Connects the points in the plot. Show Points does not have to be selected to connect points. This option is selected by default where available. This option is available only for **Line Chart**.
Add Error Bars to Mean	Adds error bars when the **Mean** statistic is selected for at least one Y variable and at least one X variable is assigned. This option is not selected by default. This option is available for **Line Chart** and **Bar Chart**, and additional options are added to the Chart launch window. See "Adding Error Bars" on page 319.

Table 11.1 Description of the Chart Launch Window *(Continued)*

Percent for quantiles	Sets the specific quantile when the **Quantiles** statistic is selected for at least one Y variable. The default value is 25. Specify a different quantile: 1. Type the value in the **Percent for quantiles** box. 2. Select a column. 3. Click **Statistics**. 4. Select **Quantiles** from the menu of statistics.

For more information about the launch window, see *Using JMP*.

After you click **OK**, the Chart report window appears. See "The Chart Report" on page 320.

Plot Statistics for Y Variables

You can plot the raw data for Y variables, or you can plot as many statistics as you want on the *y*-axis. The **Statistics** menu in the Chart launch window lists the available statistics. To specify the *y*-axis, highlight one or more numeric columns in the Select Columns list and select from the list of statistics. If all the statistics requested are counting statistics (for example, **N**) for the same column, that column is used as the category variable.

The available statistics in the Chart platform are described in the following table. They are the same as those computed by statistical platforms in the **Analyze** menu and the **Summary** command in the **Tables** menu.

Data	The value of each row in the data table when there is no categorical variable. If there is a categorical variable, **Data** produces a point plot within the variable's levels.
N	The number of nonmissing values. Also used to compute statistics when there is no column assigned as a weight variable. The Chart platform shows **N** for each level of a categorical variable.
% of Total	The percentage of the total number of rows represented by each level of the **Categories, X, Levels** variable. If summary statistics are requested on a continuous variable, then the **% of Total** equals the proportion of the sum represented by each level of the **Categories, X, Levels** variable.
N Missing	The number of missing values in each level of a categorical variable.
Min	The least value, excluding missing values, in the level of a categorical variable.
Max	The greatest value in the level of a categorical variable.
Sum Wgt	The sum of all values in a column assigned as Weight. Also used instead of **N** to compute other statistics. **Chart** shows the sum of the weight variable for each level of a categorical variable.
Sum	The sum of all values in each level of a categorical variable.

Mean	The arithmetic average of a column's values. The mean is the sum of nonmissing values divided by the number of nonmissing values.
Variance	The sample variance computed for each level of a categorical variable.
Std Dev	The sample standard deviation computed for each level of a categorical variable. It is the square root of the variance of the level values.
Std Err	The standard error of the mean of each level of a categorical variable. It is the standard deviation, **Std Dev**, divided by the square root of **N** for each level. If a column is assigned a weight variable, then the denominator is the square root of the sum of the weights.
CV	The coefficient of variation of a column's values. The CV is computed by dividing the column standard deviation by the column mean and multiplying by 100.
Range	The difference between the maximum and minimum values in each level of a categorical variable.
Median	The middle value in each level of a categorical variable. Half of the values in the level are greater than or equal to the median and half are less than the median.
Quantiles	Divides a data set so that *n*% of the data is below the nth quantile. To compute a specific quantile, enter the quantile value in the box located in the lower left of the Chart launch window *before* requesting **Quantile** from the menu.

Related Information

- "Plot a Single Statistic" on page 326
- "Plot Multiple Statistics" on page 327

Use Categorical Variables

You can assign zero, one, or two X variables whose levels are categories on the *x*-axis. The Chart platform produces a bar (or a needle, or a pie slice, and so on) for each level or combination of levels of the X variables. If you do not specify any X variable, the chart has a bar for each row in the data table.

The following table shows what type of chart to expect based on the number of X and Y variables.

X	Y	Type of Chart
none	one or more	If you do not specify a variable for categories, most statistics produce a bar (or a needle, or a pie slice, and so on) for each observation in the data table. This is useful when your data is already summarized. In that case, you usually specify **Data** as the statistic to plot. Each bar reflects the value of the Y variable. See "Plot a Single Statistic" on page 326, for an example.

X	Y	Type of Chart
one or two	none	Plots the counts for each level of the X variable. For two X variables, the counts for each level of both X variables are included (or overlaid) in a single chart. See "Plot Counts of Variable Levels" on page 328, for an example.
one or two	one or more	Plots the selected statistics for each level of the X variable. For two X variables, the selected statistics for each level of the X variables are included (or overlaid) in a single chart. See "Plot Multiple Statistics" on page 327, and "Plot Multiple Statistics with Two X Variables" on page 330, for examples.

Use Grouping Variables

If you specify one grouping variable, the result is a separate chart for each level of the grouping variable. All charts are under the same outline title. If you used the same variable as a By variable instead, the same separate charts are produced, but each chart is under its own outline title.

If you specify two or more grouping variables, the result is a matrix of charts. Each chart shows a combination of one level from each of the grouping variables.

If there are multiple statistics, the **Overlay** option is checked by default, and the Y variables (statistics) are plotted on the same chart for each level of the grouping variable. However, the levels of the grouping variable cannot be overlaid into the same plot frame. For example, if the levels of your grouping variable are Male and Female, the **Overlay** option cannot be used to combine the two Male and Female graphs into one graph. To see that type of result, use **Categories, X, Levels** instead of **Grouping** variables.

Related Information

- "Example Using Two Grouping Variables" on page 324
- "Example Using Two Grouping Variables and Two Category Variables" on page 325

Adding Error Bars

Error bars are available when the **Mean** statistic is selected for at least one Y variable, and at least one X variable is assigned. Error Bars are not available for pie charts. Selecting **Add Error Bars to Mean** causes additional options to appear in the Chart launch window.

After the option is checked, select a type of error bar from the menu that appears. Some of the types of error bar have an additional numeric field. The following table describes the different types of error bars that are available.

Range	Creates error bars based on the range of the data.
Standard Error	Creates error bars based on the standard error of the mean. You can specify the number of standard errors.

Standard Deviation	Creates error bars based on the standard deviation of the data. You can specify the number of standard deviations.
Confidence Interval	Creates error bars based on a confidence interval of the mean. The standard deviation used for the confidence interval is separate for each bar. You can specify the level of confidence.
Confidence Interval (pooled)	Creates error bars based on a confidence interval of the mean. The standard deviation used for the confidence interval is based on the pooled standard deviation. This option is not available if you have more than one category variable. You can specify the level of confidence.

The Chart Report

Follow the instructions in "Example of the Chart Platform" on page 313 to produce the report shown in Figure 11.6.

Charts can be bar charts, pie charts, line charts, needle charts, point charts, and range charts. Figure 11.6 shows a standard bar chart.

Figure 11.6 The Initial Chart Report Window

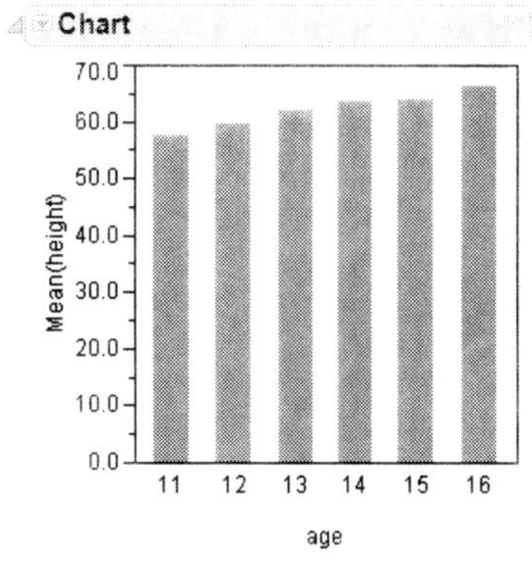

For information about additional options for the report, see "Chart Platform Options" on page 322.

Legends

Legends are shown as needed. If your chart uses different colors or markers to show levels of one or two X variables, a legend below the chart shows them. If your chart uses different colors or markers to show more than one statistic, a legend to the right of the chart shows them.

Figure 11.7 Examples of an X Legend (left) and Y Legend (right)

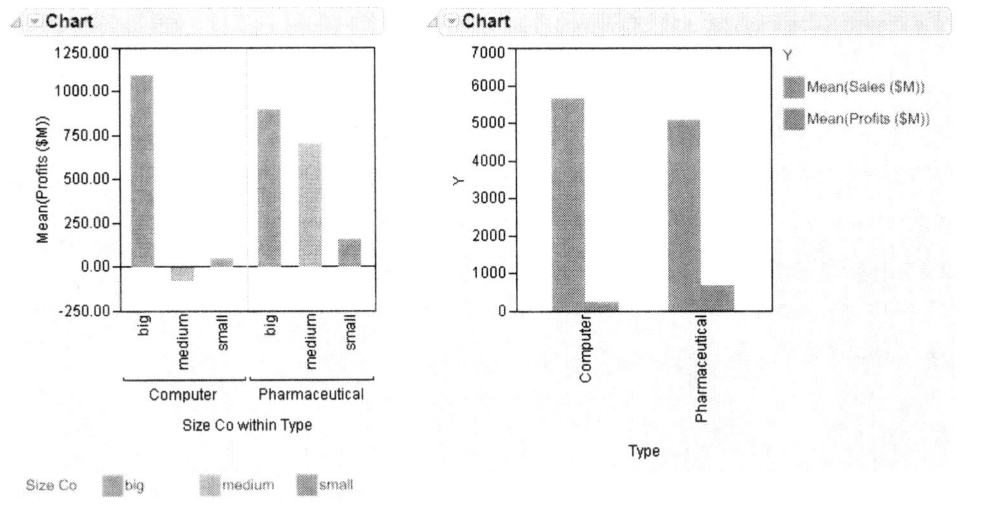

Ordering

By default, the Chart platform orders the bars using one of the common orders supported by JMP (months, days of the week, and so on). However, if the grouping column has a **Row Order Levels** column property, the levels are ordered in that order. If the grouping column has a **Value Ordering** column property, it uses that order. If both **Row Order Levels** and **Value Ordering** properties are defined, the **Value Ordering** property has precedence. With neither property in effect, bars are drawn in alphanumeric order.

Coloring Bars in a Chart

There are a few ways to color bars after the chart has been created.

Manually Set the Color of All Bars

1. Ensure that no bars are selected.
2. From the red triangle menu for Chart, select **Level Options > Colors.**
3. Select a color from the color palette that appears.

Set the Color of a Single Bar

1. Select a bar in the chart.
2. From the red triangle menu for Chart, select **Level Options > Colors.**
3. Select a color from the color palette that appears.

Note: If you assigned *both* a category variable *and* a grouping variable in your chart, all of the bars are colored even if you selected only one bar.

Automatically Assign a Color to a Level

1. Select the column in the data table.
2. Select **Cols > Column Info**.
3. Assign colors using **Value Colors** in the **Column Properties** menu. For details about the Value Colors property, see the *Using JMP* book.

Chart Platform Options

The basic Chart report is shown in Figure 11.6 on page 320.

The Chart platform has plotting options on the red triangle menu on the Chart title bar. When you select one of these options at the platform level, it affects all plots in the report if no legend levels are highlighted. If one or more plot legend levels are highlighted, the options affect only those levels. There is also a single-plot options menu for each Y variable, which appears when you highlight a Y variable legend beneath the plot and right-click.

The individual plot options are the same as those in the **Y Options** submenu at the platform level. See "Y Options" on page 323.

General Platform Options

When you select one of these options at the platform level, it affects all plots in the report if no legend levels are highlighted. If one or more plot legend levels are highlighted, the options affect only those plots.

Table 11.2 Descriptions of General Platform Options

Overlay	Displays a single overlaid chart when you have more than one Y (statistics) variable. Each statistic can be assigned a different type of chart (such as line and bar) and overlaid in a single plot. **Overlay** is selected by default. The axis notation only shows for the last chart displayed if the charts are not overlaid. When **Overlay** is not selected, the platform shows duplicate axis notation for each chart.
Vertical Chart, Horizontal Chart	Changes horizontal charts to vertical charts (**Vertical**), or vertical charts to horizontal charts (**Horizontal**). Affects all charts in the report. Pie charts are converted to bar chats.
Pie Chart	Changes a horizontal or vertical chart type to a pie chart.
Range Chart	Displays a range chart. You can change any chart that includes at least two statistics in a single plot into a range chart. See "Create a Range Chart" on page 334, for an example of a range chart.

Table 11.2 Descriptions of General Platform Options *(Continued)*

Add Error Bars to Mean	Adds error bars to charts based on means. A window opens, prompting you to select the type of error bar. If error bars already exist on a chart, you can change the error bar type. See "Adding Error Bars" on page 319 for a description of error bar types.
Stack Bars	Stacks the bars from levels of a subgroup end-to-end. To use this option, you need two **Categories, X, Levels** variables and a statistic. See "Create a Stacked Bar Chart" on page 331, for an example of stacking bars.
Y Options	Contains the options described in "Y Options" on page 323. To apply these options to a single Y variable, highlight that variable in the legend first.
Level Options	Selects colors and markers. If no levels (bars, points, or pie slices) are selected, the color or marker that you select is applied to all levels. If you select one or more levels, the color or marker that you select is applied only to the selected levels. See "Coloring Bars in a Chart" on page 321.
Label Options	Attaches labels to your plots. In the **Label Options** menu, the first two options (**Show Labels** and **Remove Labels**) turn labels on and off. The last three options (**Label by Value**, **Label by Percent of Total Values**, **Label By Row**) specify what label should appear. Only one label can be shown at a time. Label options are also available by right-clicking in the chart.
Thick Connecting Line	Toggles the connecting line in a line chart to be thick or thin.
Show Y Legend	Shows the Y legend of the plot. This option is on by default for overlaid charts.
Show Level Legend	Shows the level legend of the plot. This option is on by default when the **Show Separate Axes** option is selected.
Show Separate Axes	Duplicates the axis notation for each chart when there are multiple charts. By default, the axis notation only shows for the last chart displayed if the charts are not overlaid. This option is not available for grouped charts.
Ungroup Charts	Moves level identifiers from the right side of the charts to beneath the charts for individual charts when a grouping variable is specified.
Script	This menu contains commands that are available to all platforms. They enable you to redo the analysis or save the JSL commands for the analysis to a window or a file. For more information, see *Using JMP*.

Y Options

The following section describes the **Y Options** submenu. These commands apply to all Y variables, unless you have a legend level highlighted, then they apply to only the highlighted Y variable.

Click on the legend within a plot to highlight a Y. If you right-click on a highlighted legend level, the commands to modify that Y appear. The commands then affect only the highlighted Y.

Table 11.3 Descriptions of Y Options

Bar Chart	Displays a bar for each level of the chart variables. The default chart is a bar chart.
Line Chart	Replaces a bar chart with a line chart and connects each point with a straight line. Select the **Show Points** option to show or hide the points. **Line Chart** is also available as a platform option, which then applies to all charts at once.
Needle Chart	Replaces each bar with a line drawn from the axis to the plotted value. **Needle Chart** is also available as a platform option, which then applies to all charts at once.
Point Chart	Shows only the plotted points, without connecting them.
Show Points	Toggles the point markers on a line or needle chart on or off.
Connect Points	Toggles the line connecting points on or off.
Show Error Bars	Toggles the error bars on plots of means. Note that this option is available only for plots that involve means of variables.
Overlay Color	Assigns a color to statistics (*y*-axis) to identify them in overlaid charts.
Overlay Marker	Assigns a marker to statistics, to identify them in overlaid charts.
Pen Style	Selects a line style for connecting lines.
Label Format	**Note:** To see the impact of this change, you must turn on labels. From the red triangle menu, select **Label Options > Show Labels.** Specifies the format, field width, and number of decimals for labels. Thousands separators can be turned on or off. Enter the values in the window that appears.

Additional Examples of the Chart Platform

This section contains additional examples using the Chart platform.

Example Using Two Grouping Variables

1. Open the Car Poll.jmp sample data table.
2. Select the **Chart** command from the **Graph** menu.
3. Select age and click **Statistics**.
4. Select **Mean** from the drop-down list of statistics.
5. Select sex and click **Categories, X, Levels**.
6. Open the Additional Roles outline.

7. Select marital status and country and click **Grouping**.
8. Click **OK**.

Figure 11.8 Example of a Chart with Two Grouping Variables

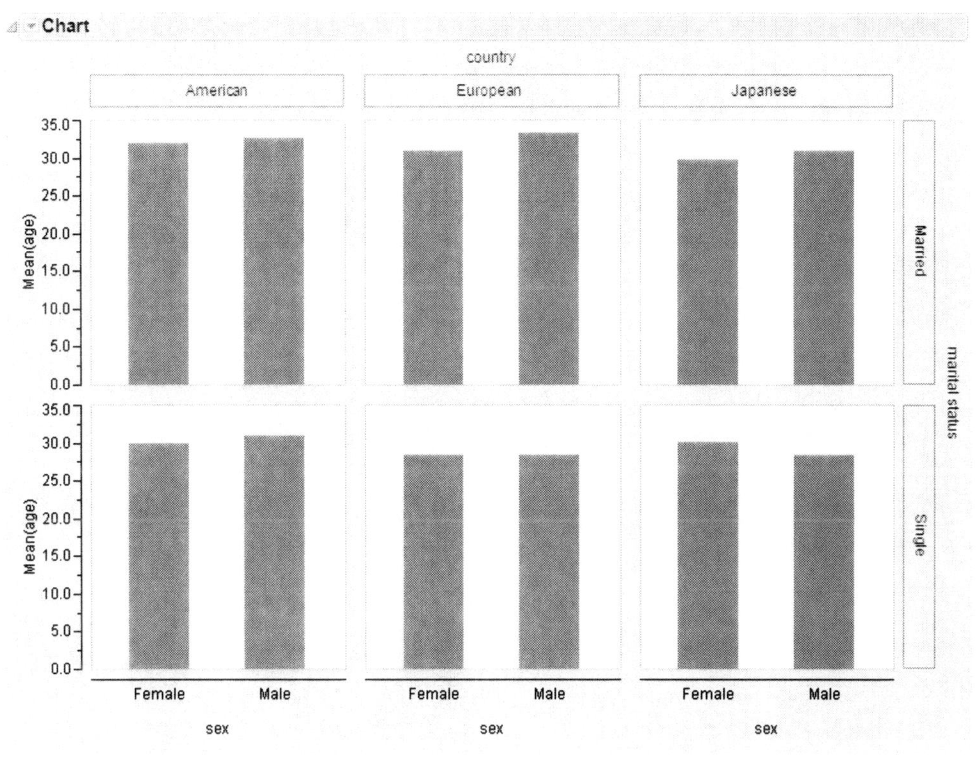

Example Using Two Grouping Variables and Two Category Variables

If you use multiple grouping and category variables, the multiple group labels appear around the borders of the charts, and the multiple X variables cause divisions within charts.

1. Open the Car Poll.jmp sample data table.
2. Select the **Chart** command from the **Graph** menu.
3. Select age and click **Statistics**.
4. Select **Mean** from the drop-down list of statistics.
5. Select sex and type and click **Categories, X, Levels**.
6. Open the Additional Roles outline.
7. Select marital status and country and click **Grouping**.
8. Click **OK**.

Figure 11.9 Example of a Chart with Two Grouping Variables and Two Category Variables

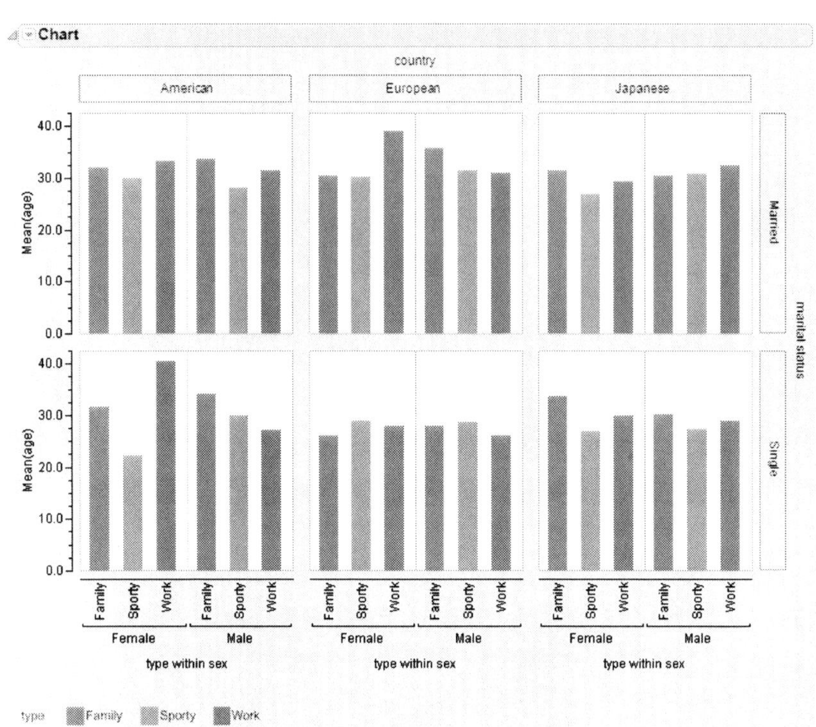

Plot a Single Statistic

If you do not specify a variable for categories, most statistics produce a bar for each observation in the data table. This is useful when your data is already summarized. In that case, you usually specify **Data** as the statistic to plot. Each bar reflects the value of the Y variable.

1. Open the Trial1.jmp sample data table.

This data table contains data from a popcorn experiment. Each row is an experiment, and the yield column is the amount of popped corn that resulted.

2. Select **Graph > Chart**.
3. Select yield and click **Statistics**.
4. Select **Data** from the menu of statistics.
5. Click **OK**.

The bar chart in Figure 11.10 shows a bar for each experiment (each row) in the data table.

Figure 11.10 Example of a Chart with One Statistic and No Categories

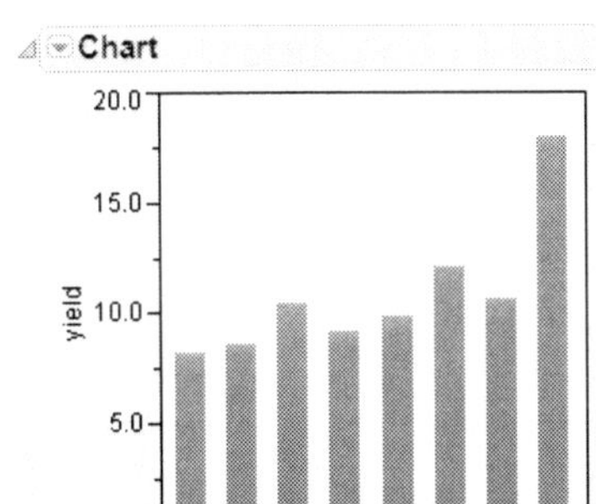

Plot Multiple Statistics

Chart more than one statistic in a single chart to compare them. If you do not assign an X variable, the chart displays a bar or point for each row in the data table.

1. Open the Financial.jmp sample data table.
2. Select **Graph > Chart**.
3. Select Type and click **Categories, X, Levels**.
4. Select Sales($M) and Assets($Mil.) and click **Statistics**.
5. Select **Mean** from the menu of statistics.
6. Click **OK**.

The bar chart in Figure 11.11 compares the mean of sales and assets for each type of company.

Figure 11.11 Example of a Chart with Two Statistics and One Category

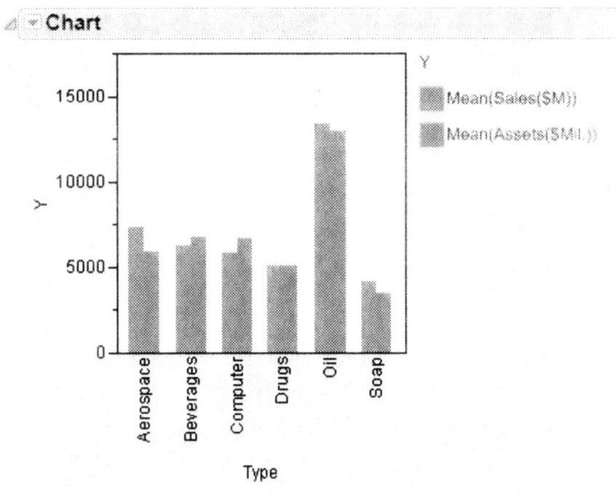

Plot Counts of Variable Levels

If you assign one or two X variables without specifying any Y variables, JMP produces bar charts that show counts for each level of the X variables.

One X Variable

1. Open the Companies.jmp sample data table.
2. Select **Graph > Chart**.
3. Select Size Co and click **Categories, X, Levels**.
4. Click **OK**.

The bar chart in Figure 11.12 shows a bar for each level of the Size Co variable.

Figure 11.12 Example of a Chart with one Category and no Statistics

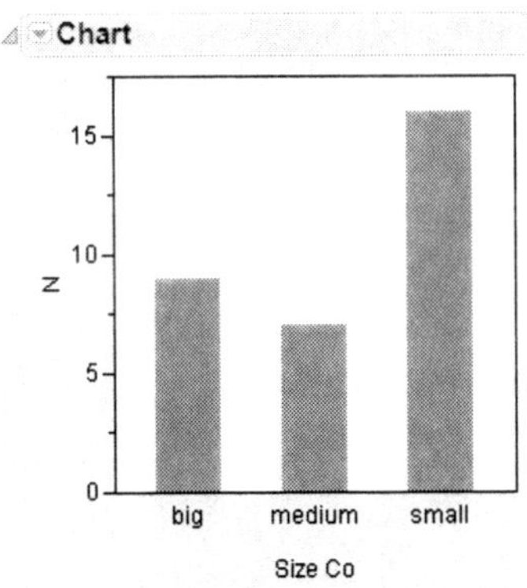

Two X Variables

If you specify two X variables (with no statistics variables), JMP divides the data into groups based on the levels in the two X variables and plots the number of members in each group.

1. Open the Companies.jmp sample data table.
2. Select **Graph > Chart**.
3. Select Type and click **Categories, X, Levels.**
4. Select Size Co and click **Categories, X, Levels.**

The order that you assign these variables is important. The levels of the second variable are nested within the levels of the first variable.

5. Click **OK.**

The bar chart on the left in Figure 11.13 shows the levels for the size of computer companies and of pharmaceutical companies. The bar chart on the right shows the results if you first selected Size Co and then Type as category variables.

Figure 11.13 Examples of Charts with Two Categories and No Statistics

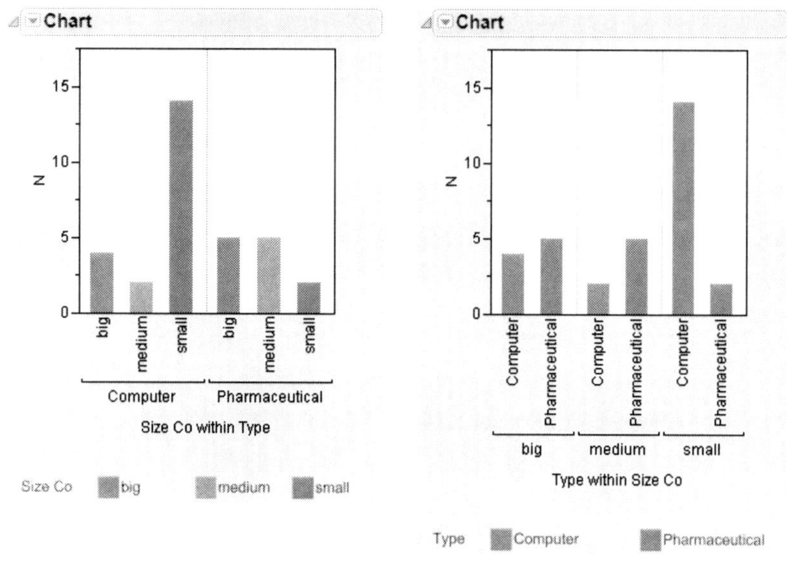

Plot Multiple Statistics with Two X Variables

When you assign two category variables, the result is a chart that shows the statistics for each level of the second category variable. This variable is nested within each level of the first category variable. The chart shows each Y side by side, using a common axis.

1. Open the Companies.jmp sample data table.
2. Select **Graph > Chart**.
3. Select Type and click **Categories, X, Levels**.
4. Select Size Co and click **Categories, X, Levels**.
5. Select Sales ($M) and Profits ($M) and click **Statistics**.
6. Select **Mean** from the menu of statistics.
7. Click **OK**.

The chart on the left in Figure 11.14 shows the result.

8. To see a separate chart for each statistic, click to deselect **Overlay** from the red triangle menu for Chart.

The chart on the right in Figure 11.14 shows the result.

Figure 11.14 Examples of Charts with Two Statistics and Two Categories

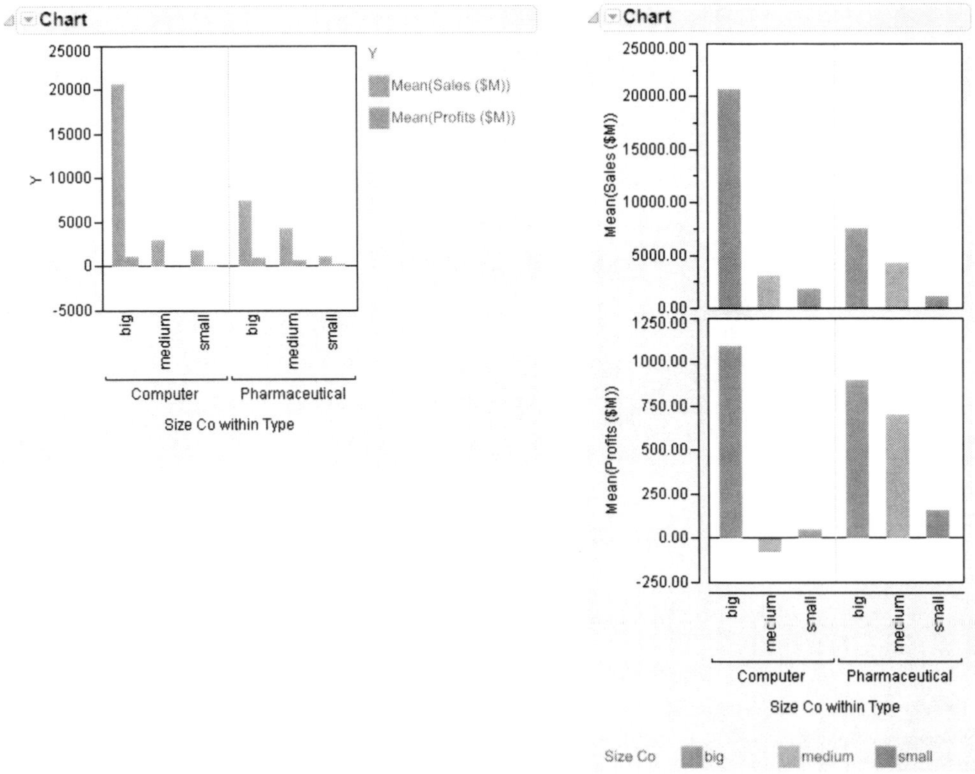

Create a Stacked Bar Chart

When you have two X levels and a single Y variable, stack the bars by selecting the **Stack Bars** command from the platform menu.

1. Open the Companies.jmp sample data table.
2. Select **Graph > Chart**.
3. Select Type and click **Categories, X, Levels**.
4. Select Size Co and click **Categories, X, Levels**.
5. Select Sales ($M) and click **Statistics**.
6. Select **Mean** from the menu of statistics.
7. Click **OK**.
8. Select **Stack Bars** from the red triangle menu for Chart.

Figure 11.15 Example of a Stacked Bar Chart

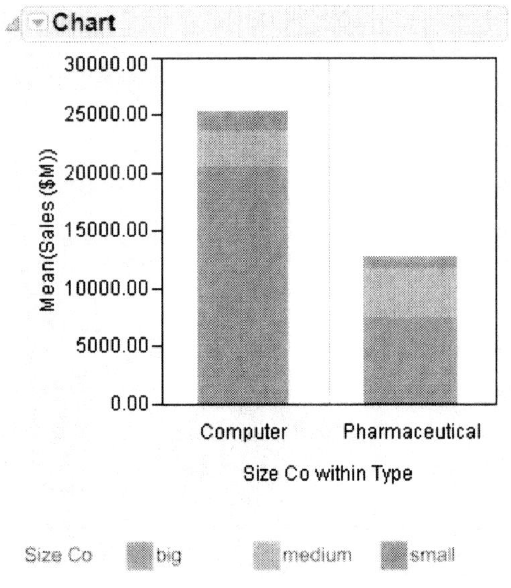

Create a Pie Chart

You can create a pie chart either in the Chart launch window or in the report window after you create another type of chart.

Create a Pie Chart in the Chart Launch Window

1. Open the Companies.jmp sample data table.
2. Select **Graph > Chart**.
3. Select Size Co and click **Categories, X, Levels**.
4. Select **Pie Chart** from the menu of chart types.
5. Click **OK**.

Change a Bar Chart to a Pie Chart

1. Open the Companies.jmp sample data table.
2. Select **Graph > Chart**.
3. Select Size Co and click **Categories, X, Levels**.
4. Click **OK**.
5. Select **Pie Chart** from the red triangle menu for chart.

   You can also right-click in the chart and select **Chart Options > Pie Chart**.

Both of the above steps produce the same pie chart shown in Figure 11.16.

Figure 11.16 Example of a Pie Chart

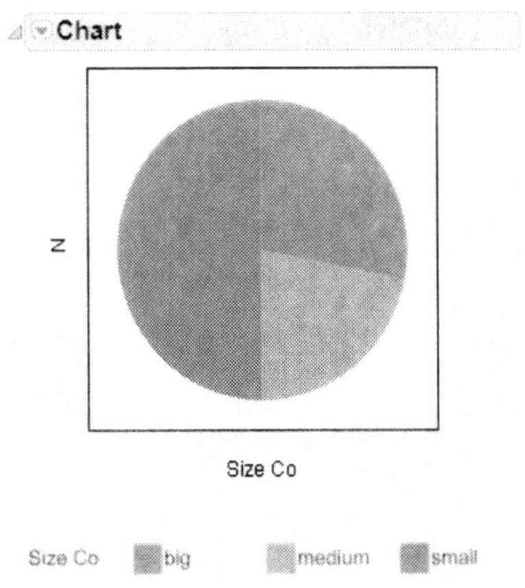

The pie chart shows that many more companies are small then are big or medium.

Show the Values or Percentages for Each Category

1. Starting with the pie chart in Figure 11.16, select **Label Options > Show Labels** from the red triangle menu for Chart.

   You can also right-click in the chart and select **Label > Show Labels.**

Label By Value is the default setting, so the pie chart now shows the number of companies that are big, medium, and small. See the chart on the left in Figure 11.16.

2. To show percentages instead, select **Label Options > Label by Percent of Total Values** from the red triangle menu for Chart.

   You can also right-click in the chart and select **Label > Label by Percent of Total Values.**

The chart on the right shows the percentage of each company size.

Figure 11.17 Example of Pie Chart Labels: Values (left) and Percentages (right)

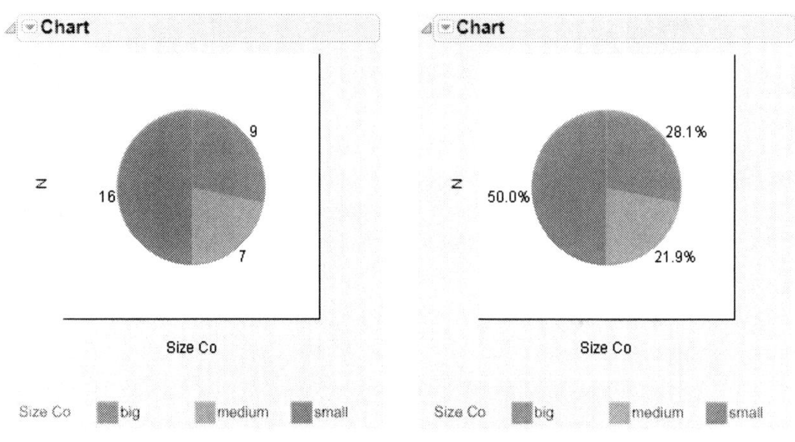

Create a Range Chart

A range chart shows the range between two values. In this example, use a range chart to compare the profits and sales of differently sized companies.

1. Open the Companies.jmp sample data table.
2. Select **Graph > Chart**.
3. Select Size Co and click **Categories, X, Levels**.
4. Select Profits ($M) and Sales ($M) and click **Statistics**.
5. Select **Mean** from the menu of statistics.
6. Click **OK**.
7. Select **Range Chart** from the red triangle menu for Chart.

Figure 11.18 Example of a Range Chart

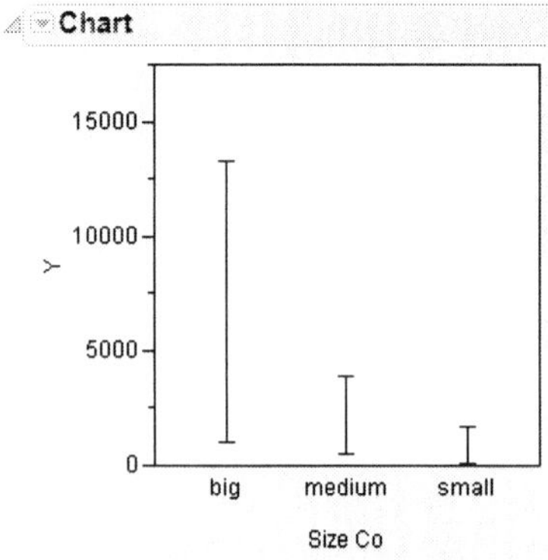

The chart shows a much larger difference between profits and sales than that for medium and small companies.

Create a Chart with Ranges and Lines for Statistics

The Stock Prices.jmp sample data table contains data for the dates and values of a stock over time. The variable YearWeek is a computed column representing the year and week in a single variable. Use a range chart to show the high, low, and average close values for each stock. For those weeks where data exists for multiple days, the average of the values is plotted.

1. Open the Stock Prices.jmp sample data table.
2. Select **Graph > Chart.**
3. Select YearWeek and click **Categories, X, Levels.**
4. Select High and click **Statistics.**
5. Select **Max** from the menu of statistics.
6. Select Low and click **Statistics.**
7. Select **Min** from the menu of statistics.
8. Select Close and click **Statistics.**
9. Select **Mean** from the menu of statistics.
10. Click **OK.**
11. Select **Range Chart** from the red triangle menu for Chart.
12. In the legend, right-click Mean(Close) and select **Connect Points.**

Figure 11.19 Example of a Combined Range and Line Chart

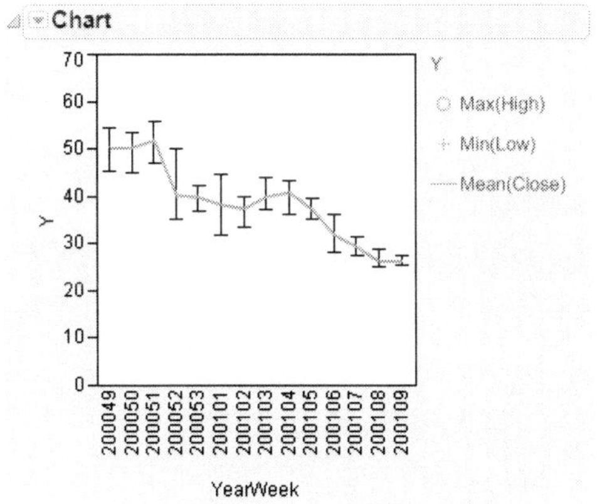

The range for each date shows the highest and lowest values the stock reached during that week. The line shows the stock's average closing price for that week.

Chapter 12

Creating Overlay Plots

Using the Overlay Plot Platform

The **Overlay Plot** command in the **Graph** menu produces plots of a single X column and one or more numeric Ys and does not accept non-numeric values for the *y*-axis. Curves can also be shown as separate plots for each Y with a common *x*-axis. Plots can be modified with range and needle options, color, log axes, and grid lines. Curves with two different scales can be overlaid on the same plot with the addition of a right axis.

Figure 12.1 Examples of Overlay Plot Graphs

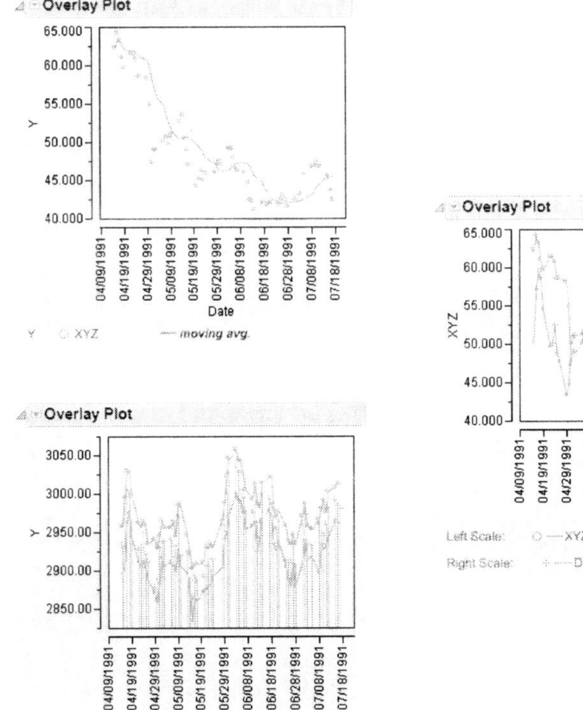

Contents

Example of an Overlay Plot

This example shows you how to plot two variables on a single *y*-axis.

1. Open the Spring.jmp sample data table.

   The table has a row for each day in the month of April. The column named April is the numeric day of the month, and the remaining columns are various weather statistics.

2. Select the **Overlay Plot** command from the **Graph** menu.
3. Select Humid1:PM and Humid4:PM and click **Y**.

   These two columns are the humidity measurements at 1:00 p.m. and 4:00 p.m.

4. Select April and click **X**.
5. Click **OK**.

   The plot shown in Figure 12.2 appears. Initially, this platform overlays all specified Y columns. The legend below the plot shows individual markers and colors that identify each Y column.

Figure 12.2 Plot with Legend

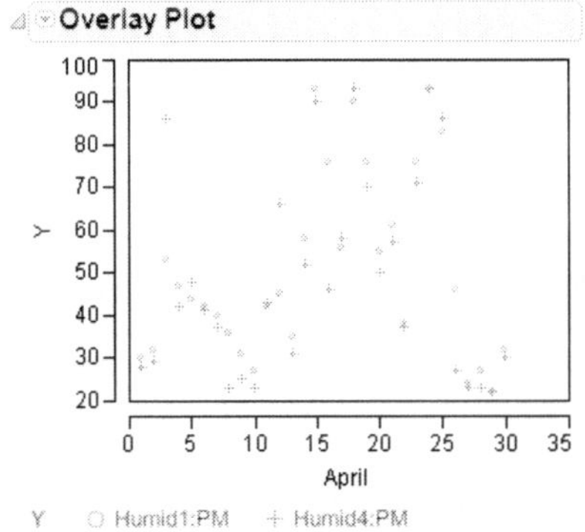

To help you quickly differentiate between the Ys, select **Y Options > Connect Points** from the Overlay Plot red triangle menu. Adjacent points are connected for each Y variable, as shown in Figure 12.3.

Figure 12.3 Plot with Connected Points

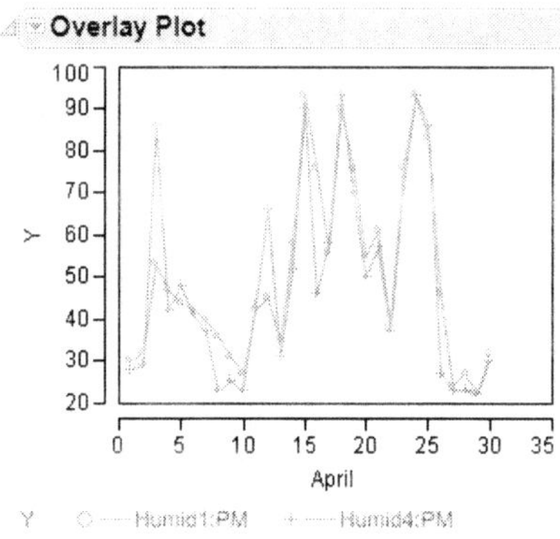

Launch the Overlay Plot Platform

Launch Overlay Plot by selecting **Graph > Overlay Plot**.

Figure 12.4 The Overlay Plot Launch Window

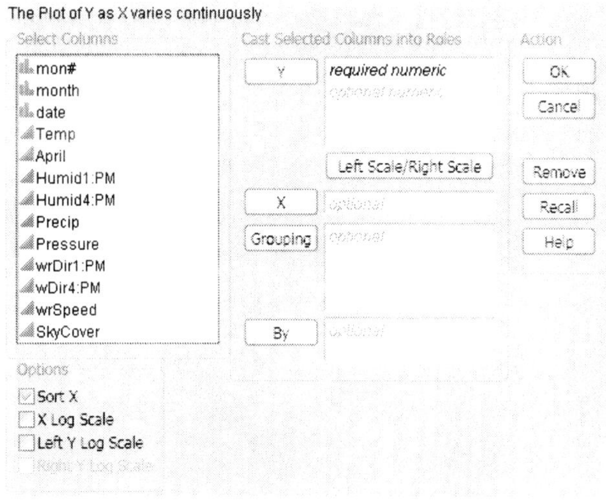

In the Overlay Plot Launch window, you assign the following:

- one X variable of any modeling type
- as many numeric Y variables as you want

Table 12.1 Description of the Overlay Plot Launch Window

Cast Selected Columns Into Roles:	
X, Y	You can graph many numeric Y variables against a single X variable.
Left Scale/Right Scale	The columns assigned to the **Y** role have a left- or right-pointing arrow to the left of the column name. This arrow designates on which vertical axis (on the left or right of the plot) the variable appears. Change the designation by highlighting the column in the **Y** list and clicking the **Left Scale/Right Scale** button.
Grouping	This option produces a matrix of graphs for each Grouping variable.
By	This option produces a separate graph for each level of the By variable. If two By variables are assigned, a separate graph for each possible combination of the levels of both By variables is produced.
Options:	
Sort X	This option causes the points to be connected in order of ascending X values. Otherwise, the points are connected in row order. This option is selected by default.
X Log Scale	This option applies a log scale to the *x*-axis.
Left Y Log Scale	This option applies a log scale to the left *y*-axis. It is available only if one or more *Y* variables are left-scaled. (See **Left Scale/Right Scale** in this table.)
Right Y Log Scale	This option applies a log scale to the right *y*-axis. It is available only if one or more Y variables are right-scaled. (See **Left Scale/Right Scale** in this table.)

For more information about the launch window, see *Using JMP*.

The Overlay Plot

Follow the instructions in "Example of an Overlay Plot" on page 339 to produce the plot shown in Figure 12.5.

Initially, this platform overlays all specified Y columns. The legend below the plot shows individual markers and colors that identify each Y column. For information about additional options for the plot, see "Overlay Plot Options" on page 342.

Figure 12.5 The Overlay Plot

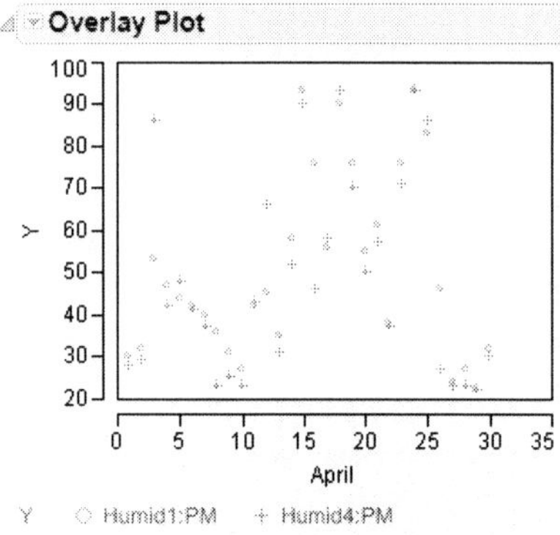

Overlay Plot Options

The basic Overlay Plot is shown in Figure 12.5.

The Overlay Plot platform has plotting options accessed from the red triangle menu on the Overlay Plot title bar. When you select one of these options at the platform level, it affects all plots in the report if no legend levels are highlighted. If one or more plot legend levels are highlighted, the options affect only those plots. There is also a single-plot options menu for each Y variable, which appears when you highlight a Y variable legend beneath the plot and right-click.

The individual plot options are the same as those in the **Y Options** submenu at the platform level. See "Y Options" on page 345.

General Platform Options

When you select one of these options at the platform level, it affects all plots in the report if no legend levels are highlighted. If one or more plot legend levels are highlighted, the options affect only those plots.

Table 12.2 Descriptions of Overlay Plot Platform Options

Overlay Plots	Contains options for overlaying: **Overlay Y's** Overlays all variables assigned to the Y role on one plot. This option is on by default and unavailable if only one Y variable is assigned. **Overlay Groups** Overlays groups and produces a legend. This option is off by default and unavailable if no grouping variables are assigned. See "Overlay Groups" on page 343. **No Overlay** Turns off overlaying for both *Y*s and groups. Creates a separate plot for each Y and each group. This option is off by default unless only one Y variable is assigned and no grouping variables are assigned. In this case, no overlaying options are available.
Separate Axes	Assigns each plot its own set of *xy*-axes. If **Separate Axes** is off, the vertical axis is shared across the same row of plots and the horizontal axis is shared on the same column of plots. The default setting is off, which results in shared axes. See "Separate Axes" on page 344.
Uniform Y Scale	Uses the same Y scale for all grouped plots. The default setting is off.
Connect Thru Missing	Connects adjacent points in the plot, regardless of missing values. The default setting is off.
Range Plot	Connects the lowest and highest points at each *x* value with a line with bars at each end. The **Needle** and **Range Plot** options are mutually exclusive.
Y Options	Contains options for the Y variables. See "Y Options" on page 345.
Ungroup Plots	Creates a separate chart for each level of a grouping variable.
Arrange Plots	Enables you to specify the number of plots in each row.
Script	This menu contains options that are available to all platforms. They enable you to redo the analysis or save the JSL commands for the analysis to a window or a file. For more information, see *Using JMP*.

Overlay Groups

Figure 12.6 shows the effect that the **Overlay Groups** option has on an overlay plot with one Y variable, one X variable, and a grouping variable. The grouping variable has two levels. The plot on the left has **Overlay Groups** turned off, so a separate graph is produced for the two levels of the grouping variable. The plot on the right has **Overlay Groups** turned on, so there is a single graph that uses colors and markers to show the two levels of the grouping variable. A legend describing the levels is added under the graph.

Figure 12.6 Overlay Groups: Off (left) and On (right)

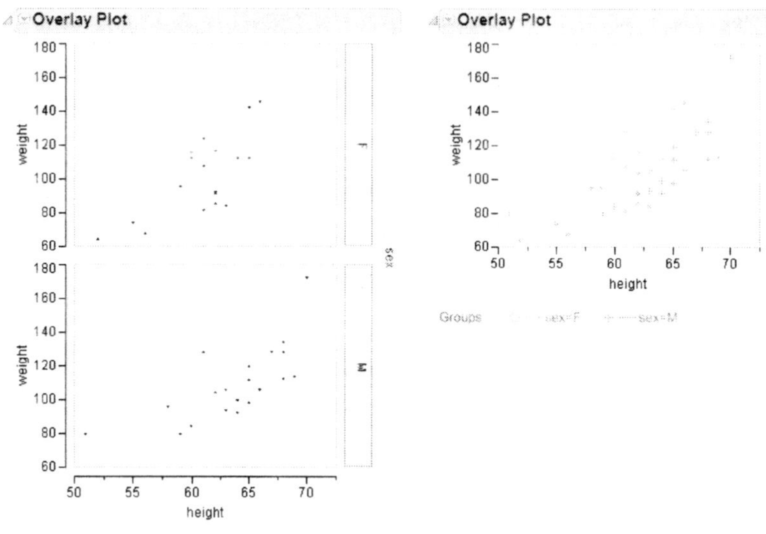

Separate Axes

Figure 12.7 shows the effect that the **Separate Axes** option has on an overlay plot with two Y variables and one X variable. The **Overlay Y's** option is turned off, so a separate plot is produced for each Y variable. The plot on the left has **Separate Axes** turned off, so the two graphs share a single *x*-axis. The plot on the right has **Separate Axes** turned on, so both graphs have their own *x*-axis.

Figure 12.7 Separate Axes: Off (left) and On (right)

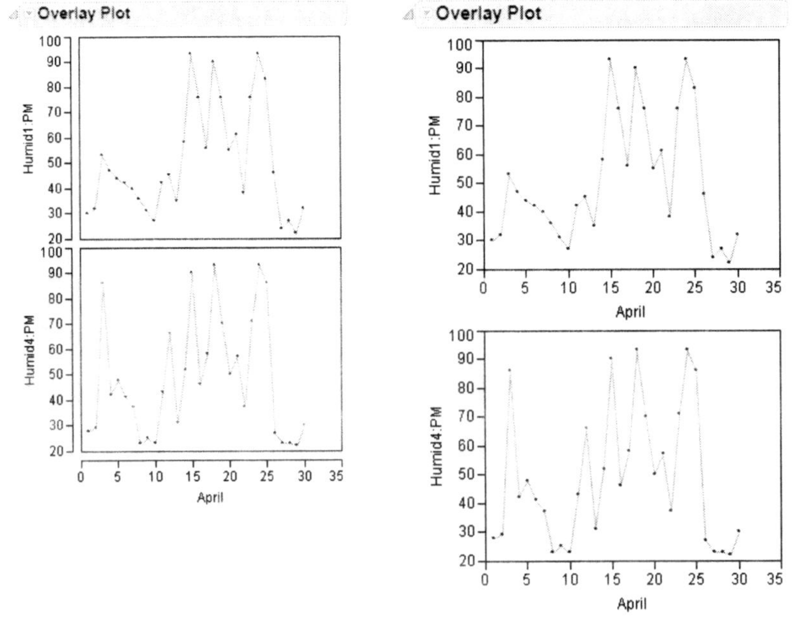

Y Options

Each Y variable is labeled in a legend beneath the plot. The Y options are available from the **Y Options** menu from the red triangle menu for Overlay Plot. You can also access the **Y Options** menu by right-clicking on any Y variable in the legend.

Note: If no Y variables are selected, any Y options that you select affect all Y variables. If one or more of the Y variables are selected, any Y options that you select affect only those you have selected.

Selecting and Deselecting Y Variables in the Legend

- Hold the SHIFT key and click to select multiple contiguous legend levels.
- Hold the CONTROL key and click to select multiple discontiguous legend levels.
- Hold the CONTROL key and click a selected legend level to deselect it.

Table 12.3 Descriptions of Y Options

Show Points	A toggle that either shows or hides points in the graph.
Connect Points	A toggle that either connects the points with lines or turns the connecting lines off. You can use **Connect Points** without showing points.
Needle	Draws a vertical line from each point to the *x*-axis.

Table 12.3 Descriptions of Y Options *(Continued)*

Step	Draws a horizontal line from each point to the *x* value of the following point, and then a vertical line to that point. You can use **Step** without showing points.
Note: The **Connect Points**, **Needle**, and **Step** options are mutually exclusive.	
Function Plot	Plots a formula (stored in the Y column) as a smooth curve. To use this function, store a formula in a column that is a function of a single X column. Assign the formula to the Y role. For an example, see "Function Plots" on page 346.
Connect Color	Displays the JMP color palette for assigning colors to connecting lines.
Overlay Marker	Displays the JMP marker palette for assigning markers to plotted points.
Overlay Marker Color	Assigns a color to all points of the selected variable.
Line Style	Enables the choice of dashed, dotted, or other line styles.
Line Width	Enables the choice of line widths.

Additional Examples of the Overlay Plot Platform

The following sections show several examples of different overlay plots.

Function Plots

Overlay Plot normally assumes you want a function plot when the Y column contains a formula. However, formulas that contain random number functions are more frequently used with simulations, where function plotting is not often wanted. Therefore, the **Function Plot** option is off by default when a random number function is present, but on for all other functions.

To See an Example of a Function Plot

1. Open the sample data file Density Compare.jmp.
2. From the **Graph** menu, select **Overlay Plot**.
3. Assign gamma1, gamma3, and gamma5 as the **Y** variables.
4. Assign Xgamma as the **X** variable.
5. Click **OK**.
6. Turn of the **Show Points** option by selecting **Y Options > Show Points** from the red triangle menu.

Figure 12.8 Function Plot

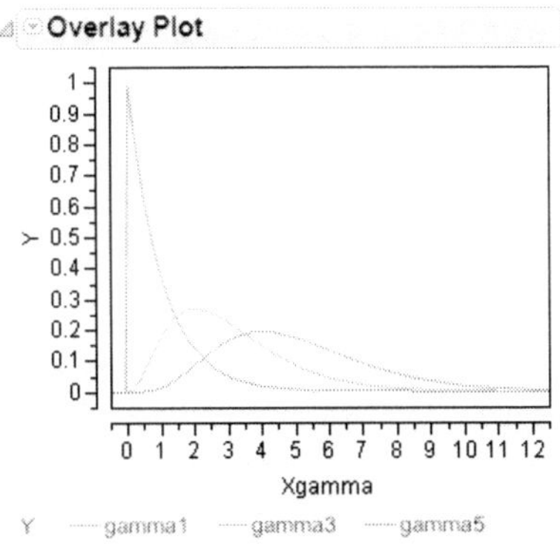

Plotting Two or More Variables with a Second Y-axis

A second *y*-axis is useful for plotting data with different scales on the same plot, such as a stock's closing price and its daily volume, or temperature and pressure. For example, consider plotting the selling price of an inexpensive stock against the Dow Jones Industrial Average.

1. Open the Stock Prices.jmp sample data table.
2. From the **Graph** menu, select **Overlay Plot**.
3. Assign High, Low, Close, and Volume as **Y** variables.
4. Select Volume in the **Y** list and click **Left Scale/Right Scale**.

   This action assigns Volume to the right axis, leaving the others on the left axis. The arrows to the left of the Y variables show you which axis each variable is assigned.
5. Assign Date as the **X** variable.
6. Click **OK**.
7. From the red triangle menu for Overlay Plot, select **Y Options > Connect Points**.

Figure 12.9 Dual Axis Overlay Plot

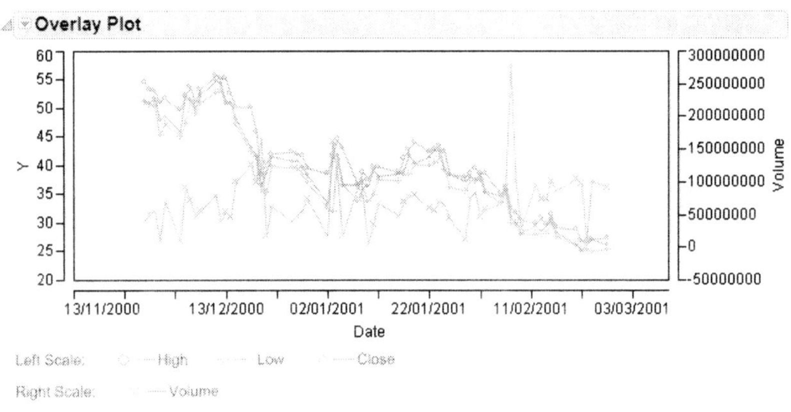

The variables **High**, **Low**, and **Close** are the stock prices of the same stock and thus are on the same scale. **Volume** is a different scale entirely, representing the trading volume of the entire Dow Jones Industrial Average.

To see why this matters, perform the same steps above without clicking the **Left Scale/Right Scale** button for **Volume**. Compare the resulting graph in Figure 12.10 to Figure 12.9.

Figure 12.10 Single Axis Overlay Plot

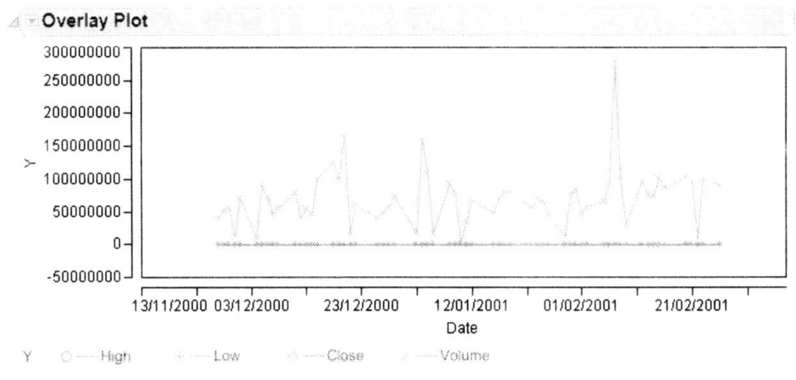

Grouping Variables

The Overlay Plot platform allows the production of several plots in one window through the use of grouping variables. With one grouping variable, a stacked vector of plots appears, with one plot for each level of the grouping variable. Two grouping variables result in a matrix of plots.

1. Open the **students.jmp** sample data table.
2. From the **Graph** menu, select **Overlay Plot**.

3. Assign weight as the **Y** variable and height as the **X** variable.
4. Assign age and sex as grouping variables.
5. Click **OK**.

   A portion of the resulting plot is shown in Figure 12.11.

Figure 12.11 Grouped Plots Without Separate Axes

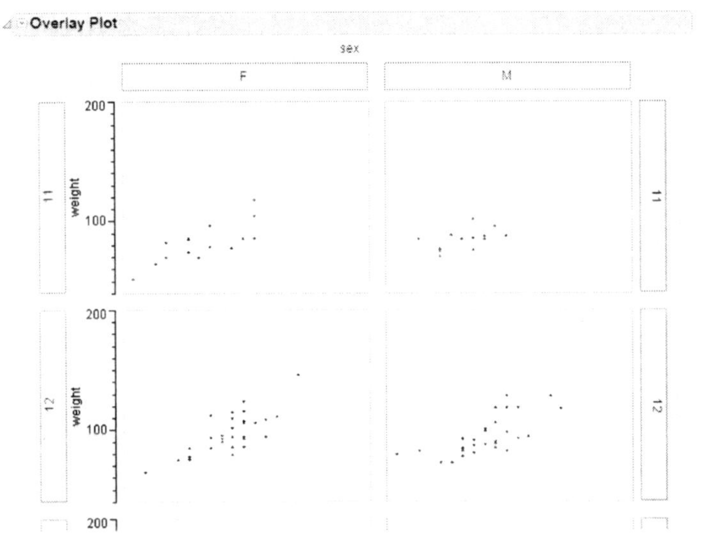

Select the **Separate Axes** option from the red triangle menu to produce plots that do not share axes. Compare Figure 12.11 to Figure 12.12.

Figure 12.12 Grouping Variables

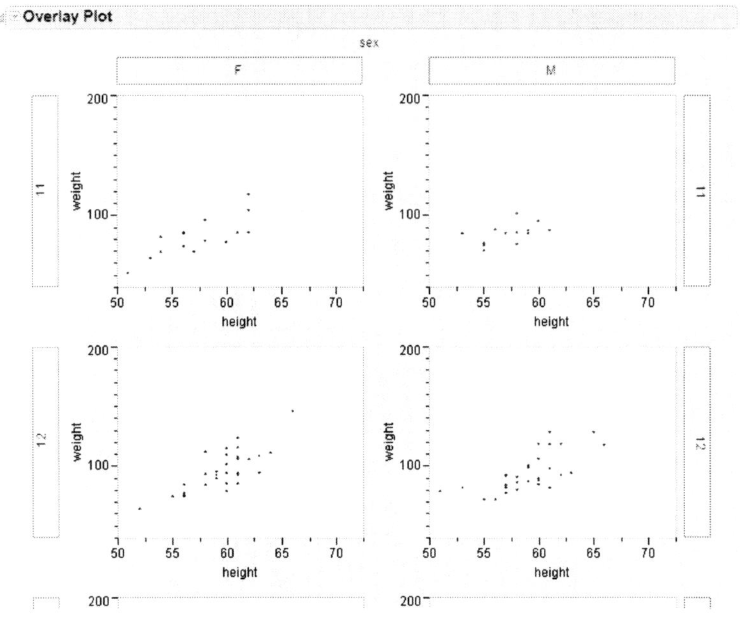

Chapter 13

Creating Three-Dimensional Scatterplots

Using the Scatterplot 3D Platform

The Scatterplot 3D platform shows the values of numeric columns in the associated data table in a rotatable, three-dimensional view. Up to three columns that you select from the associated data table are displayed at one time. See Figure 13.1.

To help visualize variation in higher dimensions, the 3D scatterplot can show a biplot representation of the points and variables when you request principal components. The most prominent directions of data are displayed on the 3D scatterplot report.

Figure 13.1 Example of a 3D Scatterplot

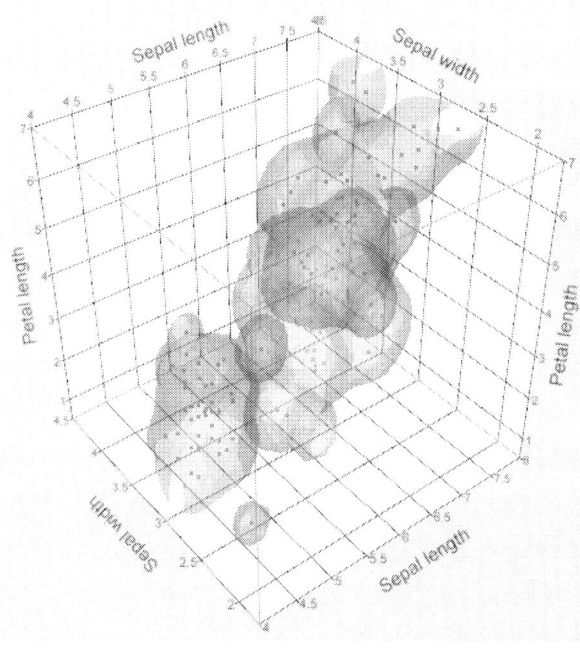

Contents

Example of a 3D Scatterplot

This example uses the Iris.jmp sample data table, which includes measurements of sepal length, sepal width, petal length, and petal width for three species of iris.

1. Open the Iris.jmp sample data table.
2. Select **Graph > Scatterplot 3D**.
3. Select **Sepal length**, **Sepal width**, and **Petal length**, and click **Y, Columns**.
4. Select **Petal width** and click **Weight**.
5. Click **OK**.

Figure 13.2 Example of an Initial 3D Scatterplot

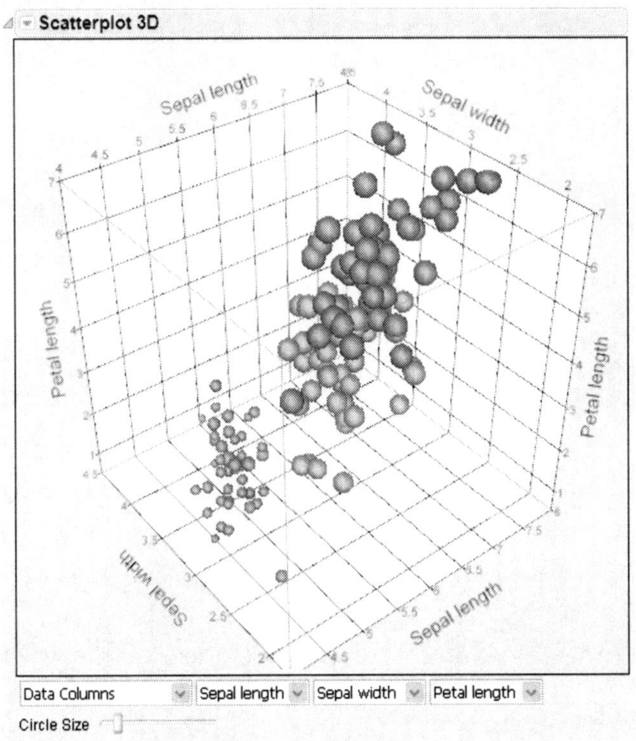

Now you can spin the 3D scatterplot to see the relationships between the variables. In this example, the data points are formatted in blue, red, and green. You might want to spin the scatterplot to see more clearly the relationships between the red and green points.

Launch the Scatterplot 3D Platform

Launch the Scatterplot 3D platform by selecting **Graph > Scatterplot 3D**.

Figure 13.3 The Scatterplot 3D Launch Window

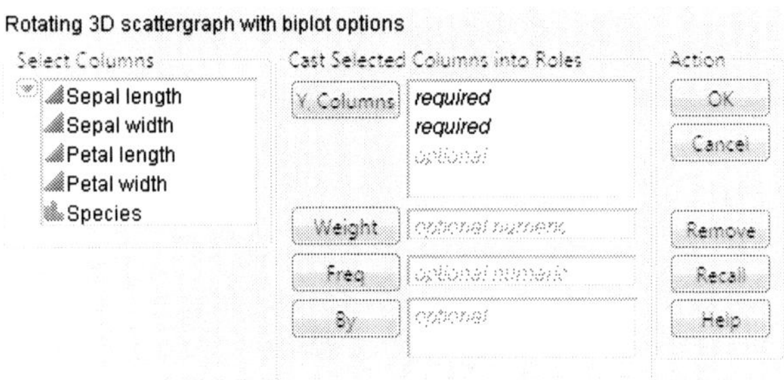

Table 13.1 Description of the Scatterplot 3D Launch Window

Y, Columns	Select the variables to plot on the 3D scatterplot. The order in which you select the variables determines where the data points appear on the axes: • The first variable appears on the *x* axis. • The second variable appears on the *y* axis. • The third variable appears on the *z* axis. You can assign the remaining variables interactively through the drop-down menus below the scatterplot.
Weight	Use the **Weight** variable to: • Assign a weight (importance or influence) to the data • Visualize a fourth variable that sizes the points **Note:** Red triangle options account for the **Weight** variable. If you do not want this variable accounted for in your analyses, remove it from the launch window. When you specify a **Weight** variable, JMP draws the points as balls. The balls are scaled so that their volume represents the weight value. You click and drag the **Circle Size** slider below the scatterplot to resize the balls.
Freq	Identifies the data table column whose variables assign a frequency to each row. This option is useful when a frequency is assigned to each row in summarized data.

Table 13.1 Description of the Scatterplot 3D Launch Window *(Continued)*

By	Produces a separate 3D scatterplot for each By variable value. When two By variables are assigned, a separate graph is produced for each combination of both By variables.

For more information about the launch window, see *Using JMP*.

The Scatterplot 3D Report

To produce the 3D scatterplot shown in Figure 13.4, follow the instructions in "Example of a 3D Scatterplot" on page 353.

The Scatterplot 3D report shows a three-dimensional spinnable view of your data. See Figure 13.4. In the launch window, you select the variables and then create the report. The variables are displayed on the 3D scatterplots' *x*, *y*, and *z* axes. Up to three variables can be displayed at a time.

Figure 13.4 Example of Information Displayed on the Scatterplot 3D Report

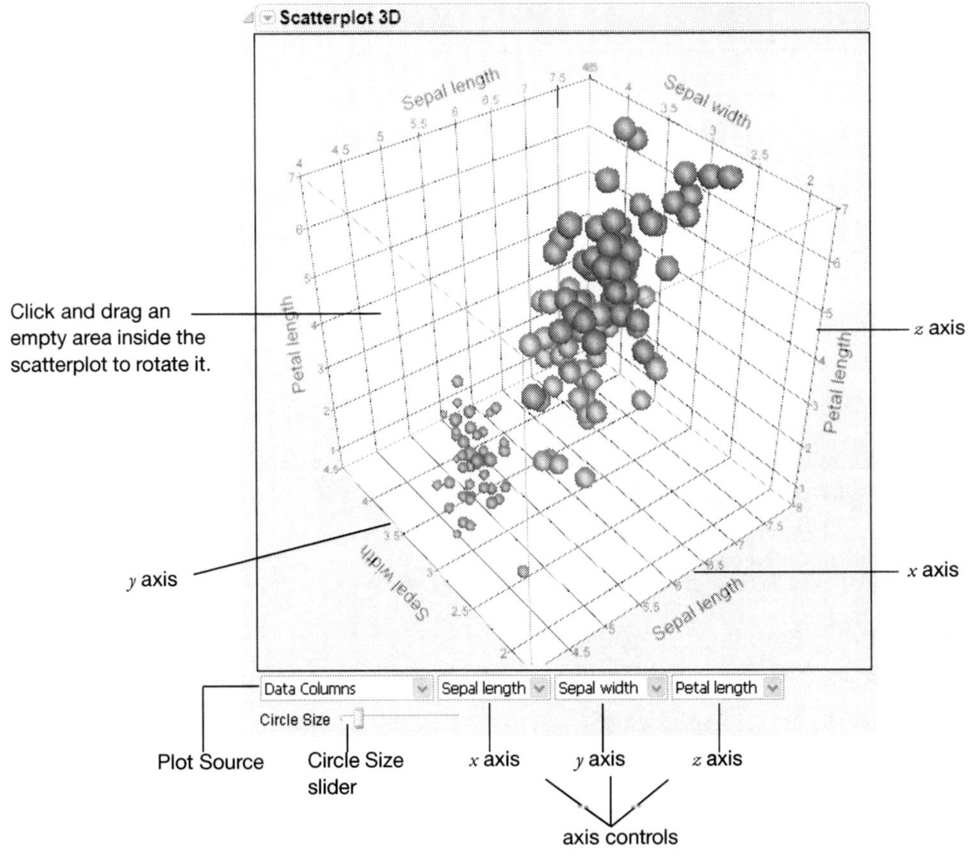

Table 13.2 Description of the Scatterplot 3D Report

Plot source	The plot source box indicates the source of the data in the plot.
Circle Size slider	**Note:** The **Circle Size** slider appears only if you have specified a **Weight** variable. Click and drag the **Circle Size** slider to resize the balls while maintaining their relative sizes.
Axis controls	Select which variable appears on each axis. Choose the **Other** option to add a new variable.
Next Axis Set (not shown)	**Note:** The **Next Axis Set** button appears only if your analysis contains more than three variables. Cycles through the axis controls for any hidden variables. See “Change Variables on the Axes” on page 357.

After you create a 3D scatterplot, you can add features such as displaying ellipses around specific data points, showing separate principal components, rotating components, connecting points, and more. See “Scatterplot 3D Platform Options” on page 359 for details.

You can also assign colors and symbols (or markers) to data points either on the 3D scatterplot itself or in the associated data table. See “Assign Colors and Markers to Data Points” on page 358 and “Assign Colors and Markers in the Data Table” on page 359.

Spin the 3D Scatterplot

You spin the 3D scatterplot report in four ways:

- Click and drag an empty area on the 3D scatterplot. The 3D scatterplot spins in the direction you dragged the mouse.

Note: Click and drag on an empty area on the 3D scatterplot, not on an axis or data point. Dragging the axis rescales the axis. Dragging a data point only selects the point.

- Slide the mouse wheel. The 3D scatterplot spins up and down only.
- Hold down an arrow key. (Before using an arrow on the number keypad, verify that NUM LOCK is turned off.)
- Hold down ESC. The 3D scatterplot spins left and right only.

In each case, the 3D scatterplot spins as long as you hold down the mouse button, arrow key, or ESC key. The spinning also continues as you slide the mouse wheel.

You can also spin the 3D scatterplot continuously as follows:

- Click and drag: Hold down SHIFT, click and drag an empty area on the plot, and release SHIFT. The faster you drag the mouse, the faster the 3D scatterplot spins.
- Mouse wheel: Hold down SHIFT, slide the wheel, and release the wheel. The 3D scatterplot spins up and down only.
- Arrow keys: Hold down SHIFT, press the arrow key, and release SHIFT.
- ESC key: Hold down SHIFT and press ESC. The 3D scatterplot spins left and right only.

In addition to automatically spinning the plot, you can oscillate the plot. Hold down SHIFT and CTRL and then click and drag the plot. The plot shakes up and down or left to right, depending on the direction in which you dragged the plot.

To stop the spinning or oscillating, click on the plot or press ESC.

Change Variables on the Axes

The variables on each axis are determined by the order in which you select the variables in the launch window. For example, the first variable that you select is displayed on the *x* axis. The second variable is displayed on the *y* axis, and the third variable is displayed on the *z* axis.

After you create a 3D scatterplot, you can change the variable assigned to an axis, plot a different set of variables, or sequence through all combinations of the variables.

1. To change the variable on a specific axis, select the axis control drop-down menu and select a different variable.
2. To add a different variable, click an axis control drop-down menu, select **Other**, select the variable, and then click **OK**.
3. To sequence through combinations of all variables, click the **Next Axis Set** button until the variables that you want to plot are displayed.

Adjust the Axes

You can manually move or rescale the axis coordinates by clicking and dragging the axis. This option shows a different set of coordinates on the 3D scatterplot. It also lets you change the space displayed between the coordinates (or the coordinate scaling).

You can also specify axis properties by double-clicking the axis and modifying settings in the specifications window.

To Move the Coordinates on the Axis

1. Place your cursor over the middle of the axis.
2. Click and drag the axis.

To Modify Coordinate Scaling

1. Place your cursor over the end of the axis.

2. Click and drag the axis.

To Rescale an Axis Precisely

1. Place your cursor over the middle of the axis (the axis, not the label).
2. Double-click the axis.

Figure 13.5 The Axis Options Window

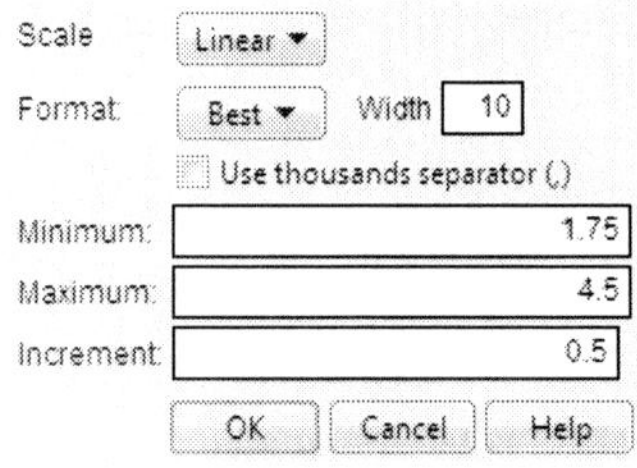

Table 13.3 Description of the Axis Options Window

Scale	Changes the scale of the axes. See the *Using JMP* book.
Format	Specifies how to display numbers on the 3D scatterplot. See the *Using JMP* book.
Use thousands separator	Displays a comma in numbers above 999. This option is available depending on which option you select in the Format drop-down menu.
Width	Controls the number of characters that appear.
Minimum	Specifies the minimum coordinate value.
Maximum	Specifies the maximum coordinate value.
Increment	Specifies the space between the coordinates.

Assign Colors and Markers to Data Points

Each point in the 3D scatterplot corresponds to a row in the associated data table. To highlight points on the 3D scatterplot, you assign colors and markers to the points. The colors and markers are then displayed on the 3D scatterplot and in the data table.

When you click a point, the following items are selected:

- the point in the 3D scatterplot
- the corresponding row in the associated data table
- the point in any other opened graph, if applicable

To select one point, click the point.

To select several points, double-click the 3D scatterplot and drag the cursor over the points. A box is displayed to indicate which points are selected.

To deselect points, double-click the 3D scatterplot.

To assign a color or marker to selected data points, proceed as follows:

1. To assign a color to the selected point, select **Rows > Colors** and then select the color.
2. To assign a marker to the selected point, select **Rows > Markers** and then select the marker.

Assign Colors and Markers in the Data Table

You can assign colors and markers to rows in the data table. The colors and markers appear next to the row number in the data table and on the 3D scatterplot. This option distinguishes points for each variable, and you can save the settings in the data table. Assigning colors and markers to specific data points (as described in "Assign Colors and Markers to Data Points" on page 358) only highlights them for the current open graphs.

See *Using JMP* for details about assigning colors and markers in the data table. For details about changing the size, quality, or transparency of markers, see "Scatterplot 3D Settings" on page 365.

Scatterplot 3D Platform Options

The red triangle menu next to Scatterplot 3D contains options to customize the display and to compute, rotate, and save principal or rotated components.

Table 13.4 Descriptions of the Scatterplot 3D Options

Show Points	Shows or hides the data points on the graph.
Show Controls	Shows or hides the source and axis controls displayed beneath the 3D scatterplot. See Figure 13.4.
Drop Lines	Draws lines from each point to the plane created by the *x* and *z* variables that you selected on the launch window.
Connect Points	Connects the points with a line. Points can be connected on the data as a whole or in groups. You can also group data by a specific variable.
Normal Contour Ellipsoids	Draws one or more normal contour ellipsoids, that is, three-dimensional ellipses that encompass a specified portion of points. You specify whether you want an ellipsoid for all of the data or for each group. You can also control the size and transparency of the ellipsoids. For details, see "Normal Contour Ellipsoids" on page 361.

Table 13.4 Descriptions of the Scatterplot 3D Options *(Continued)*

Ellipsoid Coverage	Changes the size of normal contour ellipsoids. Type a value *between* 0 and 1, where the greater the value creates a bigger the ellipsoid. The actual values "0" and "1" produce no ellipsoid, so a warning appears if you try to use those values. This option only appears after you add a normal contour ellipsoid to the 3D scatterplot.
Ellipsoid Transparency	Changes the surface of normal contour ellipsoids. The greater the value, the more opaque the ellipsoid. This option only appears after you add a normal contour ellipsoid to the 3D scatterplot.
Nonpar Density Contour	Draws nonparametric density contours, which approximately encompass a specified proportion of the points. You specify whether you want a density contour for all of the data or for each group. For details, see "Nonparametric Density Contours" on page 362.
Drop Line Thickness	Changes the width of drop lines. This option only appears after you add drop lines to the 3D scatterplot.
Principal Components	Calculates principal components on all variables. This changes the axes of the plot to have principal component scores. Biplot rays are displayed by default. You can remove them by selecting **Biplot Rays** from the red triangle menu. For details about principal components, see the *Modeling and Multivariate Methods* book.
Std Prin Components	Calculates principal components (as with the **Principal Components** option) but scales the principal component scores to have unit variance. If this option is not selected, the scores have variance equal to the corresponding eigenvalue. With standardized principal components, the correlation between the variables and the principal component scores is equal to the values in the eigenvector. This helps you quickly assess the relative importance of the variables. For details, see the *Modeling and Multivariate Methods* book. **Note:** Select this option if you want GH' rather than JK' biplots. GH' biplots try to preserve relationships between variables; JK' biplots try to preserve relationships between observations. The interpoint distance shown by GH' biplots is less meaningful, but the angles of the GH' biplot rays measure correlations better.

Table 13.4 Descriptions of the Scatterplot 3D Options *(Continued)*

Rotated Components	Specifies the number of factors that you want to rotate and the rotation method. You rotate components to better align the directions of the factors with the original variables so that the factors might be more interpretable. For details, see the *Modeling and Multivariate Methods* book.
Biplot Rays	Shows or hides biplot rays that correspond to the principal components. You must have already selected Principal Components, Std Prin Components, or Rotated Components for this option to appear.
Remove Prin Comp	Removes principal components, standardized principal components, and rotated components from the scatterplot 3D report. The 3D scatterplot reverts to its original display before principal components were selected. This option, however, does not remove any saved principal components from the data table. This option only appears after you add principal, standard, or rotated components to the 3D scatterplot.
Save Prin Components	Saves the specified number of current principal component scores as new columns in the current data table. These columns also include the formulas used for the principal components. For *n* variables in the components list, *n* principal component columns are created and named Prin1, Prin2, ... Prin*n*. This option only appears after you add principal, standard, or rotated components to the 3D scatterplot.
Save Rotated Components	Saves all rotated component scores as columns in the current data table. These columns also include the formulas that were used. If you requested *n* rotated components, then *n* rotated component columns are created and named Rot1, Rot2, ... Rot*n*. This option only appears after you add rotated components to the 3D scatterplot.
Script	Contains options that are available to all platforms. They enable you to redo or relaunch the analysis, turn off automatic recalculation, copy the script, or save the JSL commands for the analysis. For more information, see *Using JMP*.

Normal Contour Ellipsoids

A normal contour ellipsoid is a 3-dimensional ellipse that encompasses a specified portion of points. The ellipsoid is computed from a contour of the multivariate normal distribution fit to the points. The ellipsoid is a function of the means, standard deviations, and correlations of variables on the plot. See the *Modeling and Multivariate Methods* book for details about multivariate normal distributions.

When you add an ellipsoid, two formatting options are available:

- *Coverage* changes the portion of data points covered by the ellipsoid. The larger the value, the bigger the ellipsoid.
- *Transparency* changes the surface of the ellipsoid from transparent to opaque. The larger the value, the more opaque the ellipsoid.

The coverage and transparency options also appear in the red triangle menu after you add the ellipsoid.

When you add normal contour ellipsoids to a 3D scatterplot, you specify whether you want an ellipsoid for all of the data or for a specific group of data. The ellipsoid for each set of grouped data is color-coded to differentiate one group from another.

You display and remove normal contour ellipsoids by selecting and deselecting **Normal Contour Ellipsoids** from the red triangle menu.

The examples in this section use the Iris.jmp sample data table, which includes measurements of sepal length, sepal width, petal length, and petal width for three species of iris.

Related Information

- "Example of an Ungrouped Normal Contour Ellipsoid" on page 366
- "Example of Grouped Normal Contour Ellipsoids" on page 367

Nonparametric Density Contours

The nonparametric density contour shows contours that approximately encompass a specified proportion of the points. You add nonparametric density contours to see patterns in point density when the scatterplot is darkened by thousands of points.

This feature is particularly valuable when you have many points on a 3D scatterplot; the contours can be so dark that you cannot see the structure. In this situation, you remove the points so that only the contours are displayed. See "Optimizing a Dense Nonparametric Density Contour" on page 363 for details.

When you add nonparametric density contours to a 3D scatterplot, you specify whether you want a contour for all of the data or for a specific group of data. The contour for each set of grouped data is color-coded to differentiate one group from another.

You display and remove nonparametric density contours by selecting and deselecting **Nonpar Density Contours** from the red triangle menu.

Related Information

- "Example of a Grouped Nonparametric Density Contour" on page 368

Density Contour Controls

The Density Contour Controls options are displayed below the 3D scatterplot. These options let you select additional contours and change each contour's formatting.

Figure 13.6 The Density Contour Controls Window

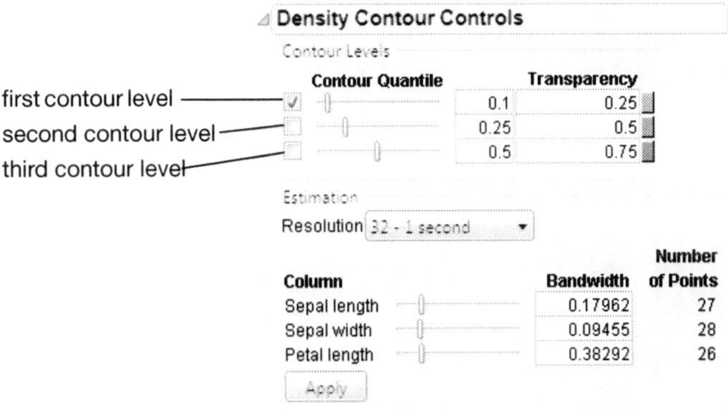

Table 13.5 Description of the Density Contour Controls Window

<table>
<tr><td>Contour Quantile</td><td>Controls which contours are shown and lets you customize the contour formatting.<ul><li>Density level represents the volume and density of the points. As the contours go from smaller to larger values, the contours cover less volume but more dense areas. A 0.9 contour represents the 10% densest part of the total, where the points are closest together. Click and drag the slider below “Contour Quantile,” or enter a value next to the slider.</li><li>Transparency changes the surface of density contours. The greater the value, the more opaque the contour. Enter a value in the box.</li><li>Color changes the color of the contour. Click the colored box and select a different color. (This option only appears for ungrouped density contours.)</li></ul>Changes to these settings take effect immediately.</td></tr>
<tr><td>Resolution</td><td>Changes the resolution of the contours. A higher resolution results in a less granular drawing of the contours but takes more time to display.</td></tr>
<tr><td>Column Bandwidth</td><td>Changes the smoothness of the fitted density. A higher bandwidth results in a smoother fitted density.

Type a new bandwidth for each variable, or click and drag the sliders. Click Apply to display your changes.</td></tr>
</table>

Optimizing a Dense Nonparametric Density Contour

When you have many points on a 3D scatterplot, the contours can be so dark that you cannot see the structure. In this situation, you remove the points so that only the contours are displayed.

To remove points from a 3D scatterplot, select **Show Points** from the red triangle menu. You can further optimize the contours by changing their size, color, and transparency. See Table 13.4 for details.

Figure 13.7 Example of Optimizing a Dense Nonparametric Density Contour

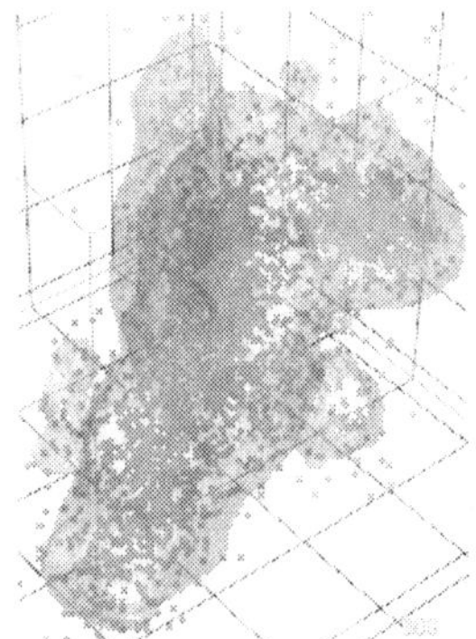

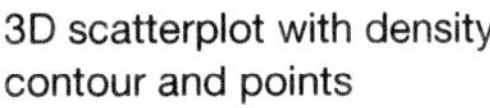
3D scatterplot with density contour and points

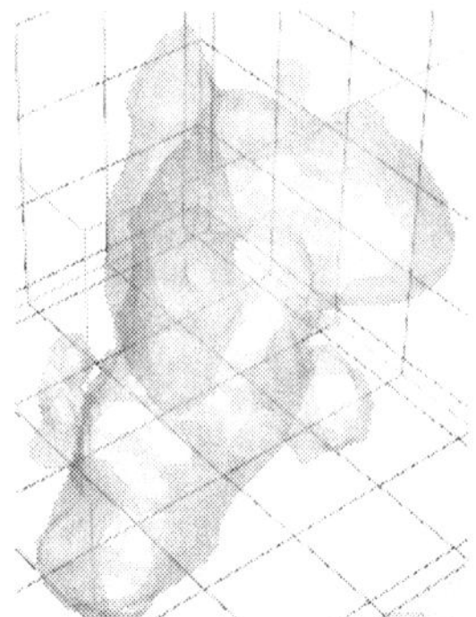
3D scatterplot with density contour and no points

Context Menu

Right-click the 3D Scatterplot to see the context menu.

Table 13.6 Descriptions of the Context Menu Options

Option	Description
Reset	Returns the orientation of the scatterplot to its original state.
Settings	Provides options to change the appearance of the 3D scatterplot. See “Scatterplot 3D Settings” on page 365.
Hide Lights Border	Shows and hides a border that displays the lights. The lights highlight different portions of the 3D scatterplot. Right-click a light to turn it on or off and to change the color.
Wall Color	Changes the color of the 3D scatterplot.
Background Color	Changes the color surrounding the 3D scatterplot.
Rows	You can color, mark, exclude, hide, and label points that correspond to rows in the associated data table. You must select the points before selecting this option. See the *Using JMP* book.
Use Hardware Acceleration	Turns hardware acceleration on or off for machines that support acceleration. This option might display the scatterplot faster. If not, try updating your graphics drivers.

Table 13.6 Descriptions of the Context Menu Options *(Continued)*

Show ArcBall	Shows and hides a globe around the 3D scatterplot. This option helps you visualize the rotation of the scatterplot. Select whether you want the ArcBall to appear always, only when you drag the scatterplot, or never.

Scatterplot 3D Settings

To customize properties such as the marker size, text size, and grid lines, right-click the 3D scatterplot and select **Settings**. The Settings window appears. As you modify the settings, a preview appears on the 3D scatterplot.

Figure 13.8 The Scatterplot 3D Settings Window

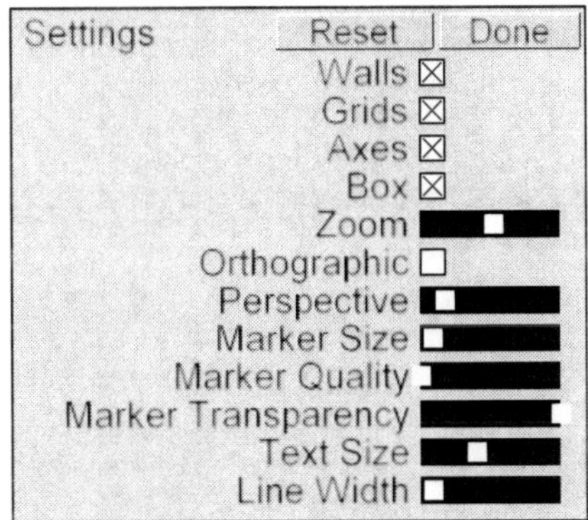

Note the following:

- Move the sliders left to decrease the selected property or to the right to increase the selected property.
- To move the Settings window around the scatterplot, click and drag the top portion of the window.

Table 13.7 Descriptions of the Scatterplot 3D Settings Window Options

Reset	Resets the default settings.
Done	Closes the window.
Walls	Adds or removes the 3D scatterplot walls. Without walls, the background color of the 3D scatterplot is displayed.
Grids	Shows or hides the coordinate lines.
Axes	Shows or hides the variable names that appear above each axis.
Box	Shows or hides the box. Without the box, the 3D scatterplot is displayed as an open plot.

Table 13.7 Descriptions of the Scatterplot 3D Settings Window Options *(Continued)*

Zoom	Enlarges or shrinks the 3D scatterplot.
Orthographic	Changes the view of the scatterplot from 3-dimensional to an orthographic projection. In the orthographic view, the walls of the scatterplot do not converge to a vanishing point. This means that you can compare near and far distances and see the structure between data points. **Note:** If you turn off orthographic view and completely decrease the perspective, the walls of the scatterplot do not converge. This is the same effect that you get when you turn on orthographic view.
Perspective	Increases or decreases the perspective. Large values create a view that is unnaturally large and visually disorienting. In this case, you need to resize the scatterplot window to show the entire plot.
Marker Size	Increases or decreases the size of the data point markers.
Marker Quality	Increases and decreases the data marker quality. For example, when you increase the marker quality, some markers have an opaque center. Other symbol markers are formatted in bold. Increase the zoom to see these changes in quality.
Marker Transparency	Increases or decreases the transparency of the data markers.
Text Size	Increases or decreases the text size.
Line Width	Changes the width of the coordinate and axes lines.

Additional Examples of the Scatterplot 3D Platform

This section contains additional examples using 3D scatterplots.

Example of an Ungrouped Normal Contour Ellipsoid

This example shows how to add a normal contour ellipsoid to more than 75% of the data points. The ellipsoid is 25% transparent.

1. Open the Iris.jmp sample data table.
2. Select **Graph > Scatterplot 3D**.
3. Select **Sepal length**, **Sepal width**, **Petal length**, and **Petal width** and click **Y, Columns**.
4. Click **OK**.
5. From the red triangle menu, select **Normal Contour Ellipsoids**. Notice that **Ungrouped** is already selected.
6. Type 0.75 next to Coverage.

7. Type 0.25 next to Transparency.
8. Click **OK**.

Figure 13.9 Example of an Ungrouped Normal Contour Ellipsoid

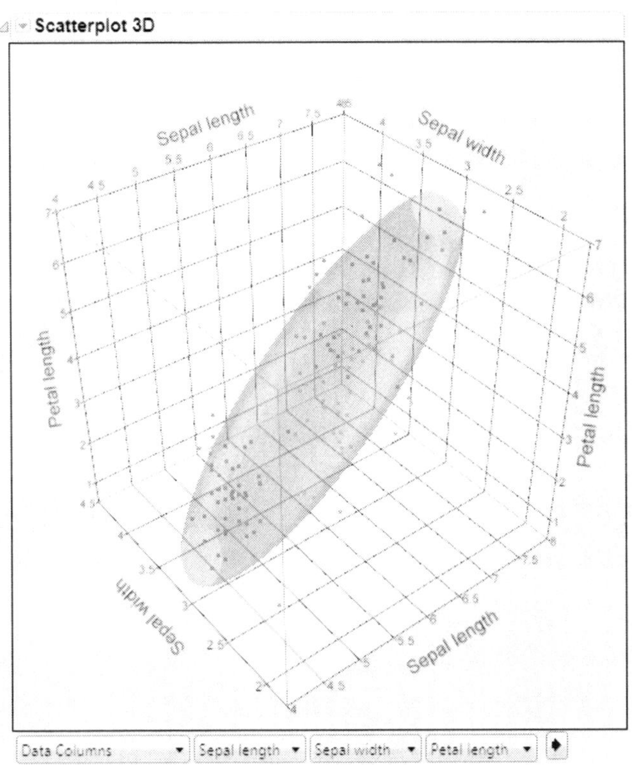

Example of Grouped Normal Contour Ellipsoids

This example shows how to group measurements by species and to format each group with a normal contour ellipsoid. The ellipsoids cover 75% of the data points and are 50% transparent. The contours are color-coded based on species.

1. Open the Iris.jmp sample data table.
2. Select **Graph > Scatterplot 3D**.
3. Select **Sepal length**, **Sepal width**, **Petal length**, and **Petal width** and click **Y, Columns**.
4. Click **OK**.
5. From the red triangle menu, select **Normal Contour Ellipsoids**.
6. Select **Grouped by Column**.
7. Select **Species**.

8. Type 0.75 next to Coverage.
9. Type 0.5 next to Transparency.
10. Click **OK**.

Figure 13.10 Example of Grouped Normal Contour Ellipsoids

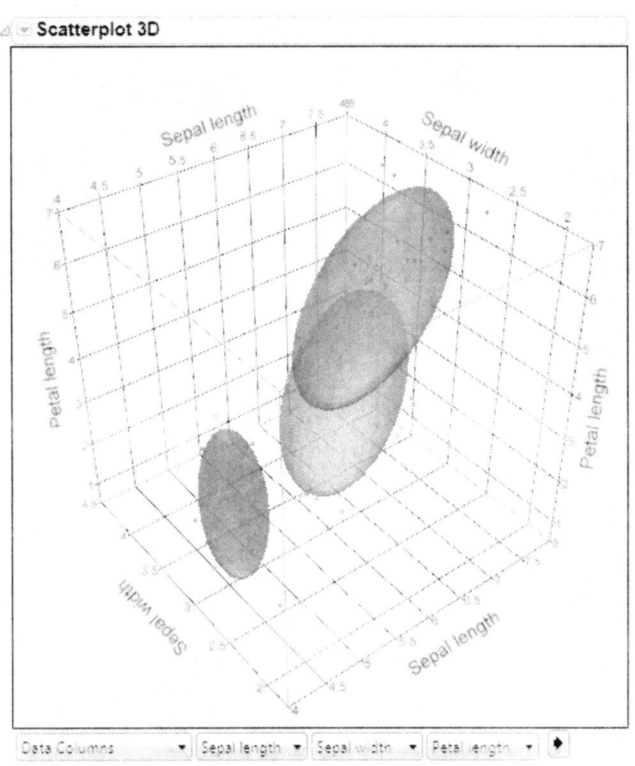

Example of a Grouped Nonparametric Density Contour

This example shows how to group data points and format each nonparametric density contour.

1. Open the Iris.jmp sample data table.
2. Select **Graph > Scatterplot 3D**.
3. Select **Sepal length**, **Sepal width**, **Petal length**, and **Petal width** and click **Y, Columns**.
4. Click **OK**.
5. From the red triangle menu, select **Nonpar Density Contour**.
6. Select **Grouped by Column**.
7. Select **Species** and click **OK**. A different colored contour is displayed for each of the three species.

8. Type 0.25 in the first Contour Quantile box. 25% of the data points appear outside the contour surfaces, which results in smaller contours.
9. Type 0.15 in the first Transparency box. The contours are 15% opaque.

Figure 13.11 Changing the Nonparametric Density Contour Transparency and Density

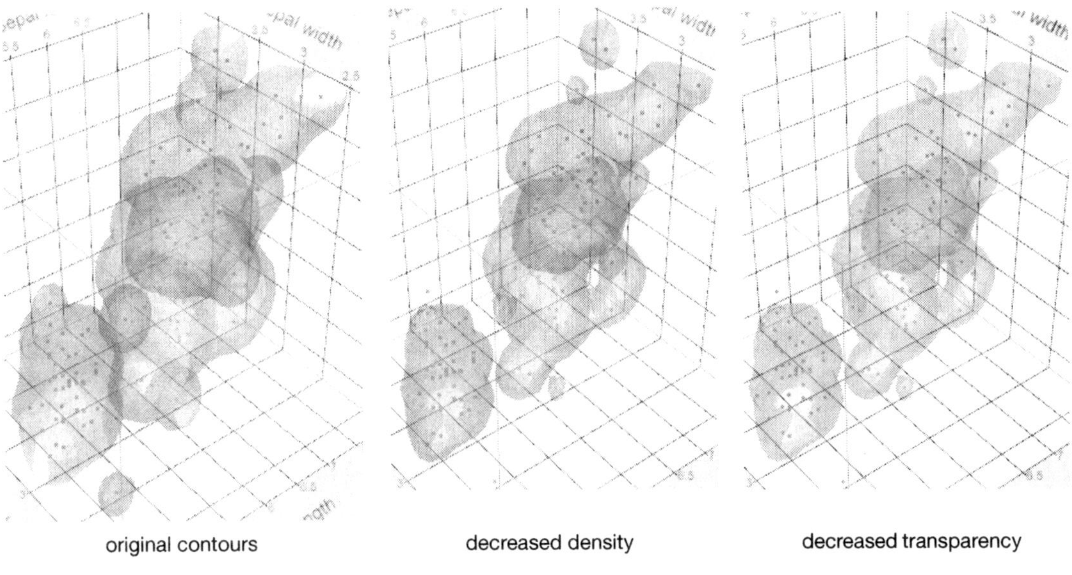

10. Select the second check box. The contour quantiles are the same (.25), so the new contours overlap the first ones.
11. Type .5 in the second Contour Quantile box. 50% of the data points appear outside the contour surfaces. A second set of contours appears within the first, to further illustrate the density of the data points.

Figure 13.12 Adding a Second Nonparametric Density Contour

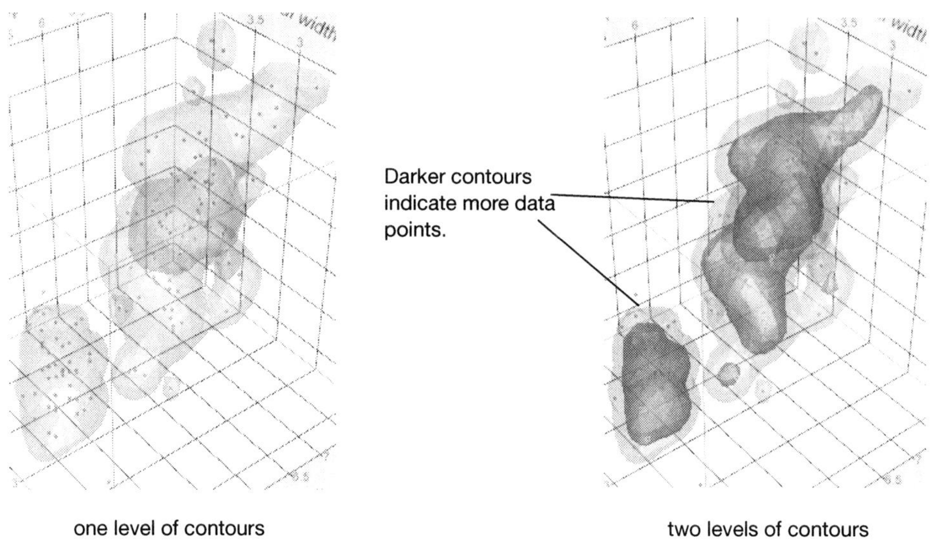

You can now format the second levels of contours and turn on the third level of contours.

The options for formatting the grouped and ungrouped nonparametric density contours are similar. The only difference is that you cannot change the color of each *grouped* nonparametric density contour. See Table 13.4 for options.

Chapter 14

Creating Contour Plots

Using the Contour Plot Platform

The **Contour Plot** command in the **Graph** menu constructs contours of a response in a rectangular coordinate system. A contour plot shows a three-dimensional surface in two dimensions. Contours delineate changes in the third dimension.

Here are some of the options available with the Contour platform:

- specify the number of contour levels
- choose to plot contour lines or filled contours
- show or hide data points
- label contours with response values
- define and use a custom coloring scheme

Figure 14.1 Examples of Contour Plots

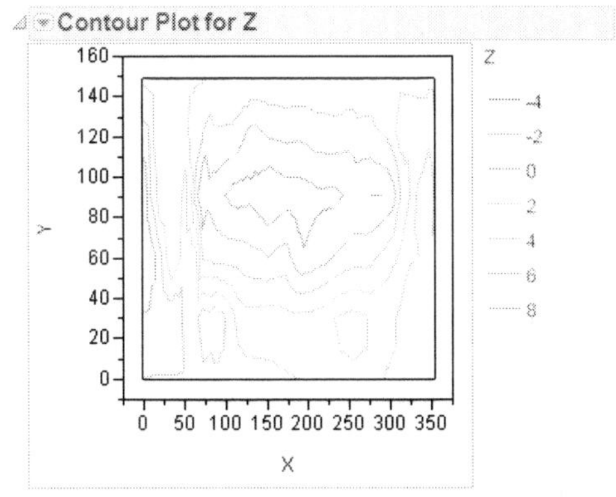

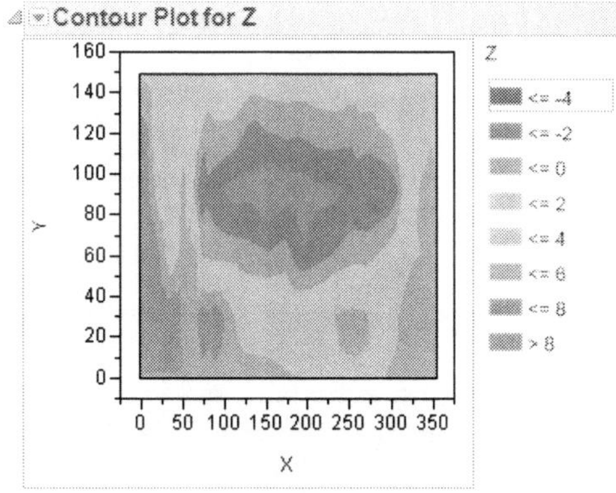

Contents

Example of a Contour Plot

To create a contour plot, you need two variables for the *x*- and *y*-axes and at least one more variable for contours. You can also use several *y*-variables. This example uses the Little Pond.jmp sample data table. X and Y are coordinates of a pond. Z is the depth.

1. Open the Little Pond.jmp sample data table.
2. Select **Graph > Contour Plot**.
3. Select the X and Y coordinates and click **X**.
4. Select the depth, Z, and click **Y**.

Note: In a contour plot, the X1 and X2 roles are used for the X and Y axes.

5. Click **OK**.

Figure 14.2 Example of a Contour Plot with Legend

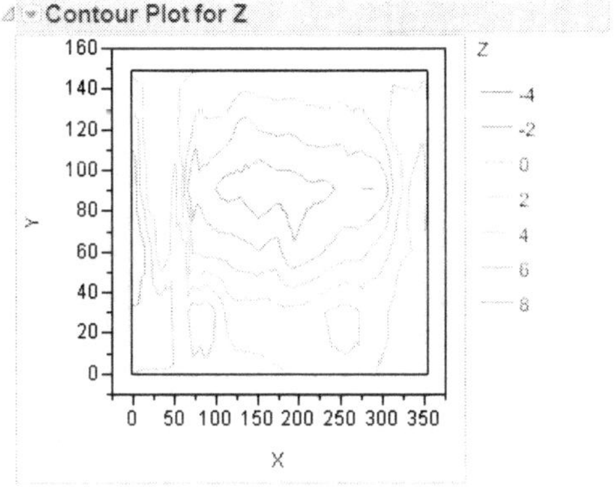

The *x*- and *y*-axes are coordinates and the contour lines are defined by the depth variable. This contour plot is essentially a map of a pond showing depth.

Launch the Contour Plot Platform

Launch the Contour Plot platform by selecting **Graph > Contour Plot**.

By default, the contour levels used in the plot are values computed from the data. You can specify your own number of levels and level increments in the Launch window before you create the plot. You can also do so in the red triangle menu for Contour Plot after you create the plot. You can use a column formula to compute the contour variable values.

Figure 14.3 The Contour Plot Launch Window

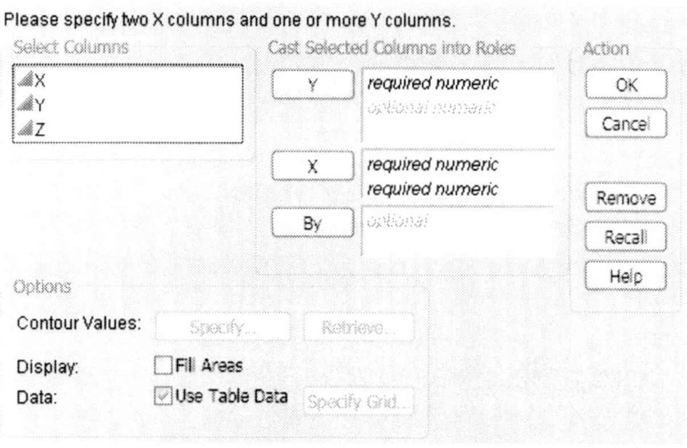

Table 14.1 Description of the Contour Plot Launch Window

Y	Columns assigned to the **Y** role are used as variables to determine the contours of the plot. You must specify at least one, and you can specify more than one. You can also assign a column with a formula to this role. If you do so, the formula should be a function of exactly two variables. Those variables should be the *x* variables entered in the Launch window.
X	Columns assigned to the **X** role are used as the variables for the *x*- and *y*-axes. You must specify exactly two columns for **X**.
By	This option produces a separate graph for each level of the By variable. If two By variables are assigned, a separate graph for each possible combination of the levels of both By variables is produced.
Options:	
Contour Values	Specify your own number of levels and level increments. See "Contour Specification" on page 377.
Fill Areas	Fill the areas between contour lines using the contour line colors.

Table 14.1 Description of the Contour Plot Launch Window *(Continued)*

Use Table Data **Specify Grid**	Most often, you construct a contour plot for a table of recorded response values. In that case, **Use Table Data** is selected and the **Specify Grid** button is unavailable. However, if a column has a formula and you specify that column as the response (**Y**), the **Specify Grid** button becomes available. When you click **Specify Grid**, you can define the contour grid in any way, regardless of the rows in the existing data table. This feature is also available with table templates that have one or more columns defined by formulas but no rows. See "Use Formulas for Specifying Contours" on page 379.

For more information about the launch window, see *Using JMP*.

The Contour Plot

Follow the instructions in "Example of a Contour Plot" on page 373 to produce the plot shown in Figure 14.4.

The legend for the plot shows individual markers and colors for the *Y* variable. Replace variables in the plot by dragging and dropping a variable, in one of two ways: swap existing variables by dragging and dropping a variable from one axis to the other axis; or, click on a variable in the Columns panel of the associated data table and drag it onto an axis.

For information about additional options for the report, see "Contour Plot Platform Options" on page 376.

Figure 14.4 The Contour Plot Report

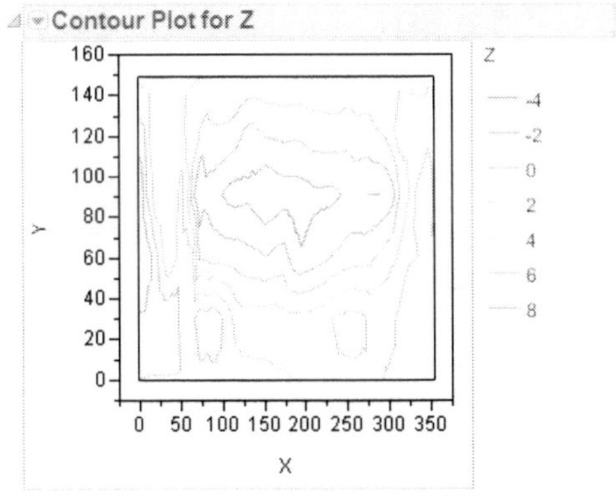

Contour Plot Platform Options

Using the options in the red triangle menu next to Contour Plot, you can tailor the appearance of your contour plot and save information about its construction.

Table 14.2 Descriptions of Contour Plot Platform Options

Show Data Points	Shows or hides (*x*, *y*) points. The points are hidden by default.
Show Missing Data Points	Shows or hides points with missing *y* values. Only available if **Show Data Points** is selected.
Show Contours	Shows or hides the contour lines or fills. The contour lines are shown by default.
Show Boundary	Shows or hides the boundary of the total contour area. The boundary is shown by default.
Fill Areas	Fills the areas between the contours with a solid color. It is the same option that is available in the Launch window. If you leave it deselected in the Launch window, you can see the line contours before filling the areas. See "Fill Areas" on page 376.
Label Contours	Shows or hides the label (*z*-value) of the contour lines.
Color Theme	Select another color theme for the contours.
Reverse Colors	Reverses the order of the colors assigned to the contour levels.
Change Contours	Set your own number of levels and level increments. See "Contour Specification" on page 377.
Save	This menu has options to save information about contours, triangulation, and grid coordinates to data tables. See "Contour Plot Save Options" on page 379.
Script	This menu contains commands that are available to all platforms. They enable you to redo the analysis or save the JSL commands for the analysis to a window or a file. For more information, see *Using JMP*.

Fill Areas

If you select **Fill Areas**, the areas between contour lines are filled with the contour line colors. This option is available in the Launch window and in the red triangle menu for Contour Plot. Figure 14.5 shows a plot with contour lines on the left and a plot with the contour areas filled on the right.

Figure 14.5 Comparison of Contour Lines and Area Fills

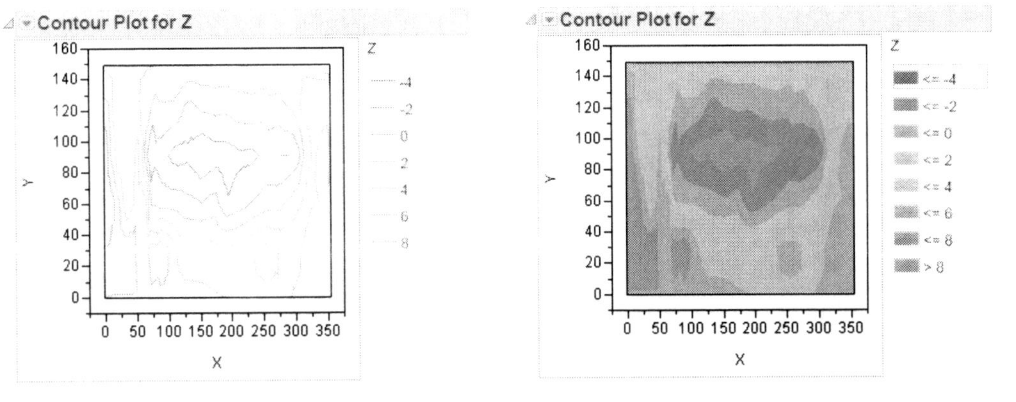

Areas are filled from low to high values. An additional color is added in the filled contour plot for the level above the last, and highest, contour line.

Contour Specification

If you do not select options in the Launch window, the default plot spaces the contour levels equally within the range of the *Y* variable. The default colors are assigned via the Continuos Color Theme in Preferences. You can see the colors palette with the **Colors** command in the **Rows** menu or by right-clicking on an item in the Contour Plot legend.

You can specify contour levels either in the Launch window (the **Specify** button) or in the report window from the red triangle menu for Contour Plot (the **Specify Contours** option).

Figure 14.6 Example of Contour Specification: Launch Window (on the left) and Menu (on the right)

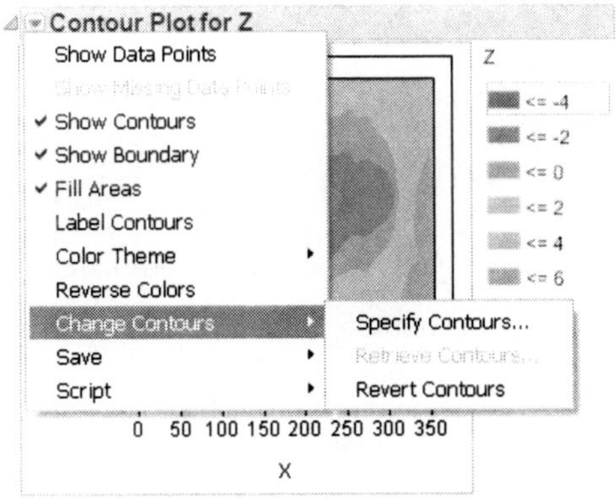

Specify

This option is both in the Launch window and on the red triangle menu for Contour Plot (the **Specify Contours** option).

Selecting this option displays the Contour Specification window. See Figure 14.7. Using this window, you can do the following:

- change the number of contours
- specify minimum and maximum values to define the range of the response to be used in the plot
- change the increment between contour values

You supply any three of the four values, and the remaining value is computed for you. Click on the check box to deselect one of the numbers and automatically select the remaining check box.

Figure 14.7 The Contour Specification Window

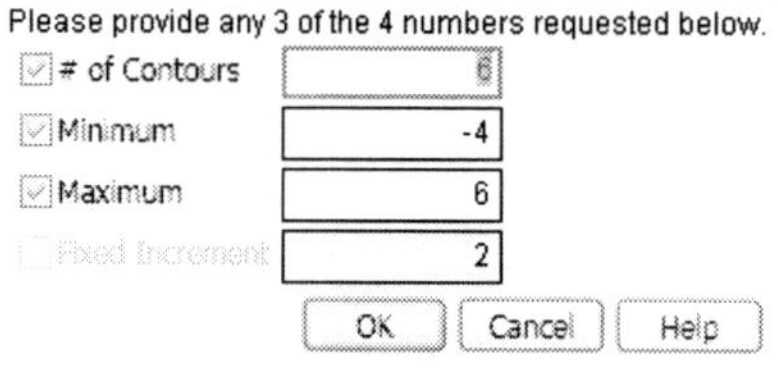

Colors are automatically assigned and are determined by the number of levels in the plot. After the plot appears, you can right-click (press CONTROL and click on the Macintosh) on any contour in the plot legend and choose from the JMP color palette to change that contour color.

Retrieve

This option is both in the Launch window (the **Retrieve** button) and on the red triangle menu for Contour Plot (the **Retrieve Contours** option).

Note: Neither the button nor the menu option are active unless there is an open data table *in addition to* the table that has the contour plotting values. When you click **Retrieve** or select **Retrieve Contours**, a window with a list of open data tables appears.

Using this option, you can retrieve the following from an open JMP data table:

- the number of contours
- an exact value for each level
- a color for each level

From the list of open data tables, select the data table that contains the contour levels.

For level value specification, the Contour Plot platform looks for a numeric column with the same name as the response column that you specified in the Launch window. The number of rows in the data table defines the number of levels.

If there is a row state column with color information, those colors are used for the contour levels. Otherwise, the default platform colors are used.

Revert Contours

This option appears only on the red triangle menu for Contour Plot.

If you have specified your own contours, selecting this option reverts your Contour Plot back to the default contours.

Contour Plot Save Options

This menu has options to save information about contours, triangulation, and grid coordinates to data tables.

Save Contours creates a new JMP data table with columns for the following:

- the *x*- and *y*-coordinate values generated by the Contour platform for each contour
- the response computed for each coordinate set
- the curve number for each coordinate set

The number of observations in this table depends on the number of contours you specified. You can use the coordinates and response values to look at the data with other JMP platforms. For example, you can use the Scatterplot 3D platform to get a three-dimensional view of the pond.

Generate Grid displays a window that prompts you for the grid size that you want. When you click **OK**, the Contour platform creates a new JMP data table with the following:

- the number of grid coordinates you requested
- the contour values for the grid points computed from a linear interpolation

Save Triangulation creates a new JMP data table that lists coordinates of each triangle used to construct the contours.

Use Formulas for Specifying Contours

Most often you construct a contour plot for a table of recorded response values such as the **Little Pond** data table. In that case, in the launch window, **Use Table Data** is checked and the **Specify Grid** button is unavailable. However, if a column has a formula and you specify that column as the response (**Y**), the **Specify Grid** button becomes active.

When you click **Specify Grid**, the window shown in Figure 14.8 appears.

Figure 14.8 Example of the Contour Specification for Formula Column

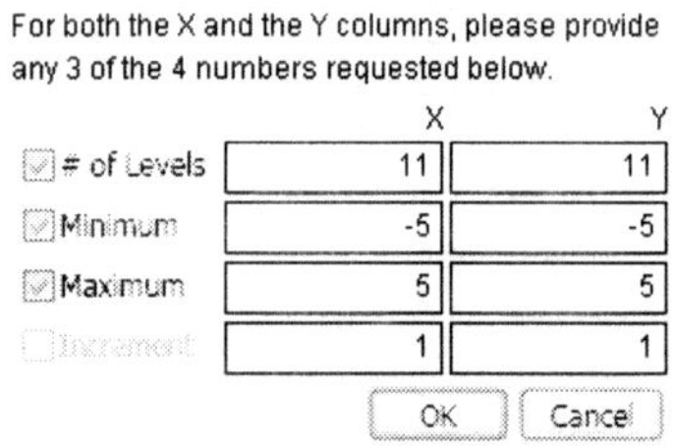

For both the X and the Y columns, please provide
any 3 of the 4 numbers requested below.

	X	Y
# of Levels	11	11
Minimum	-5	-5
Maximum	5	5
Increment	1	1

OK Cancel

You can complete the Specify Grid window and define the contour grid in any way, regardless of the rows in the existing data table. This feature is also available with table templates that have one or more columns defined by formulas but no rows.

Chapter 15

Creating Bubble Plots

Using the Bubble Plot Platform

A bubble plot is a scatter plot that represents its points as circles, or bubbles. Bubble plots can be dynamic (animated over time) or static (fixed bubbles that do not move). Use bubble plots to:

- dynamically animate bubbles using a time variable, to see patterns and movement across time
- use size and color to clearly distinguish between different variables
- aggregate data (rows) into a single bubble, simplifying the bubble plot

Because you can see up to five dimensions at once (x position, y position, size, color, and time), bubble plots can produce dramatic visualizations and readily show patterns and trends.

Note: Dynamic bubble plots were pioneered by Hans Rosling, Professor of International Health, Karolinska Institutet, and the people involved in the Gapminder.org project.

Figure 15.1 Example of a Bubble Plot

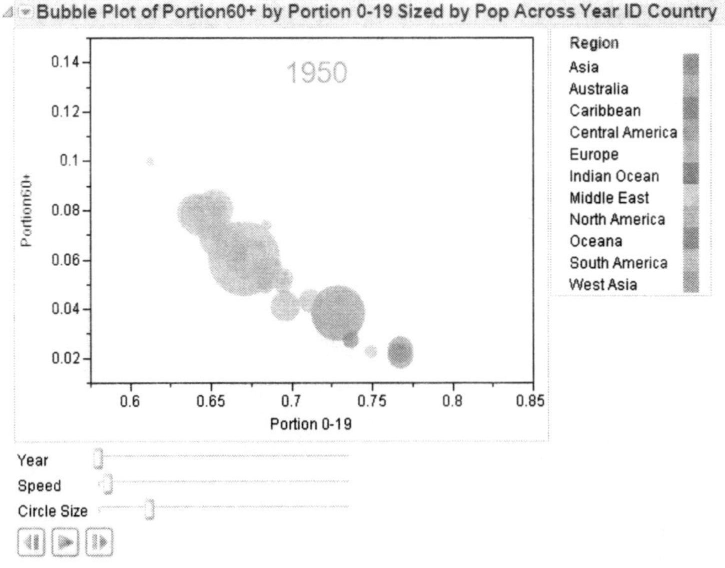

Contents

Example of a Dynamic Bubble Plot

This example uses the PopAgeGroup.jmp sample data table, which contains population data for countries and regions around the world. Examine the relationship between the proportion of younger and older people in the sample populations.

1. Open the PopAgeGroup.jmp sample data table.
2. Select **Graph > Bubble Plot**.

   The launch window appears.

Figure 15.2 The Bubble Plot Launch Window

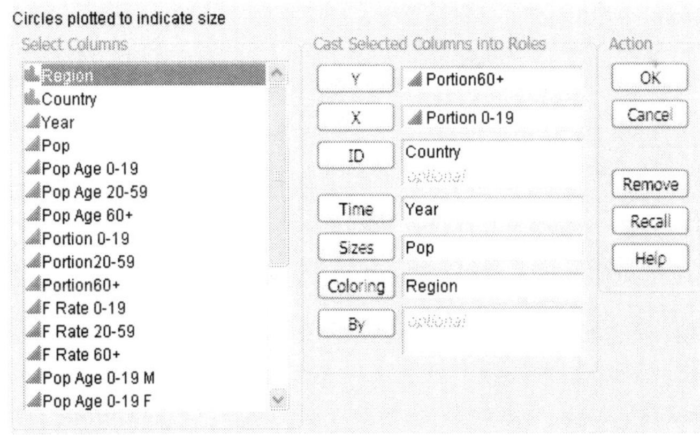

3. Select Portion60+ and click **Y**.

   The portion of the population that are 60 years or older becomes the *y* coordinate.
4. Select Portion 0-19 and click **X**.

   The portion of the population that are 0-19 years becomes the *x* coordinate.
5. Select Country and click **ID**.

   All the rows for each country are aggregated into a single bubble.
6. Select Year and click **Time**.

   The bubble plot shows a unique plot for each year's data.
7. Select Pop and click **Sizes**.

   The sizes of the bubbles reflect the overall population values.
8. Select Region and click **Coloring**.

   The bubble for each region is colored differently.
9. Click **OK**.

   The report window appears.

Figure 15.3 The Bubble Plot Report Window

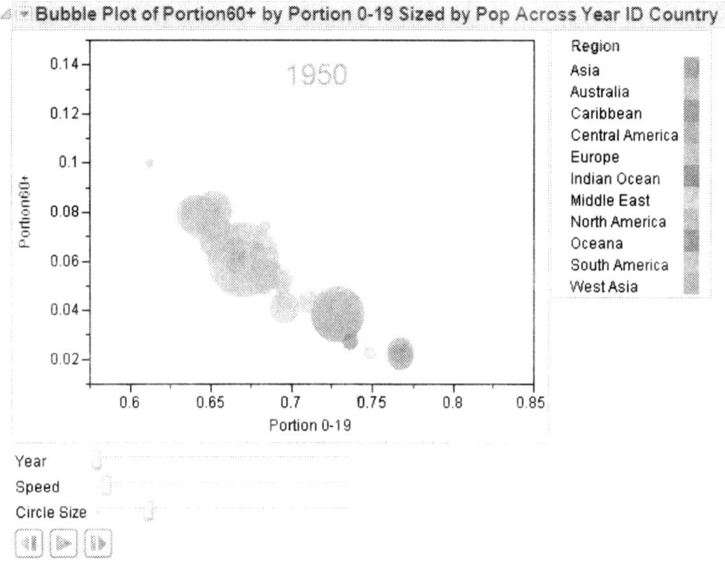

10. Click the play button to see the animated, dynamic report. Alternatively, you can click the previous button to move forward by one year.
11. (Optional) To view a legend that identifies each color with its region, select **Legend** from the red triangle menu.

As time progresses, you can see that the portion of the population that is 0-19 years decreases, and the portion of the population that is 60 years or more increases.

Launch the Bubble Plot Platform

Launch the Bubble Plot platform by selecting **Graph > Bubble Plot**.

Figure 15.4 The Bubble Plot Launch Window

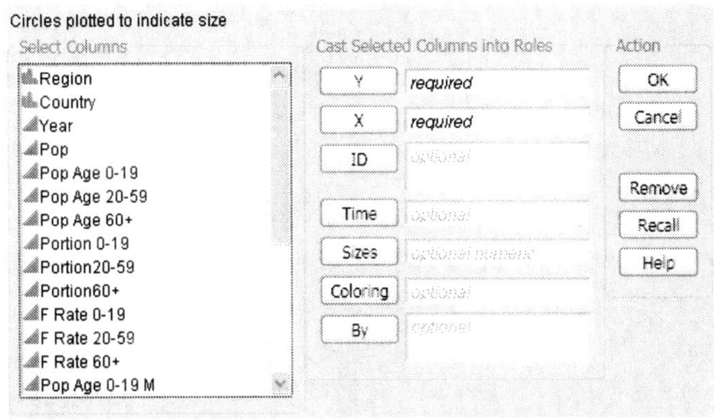

Table 15.1 Description of the Bubble Plot Launch Window

Y, X	The Y and X columns become the *y* and *x* coordinates of the bubbles in the plot. These values can be continuous or categorical (nominal or ordinal).
ID	(Optional) ID variables identify rows that should be aggregated and shown as a single bubble. The default coordinates of each bubble are the averaged *x* and *y* values, and the default size of each bubble is the sum of the sizes of all aggregated members. See "Specifying Two ID Variables" on page 386.
Time	Maintains separate coordinates, sizes, and colors for each unique time period. The bubble plot shows these values for a single time period. For example, if the **Time** column contains years, the bubble plot is updated to show data by each year. See "Specifying a Time Variable" on page 386.
Sizes	Controls the size of the bubbles. The area of the bubbles is proportional to the **Size** value. There is a minimum bubble size, to keep bubbles visible, even if the size value is zero. If **Size** is left blank, the default bubble size is proportional to the number of rows in that combination of **Time** and **ID**.
Coloring	Colors the bubbles according to the selected variable. If the selected variable is categorical (nominal or ordinal), each category is colored distinctly. If the selected variable is continuous, a gradient of colors is used.
By	Place a column here to produce a separate bubble plot for each level of the variable.

For more information about the launch window, see *Using JMP*.

Specifying Two ID Variables

Specifying a second **ID** variable provides a hierarchy of categories, but the bubbles are not split by the second category until they are selected and split interactively. In the launch window, if you specify a second **ID** variable, **Split** and **Combine** buttons appear in the report window.

For example, you might specify a country as the first **ID** variable, resulting in a separate aggregated bubble for each country. A second **ID** variable, perhaps designating regions within each country, would further split each country when the interactive **Split** button under the graph is pressed.

Specifying a Time Variable

Maintains separate coordinates, sizes, and colors for each unique time period. The bubble plot shows these values for a single time period. For example, if the **Time** column contains years, the bubble plot is updated to show data by each year.

To move the time label on the plot, click and drag the label.

If data is missing within a time period, the value is linearly interpolated. If data is missing for the first or last time period, the value is not estimated, but left as missing.

Related Information

- "Control Animation for Dynamic Bubble Plots" on page 386
- "Example of Specifying Only a Time Variable" on page 392

Interact with the Bubble Plot

Use the Bubble Plot platform in one of two modes:

- *Static mode*, where the bubbles are fixed and do not animate over time (no **Time** variable is specified). See "Example of a Static Bubble Plot" on page 396.
- *Dynamic mode*, where the bubbles are animated over time (a **Time** variable is specified). See "Example of a Dynamic Bubble Plot" on page 383.

You interact with both static and dynamic bubble plots in different ways.

Control Animation for Dynamic Bubble Plots

Use sliders and buttons to control the animation of dynamic bubble plots.

Figure 15.5 Animation Controls

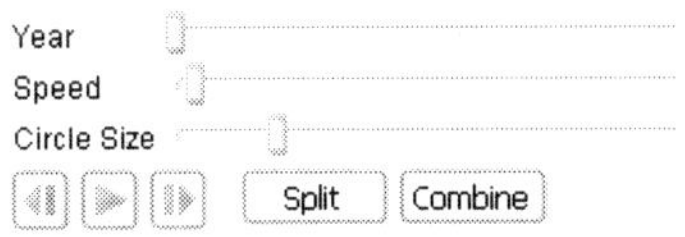

Table 15.2 Descriptions of the Animation Controls

Slider or Button	Description	Visibility
<Time variable>	Controls which time values appear in the bubble plot. You manually drag the slider to see a progression of time. Click and drag on the time variable in the bubble plot to move its position.	Only appears if you have specified a variable for **Time.**
Speed	Adjusts the speed of the animation.	Only appears if you have specified a variable for **Time.**
Circle Size	Adjusts the size of the bubbles. The bubbles maintain their relative size, but their absolute size can be adjusted.	Appears on all bubble plots.
	Adjusts the time value by one unit and shows the previous time value.	Only appears if you have specified a variable for **Time.**
	Press play to animate the bubble plot. Moves through all of the time values in order, and loops back to the beginning when the last time period is reached. Press pause to stop the animation.	Only appears if you have specified a variable for **Time.**
	Adjusts the time value by one unit and shows the next time value.	Only appears if you have specified a variable for **Time.**
Split	Separates the bubble represented by the first, larger **ID** variable into its smaller constituent parts, which are defined by the second, smaller **ID** column. Select the bubble and click **Split**. See Figure 15.10.	Only appears if you have specified two **ID** variables.
Combine	Reverses the action of the **Split** button by recombining the smaller bubbles back into their original bubble. Select any of the smaller bubbles in the group and click **Combine.**	Only appears if you have specified two **ID** variables.

Select Bubbles

Click on a bubble to select it. Note the following:

- Visually, selected bubbles become darker or brighter, and non-selected bubbles are more transparent.
- If the bubble was not filled initially, selection fills it.
- If no bubbles are selected, all of the bubbles are semi-transparent.

When you select a bubble, all of the rows in the data table that correspond to the selected bubble are highlighted. Note the following:

- If the bubble is an aggregate based on an **ID** column, all of the rows for that **ID** are highlighted. Otherwise, the one row represented by that bubble is highlighted.
- If you specify an **ID** and a **Time** variable, selecting a bubble highlights all of the rows for that **ID**, across all of the **Time** levels.

If you select a row from the data table, it is selected in the associated bubble plot. Note the following:

- If you have not specified a **Time** variable, selecting one row from the data table highlights the corresponding bubble in the plot.
- If you have specified a **Time** variable, selecting one row from the data table highlights the corresponding bubble for only that time period in the dynamic bubble plot.

Use the Brush Tool

Use the brush tool to temporarily select bubbles and obtain more information about the selected bubbles. For dynamic bubble plots, turn on the **Trail Bubbles** or the **Trail Lines** option to show the selected bubbles during animation. When you select bubbles with the brush tool, the corresponding rows are highlighted in the associated data table.

Note: For a more granular examination of the highlighted rows, use the **Tables > Subset** command or the Row Editor. See the *Using JMP* book.

Bubble Plot Platform Options

The following table describes the options in the Bubble Plot red triangle menu.

Draw	• Select **Filled** to fill all of the bubbles. • Select **Outlined** to outline all of the bubbles. • Select **Filled** and **Outlined** to fill and outline all of the bubbles.

Set Shape	Change the shape of the bubble. You can create a custom shape using JSL. The **Custom** option opens the custom shape. If no custom shape has been created, the **Custom** option uses the default circle shape. For more information about creating custom shapes, see the *Scripting Guide*.
Orient Shapes	Orients the shapes as they move in particular directions over time, following the shape of the data. This option appears only if you have specified a variable for **Time**.
Trail Bubbles	Shows the past history of bubbles as a semi-transparent trail. You must click on a bubble before you can see trail bubbles. See "Example of Specifying Only a Time Variable" on page 392. Note the following: • This option appears only if you have specified a variable for **Time**. • If you do not want to see the bubble labels, select the **Label > None** option.
Trail Lines	Shows the past history of bubbles as connected line segments. You must click on a bubble before you can see trail lines. See "Example of Specifying Only a Time Variable" on page 392. Note the following: • This option appears only if you have specified a variable for **Time.** • If you do not want to see the bubble labels, select the **Label > None** option.
Label	• Select **None** to label none of the bubbles in the plot. • Select **All** to label all of the bubbles in the plot. • Select **Selected** to label bubbles only when you select them. **Note:** Click and drag on a label to move it.
Color Theme	Change the colors representing the high, middle, and low values of the color variable. This option appears only if you have specified a variable for **Coloring.**
Revert Color Theme	Reverts back to the original color theme. This option appears only if you have applied a color theme.
Legend	Shows a legend that describes the colors in the bubble plot. This option appears only if you have specified a variable for **Coloring.**

Selectable Across Gaps	If a bubble is selected, this option keeps the bubble selected during time periods where data is missing. Otherwise, the bubble is not selected during time periods where data is missing.
Show Roles	Shows the variables that are used in the bubble plot. You can change and delete the variables. See "Show Roles" on page 390.
Split All	Splits all bubbles into their constituent parts. Unlike the **Split** button, the bubbles do not have to be selected. This option appears only if you have specified two **ID** variables.
Combine All	Combines all constituent bubbles within a group into their larger bubble. Unlike the **Combine** button, the bubbles do not have to be selected. This option appears only if you have specified two **ID** variables.
Lock Scales	Prevents axis scales and gradient legend scales from automatically adjusting in response to data or filtering changes.
Aggregation Options	Alters how the **X**, **Y**, and **Sizes** roles are computed. By default, the values are calculated using means for **X** and **Y**, and sums for **Sizes**. • Selecting **X as Sum** or **Y as Sum** computes the **X** and **Y** values using sums. • Deselecting **Size as Sum** computes **Size** values using means. • Selecting **Color as Sum** computes the sum of the data values and maps to a color. This option appears only for continuous variables.
Save for Adobe Flash platform (.SWF)	Saves the bubble plot as .SWF files that are Adobe Flash player compatible. You can use these files in presentations and in Web pages. An HTML page is also saved that shows you the correct code for using the resulting .SWF file. For more information about this option, go to http://www.jmp.com/support/swfhelp/en.
Script	This menu contains options that are available to all platforms. They enable you to redo the analysis or save the JSL commands for the analysis to a window or a file. For more information, see *Using JMP*.

Show Roles

Using the **Show Roles** option in the red triangle menu, you can make changes to your existing variables without having to relaunch the platform and start your analysis over.

Follow the instructions in "Example of a Dynamic Bubble Plot" on page 383 to produce the report window shown in Figure 15.6.

Figure 15.6 Example of Bubble Plot with Show Roles Selected

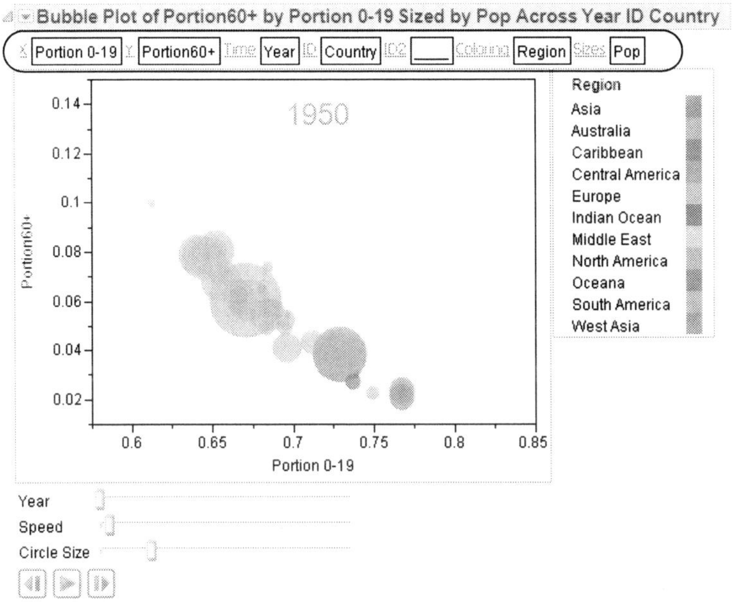

Change the Variable Assigned to a Role

To change the variable assigned to a role, click on a blue underlined role name. For example, in Figure 15.6, to change the **Coloring** variable from Region to Country, proceed as follows:

1. Click on the Coloring link.

   The Select column for Coloring window appears.

2. Click on Country.
3. Click **OK**.

   Country now replaces Region as the **Coloring** variable in the bubble plot.

Remove a Variable

To remove an existing variable from the bubble plot, make sure that nothing is selected in the Select column for <*Role*> window, and click **OK**. For example, in Figure 15.6, to remove the **Sizes** variable (Pop), proceed as follows:

1. Click on the Sizes link.

   The Select column for Sizes window appears.

2. Ensure that nothing is selected. If a variable is selected, deselect it by holding down the CTRL key and clicking on the variable.
3. Click **OK**.

   The Sizes role now appears with an empty box.

Note: The X and Y variables can be changed only and cannot be removed.

Add a Variable

Once you have removed an existing variable from the bubble plot, there are two ways to add a new variable:

- Click on the blue underlined role name. See "Change the Variable Assigned to a Role" on page 391.
- In the data table, click on the variable in the column panel, and drag it into the empty role box.

Additional Examples of the Bubble Plot Platform

The following examples further illustrate the features of the Bubble Plot platform.

Example of Specifying Only a Time Variable

For dynamic bubble plots, you might specify only a **Time** variable and no **ID** variable. The resulting bubble plot contains a single moving bubble that tracks the series as the **Time** value changes.

1. Open the PopAgeGroup.jmp sample data table.
2. Select **Graph > Bubble Plot**.
3. Select Portion60+ and click **Y**.
4. Select Portion 0-19 and click **X**.
5. Select Year and click **Time**.
6. Select Region and click **Coloring**.
7. Click **OK**.

   The initial report window appears.

Figure 15.7 The Initial Report Window with a Time Variable

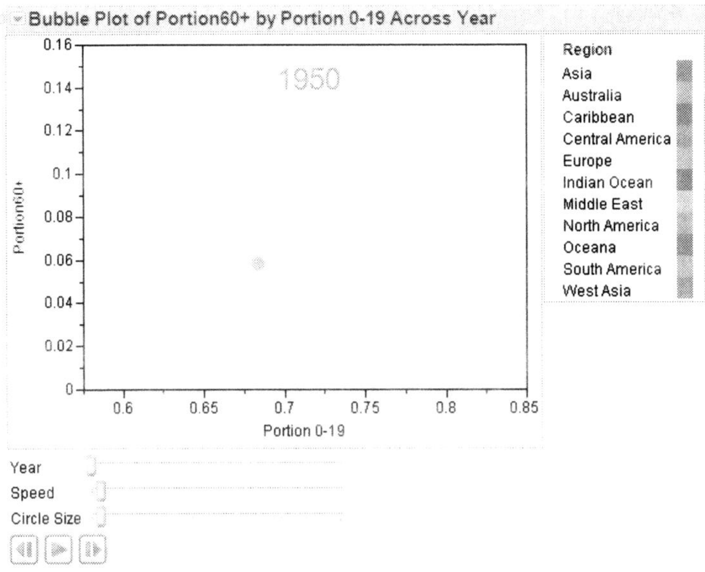

8. Click on the bubble to select it.

   All rows in the data table are also highlighted.

9. From the red triangle menu, select **Trail Bubbles** and **Trail Lines**.

10. Click the play button.

    The bubble plot animates, showing a trail for the single bubble.

Figure 15.8 Animated Bubble Plot over Time

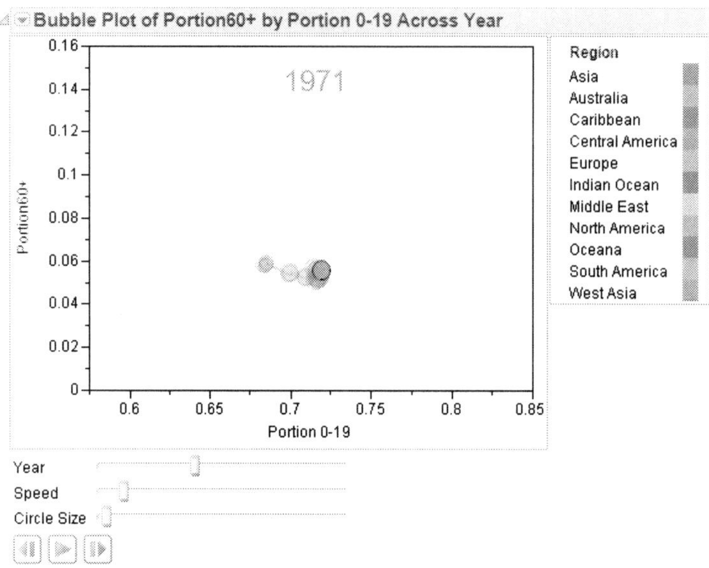

Example of Specifying Only ID Variables and Splitting a Bubble

For static bubble plots, you might specify one or two **ID** variables and no **Time** variable. The resulting bubble plot contains a bubble at each **ID** value. Note that although this bubble plot is static, you can perform splitting on bubbles.

1. Open the PopAgeGroup.jmp sample data table.
2. Select **Graph > Bubble Plot**.
3. Select Portion60+ and click **Y**.
4. Select Portion 0-19 and click **X**.
5. Select Region and Country and click **ID**.
6. Select Region and click **Coloring**.
7. Click **OK**.

   The initial report window appears.

Figure 15.9 Initial Report Window with ID Variables

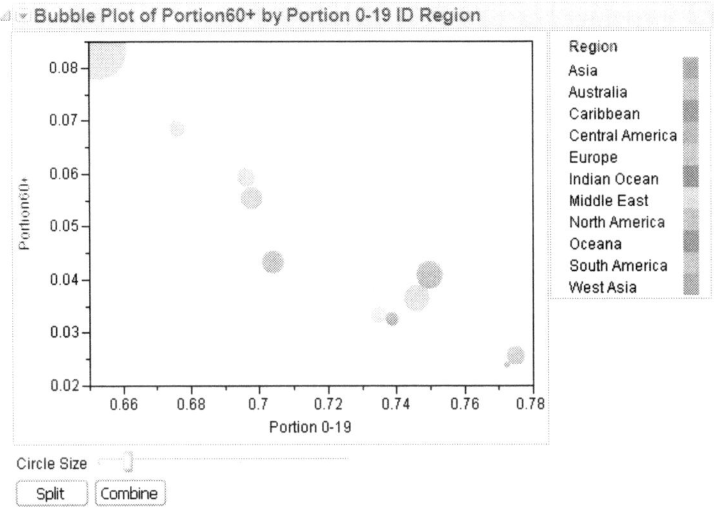

Split the bubble representing the region of North America into countries.

8. Click on the bubble representing North America (hover over a bubble to see its label, or use the legend to find the color of North America.)
9. Click **Split**.

   You see that the North America bubble has split into three bubbles, representing the countries within the region of North America (the United States of America, Canada, and Mexico).

Figure 15.10 Splitting the North America Bubble

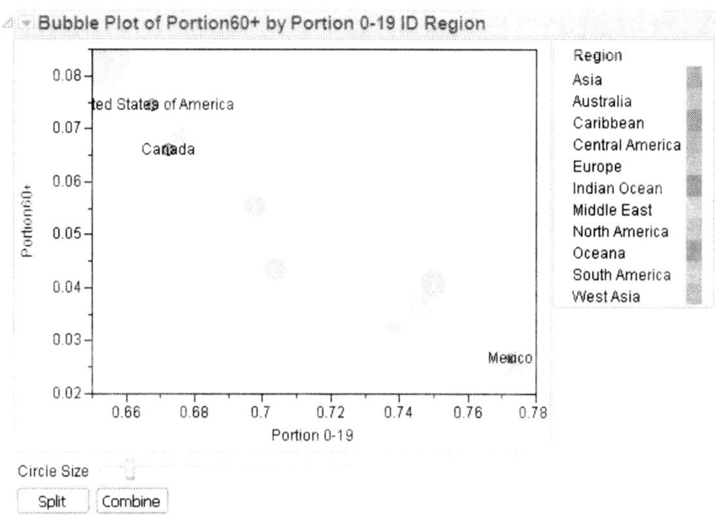

Example of a Static Bubble Plot

This example uses the SATByYear.jmp sample data table, which contains SAT verbal and math test scores for a selection of the US population in 2004.

1. Open the SATByYear.jmp sample data table.
2. Select **Graph > Bubble Plot**.
3. Select SAT Verbal and click **Y**.
4. Select SAT Math and click **X**.
5. Select State and click **ID**.
6. Select % Taking (2004) and click **Sizes**.
7. Click **OK**.

   The report window appears.

Figure 15.11 The Static Bubble Plot Report Window

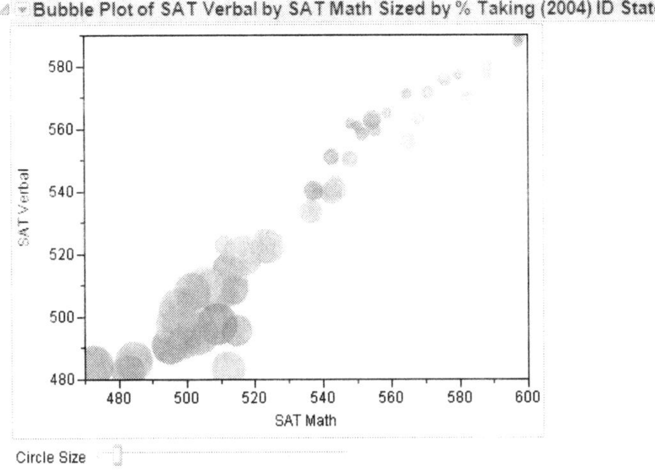

From Figure 15.11, you draw the following conclusions:

- Higher verbal scores appear to be associated with higher math scores, since the two track very closely in the bubble plot. This signifies a correlation between verbal and math scores.
- The larger bubbles represent the US states that have a high percentage of individuals taking the SAT test in 2004. These larger bubbles are all grouped together in the lower left of the graph. This shows that when a state has a high percentage of individuals taking the test, both the math and verbal scores are low.

Instead of grouping the bubbles primarily by state, group the bubbles primarily by region as follows:

1. From the red triangle menu, select **Show Roles**.
2. Click on the **ID** link.
3. Select Region, and click **OK**.

   Region is now the primary **ID** variable.
4. Click on the **ID2** link.
5. Select State, and click **OK**.

   State is now the secondary **ID** variable.

Figure 15.12 Example of Bubble Plot Grouped by Region and State

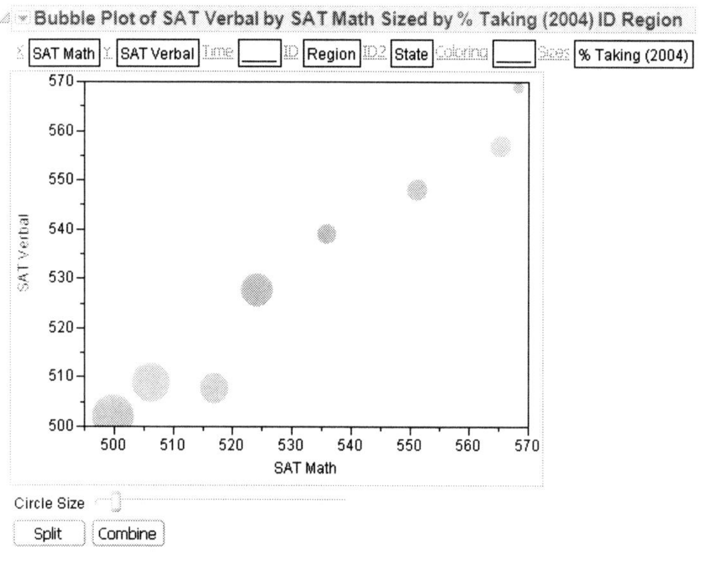

6. Click on the bubble that represents the Southwest region (hover over a bubble or click on it to see its label).
7. Click **Split**.

   Now the bubbles are split by the secondary **ID** variable, which is State. You now see each state within the Southwest region.

Figure 15.13 Example of Southwest Region Split by State

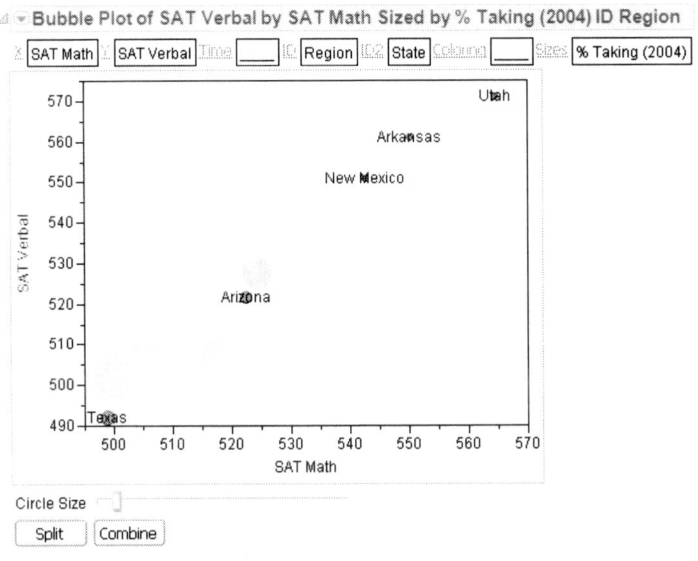

From Figure 15.13, you see that there is significant variation between the scores from the Southwest states.

8. Click **Combine** to combine the southwest states again.
9. To do a comparison, click on the New England bubble (hover over a bubble or click on it to see its label).
10. Click **Split**.

Figure 15.14 Example of New England Region Split by State

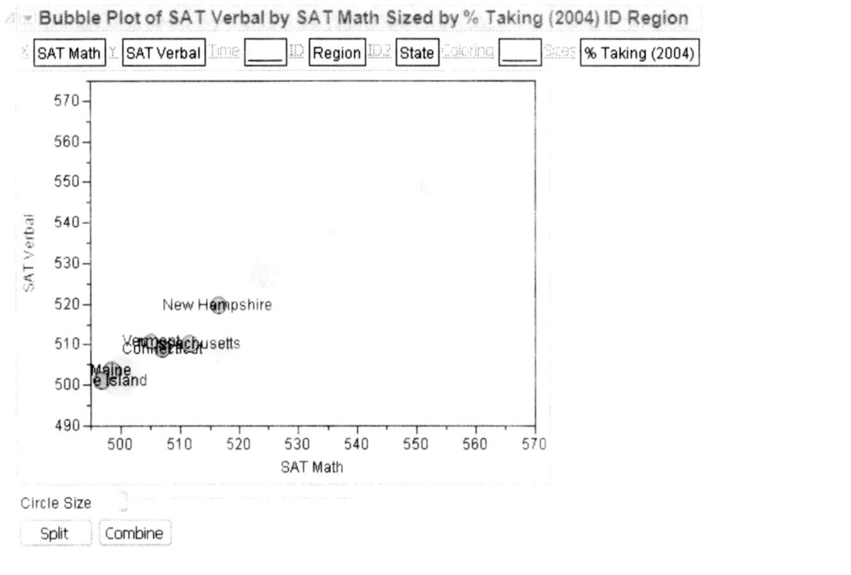

You see that the New England states do not have as much variation as the Southwest states.

Example of a Bubble Plot with a Categorical Y Variable

All of the examples shown so far use continuous Y variables. If you use a categorical (nominal or ordinal) Y variable, the bubble plot appears differently.

This example uses the blsPriceData.jmp sample data table, which shows the price of commodities over several years. Because the value of the US dollar changes over time, a column named Price/Price2000 shows the ratio of a commodity's price at any given time to the price in the year 2000.

1. Open the blsPriceData.jmp sample data table.
2. Select **Graph > Bubble Plot.**
3. Select Series and click **Y.**
4. Select Price/Price2000 and click **X.**
5. Select date and click **Time.**
6. Click **OK.**

   The report window appears.

This produces a bubble plot that, when animated by clicking the play button, shows the price bubbles moving side to side according to their price ratio.

Figure 15.15 Static Example of Animated Bubbles

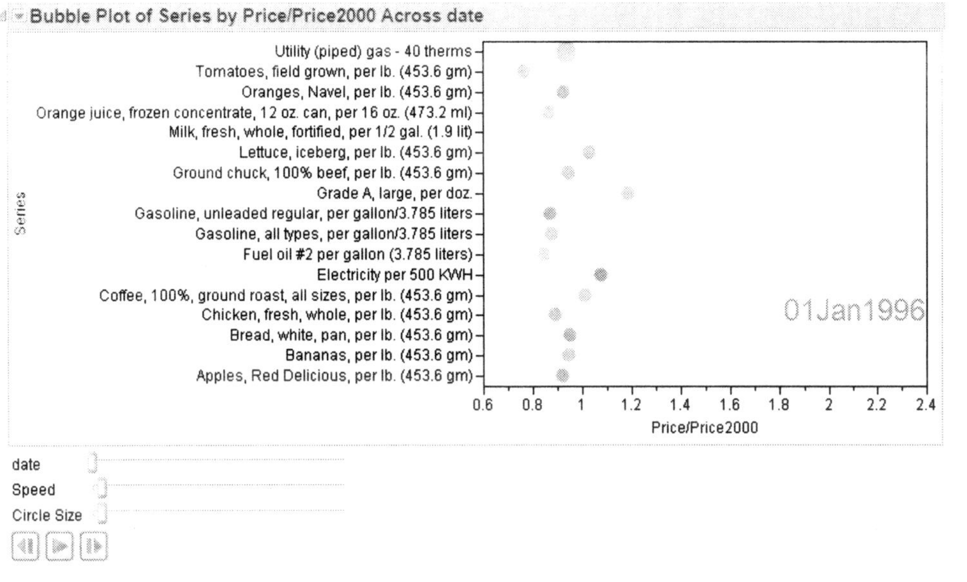

For easier readability, add grid lines as follows:

7. Double-click on the categorical axis.
8. In the Y Axis Specification window, select the box under Gridline next to Major.
9. Click **OK**.

To animate the bubble plot, click the play button. The price bubbles move side to side, according to their price ratio.

Chapter 16

Creating Maps

Using Map Functions

JMP transforms numbers and geographic data into compelling images, and turns simple tables of numbers into captivating pictures that bring the story in your data to life. JMP can help you display your data on geographical maps. Choose from built-in high-quality images, or use a Web Map Service to get custom map images from the Internet. JMP includes shape files for borders or many geographic regions and lets you add your own custom shapes, such as for a manufacturing plant or campus.

Figure 16.1 Example of a Map

Contents

Overview of Mapping

There are two types of map support in JMP: one where a map shows the data (Graph Builder) and one where a map provides context for the data (Background Maps). You can also create your own maps.

Graph Builder

You can interact with Graph Builder to create compelling visualizations of your data. JMP includes graphical support to display analyses using background maps and shape files. You can add color and geographical boundaries to maps through the following zones:

- The **Map Shape** zone assigns geographical boundaries to a map based on variables in the data table. The map shape value determines the x and y axes.

  Boundaries such as U.S. state names, Canadian provinces, and Japanese prefectures are installed with JMP. You can also create your own boundaries (geographical or otherwise) and specify them as a **Map Role** column property in the data table.
- The **Color** zone applies color based on a variable to geographical shapes.
- The **Size** element scales map shapes according to the size variable, minimizing distortion.

Background Maps

You can add background maps to any JMP graph through the Set Background Map window. You can use built-in background maps or connect to a Web Map Service (WMS) to display specialty maps like satellite images, radar images, or roadways. Right-click in a graph and select Background Map to choose from the following images and boundaries:

- **Simple Earth** and **Detailed Earth** maps are installed with JMP.
- **NASA server** provides maps using a WMS to show their most up-to-date maps.
- **Web Map Service** lets you enter the URL for a Web site that provides maps using the WMS protocol. You can also specify the map layer.
- Boundaries for various regions.

Example of Creating A Map in Graph Builder

This example uses the Crime.jmp sample data table, which contains data on crime rates for each US state.

1. Open the Crime.jmp sample data table (C:\Program Files\SAS\JMP\<Version Number>\Samples\Data).
2. Select **Graph > Graph Builder**.
3. Drag and drop State into the **Map Shape** zone.
4. Drag and drop Burglary into the **Color** zone.

Figure 16.2 Example of Burglary by State

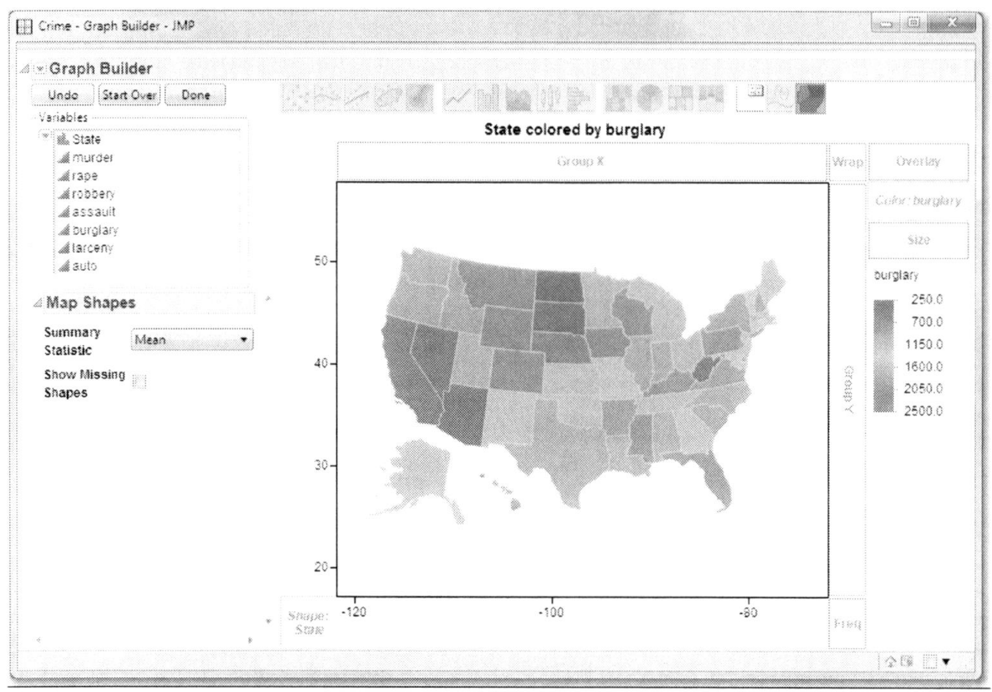

Note the following:

- The latitude and longitude appear on the Y and X axes.
- The legend shows the colors that correspond to the burglary rates. Since **Burglary** is a continuous variable, the colors are on a gradient.
- The map is projected so that relative areas are not distorted (the 49th parallel across the top of the US is not a straight line).

Graph Builder

Open a data table that contains geographic data. Launch Graph Builder by selecting **Graph** > **Graph Builder.** The primary element in the Graph Builder window is the graph area. The graph area contains drop zones (Map Shape, Color and Size), and you can drag and drop variables into the zones. From here you can map shapes for data tables that include place names.

Figure 16.3 The Graph Builder Window

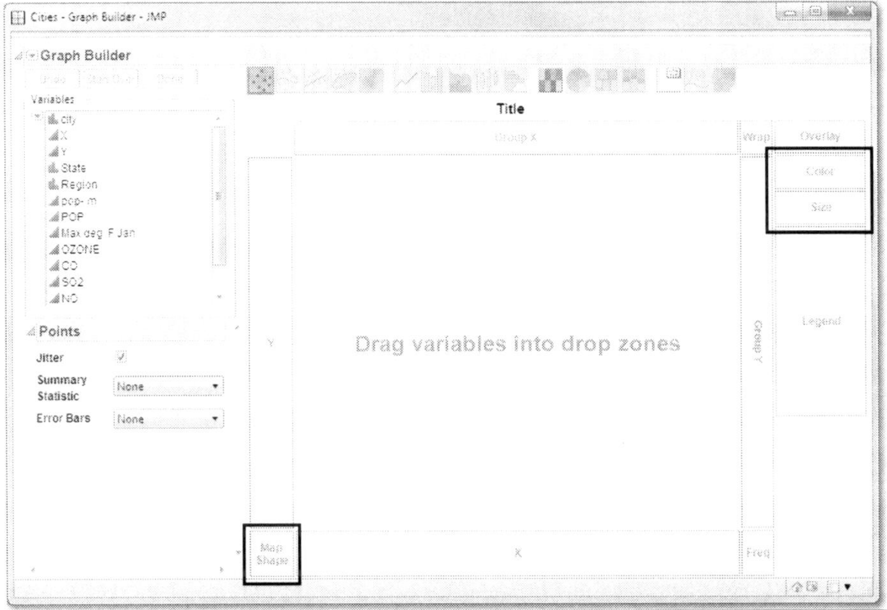

Map Shape

When a column contains the names of geographical regions (such as countries, regions, states, provinces, counties), you can assign the column to the **Map Shape** zone. When a variable is dropped in **Map Shape**, Graph Builder looks for map shapes that correspond to the values of the variable and draws the corresponding map. The variable can have a column property that tells JMP where to find the map data. If not, JMP looks through all known map files. If you have a variable in the Map Shape zone, the X and Y zones disappear. The Map Shape zone is positional and influences the types of graph elements that are available.

Figure 16.4 Example of Cities.jmp after Dragging State to Map Shape

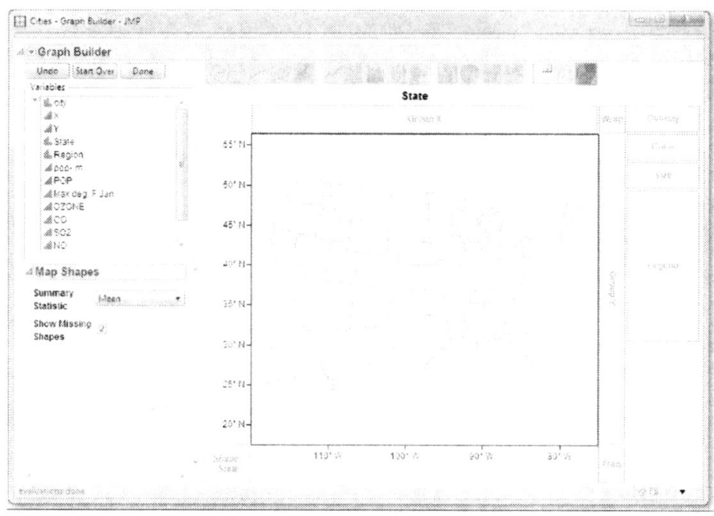

For each map there are two .jmp files; one for the name data (one row per entity) and one for coordinate data (many rows per entity). They are paired via a naming convention; xxx-Name.jmp and xxx-XY.jmp, where "xxx" is some common prefix. Some examples of sample files that are shipped with the product are:

- World-Name.jmp
- World-XY.jmp
- US-State-Name.jmp
- US-State-XY.jmp

Map Name Files

Each xxx-Name.jmp can contain any number of shape name columns, which are identified with a column property. Multiple name columns support localizations and alternate names styles (such as abbreviations), but a given graph usage uses only one column of names. The first column of the Name file must contain unique Shape ID numbers in ascending order.

Figure 16.5 Example of US-State-Name.jmp

	Shape ID	Name	State FIPS	USPS Code	Abbr	Max Grow Scale	Max Grow X Offset	Max Grow Y Offset
1	1	Alabama	01	AL	Ala.	1	0	0
2	2	Alaska	02	AK	Alaska	1	0	0
3	3	Arizona	04	AZ	Ariz.	1	0	0
4	4	Arkansas	05	AR	Ark.	1	0	0
5	5	California	06	CA	Calif.	3	-1.7	-3
6	6	Colorado	08	CO	Colo.	1	0	0
7	7	Connecticut	09	CT	Conn.	4	0.8	-1.1
8	8	Delaware	10	DE	Del.	3	0.5	-0.3
9	9	District of Columbia|Washi	11	DC	D.C.	70	2	-2.5
10	10	Florida	12	FL	Fla.	4	2	-2
11	11	Georgia	13	GA	Ga.	1	0	0
12	12	Hawaii	15	HI	Hawaii	20	0	0
13	13	Idaho	16	ID	Idaho	1	0	0

Map XY Files

Each xxx-XY.jmp file has four columns. Each row is a coordinate in some shape. Each shape is made of one or more parts. Each part is a closed polygon. The first column is the same Shape ID as in the xxx-Name file. The second column is the Part ID. The next two columns are X and Y.

Figure 16.6 Example of US-State-XY.jmp

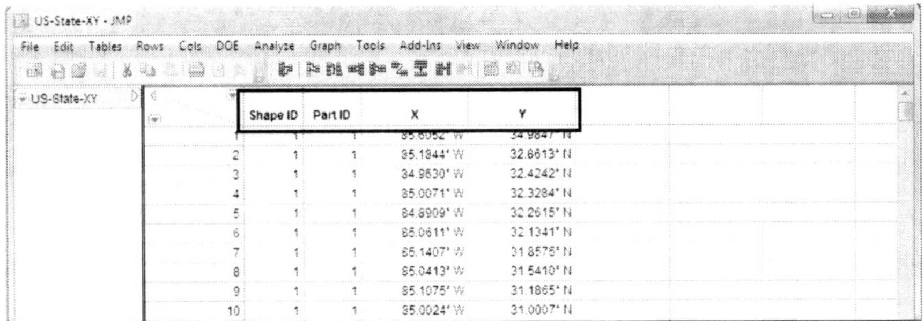

	Shape ID	Part ID	X	Y
1	1	1	85.6052° W	34.9847° N
2	1	1	85.1844° W	32.8613° N
3	1	1	84.9630° W	32.4242° N
4	1	1	85.0071° W	32.3284° N
5	1	1	84.8909° W	32.2615° N
6	1	1	85.0611° W	32.1341° N
7	1	1	85.1407° W	31.8575° N
8	1	1	85.0413° W	31.5410° N
9	1	1	85.1075° W	31.1865° N
10	1	1	85.0024° W	31.0007° N

Color

You can add color to geographic regions on a map to show statistical differences across geography. This type of map is called a choropleth map. Choropleth maps shade or give patterns to areas in proportion to the measurement of the statistical variable being displayed on the map. Such maps let you visualize differences across a geographic area. The Graph Builder platform lets you quickly create this type of map.

Drag a column containing geographic place-names, like countries, regions, states, or provinces, into the **Map Shape** zone and create a map. Then drag a column to the **Color** zone to color the map by that column. The categorical or continuous color theme selected in your Preferences is applied to each shape.

Figure 16.7 Example of SAT.jmp after Dragging 2004 Verbal to Color

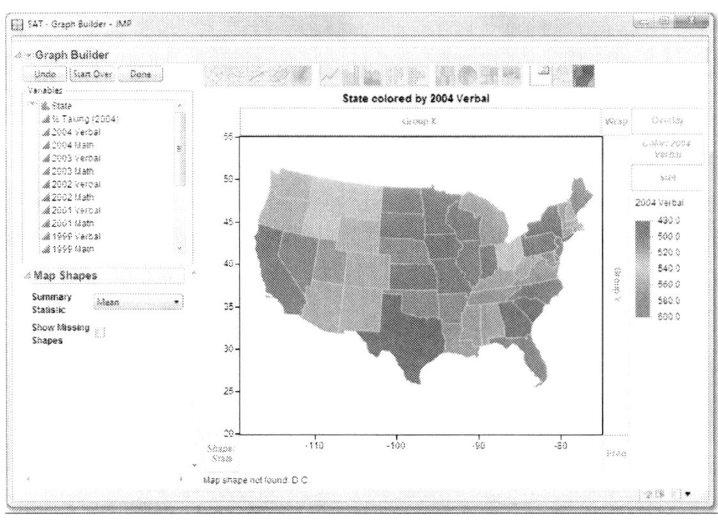

Size

Use the **Size** element to scale map shapes according to the size variable, minimizing distortion.

Figure 16.8 Example of SAT.jmp after Dragging Population to Size

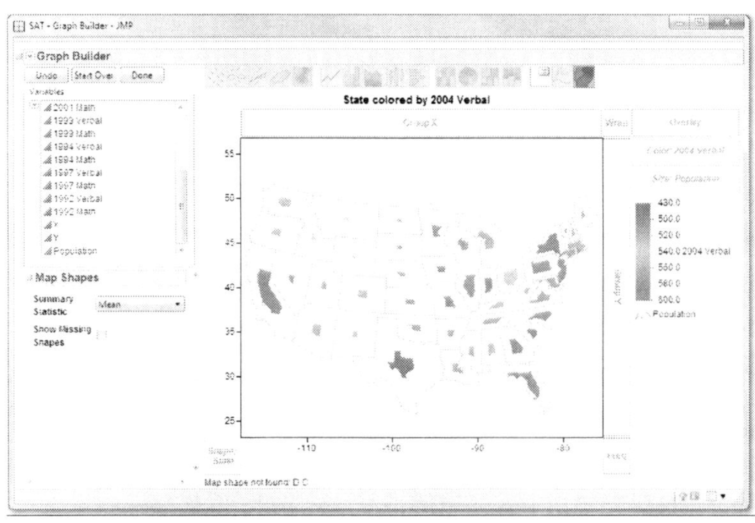

Customizing Graphs

To change colors and transparency for a map, right-click on the color bar in the legend. The right-click options vary, depending on whether the color variable is continuous or categorical (nominal or ordinal). However, for both types of variables, you can change the transparency.

To change the transparency of a graph:

1. Right-click on the color of the variable level on the color bar that you want to change and select **Transparency**.
2. Specify the transparency between 0 (clear) and 1 (opaque).
3. Click **OK**.

You can also change the transparency of images (for example, Simple Earth and Detailed Earth). To set the transparency, right-click over the graph and select **Customize...**. This brings up the Customize Graph window, where you can select the Background Map and assign a value for transparency. A valid value for transparency goes from 0.0 (completely transparent) to 1.0 (completely opaque). Within Graph Builder, you can also right-click over the graph and select Graph > Transparency.

Categorical (nominal or ordinal) variables use a singular coloring system, where each level of the variable is colored differently.

To change the color of one of the variable levels:

1. Right-click on the color of the variable level that you want to change and select **Fill Color**.
2. Select the new color.

Continuous variables use a color gradient.

To change the color theme:

1. Right-click on the color bar and select **Gradient**.
2. In the Gradient Settings window, select a different **Color Theme**.

Graphs consist of markers, lines, text, and other graphical elements that you can customize. If you right-click an image, there are several options for working with the graph. The options differ based on what you clicked. For more information, see the "Interactive Data Visualization" chapter on page 255 and *Using JMP*. Below are a few options.

Figure 16.9 Right-click Menu for Graphics

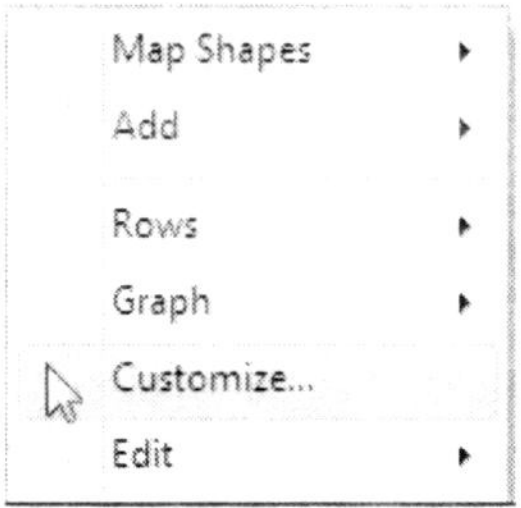

- Map Shapes:
  - Change To: Caption Box - a summary statistic value for the data
  - Summary Statistics - provides options for changing the statistic being plotted
  - Show Missing Shapes - Shows or hides missing data from a map (turned off by default). Missing Shape means that there are some shape names that exist in the map file but not in the data table for analysis.
  - Remove
- Customize - You can change the properties of the graph such as contents, grid lines, or reference lines. The graphical elements that you can customize differ for each graph. Select **Background Map** to change the transparency of a background map or **Map Shape** to change the line color, line style and width, fill color, missing shape fill or missing value fill. Click **Help** in the Customize Graph window for a more detailed explanation of the customize options.

Built-in Map Files

When you use the **Map Shape** zone, JMP searches through its built-in map files for matching names. Built-in map files normally include the following:

- World countries and regions
- States and counties in the United States
- First-level divisions for Canada, China, the United Kingdom, Spain, France, Italy, Japan, and Germany

JMP installs built-in map files in the following directory:

- Windows: C:\Program Files\SAS\JMP\<Version Number>\Maps
- Macintosh: /Library/Application Support/JMP/<Version Number>/Maps

Each map consists of two JMP data tables with a common prefix. Each data table has the same base name with one of these two suffixes:

- -Name file that contains the unique names for the different regions
- -XY file that contains the latitude and longitude coordinates of the boundaries

The two files are implicitly linked by a Shape ID column. For example:

- The World-Name.jmp file contains the exact names and abbreviations for different countries and regions throughout the world
- The World-XY.jmp file contains the latitude and longitude numbers for each country, by Shape ID

Note: If JMP does not recognize the names in your data, check the built-in map files for the spelling that JMP recognizes. For example, United Kingdom is a country name, but not Great Britain.

Custom Map Files

You can create your own map files by following the same pattern as the built-in files. To add your own map files, you need two things: a series of XY coordinates for the vertices of the polygons that describe the shape, and a set of names for each polygon. Data and shape attributes are required to map custom shapes so you can add your own shapes to JMP. There are two common sources for data like this: ESRI® shapefiles and SAS/GRAPH® map data sets.

In order for JMP to automatically find your files, place them in the following directory:

- On Windows: C:\Users\<user name>\AppData\Local\SAS\JMP\Maps
- On Mac: /Users/<user name>/Library/Application Support/JMP/Maps

Or, you can link the map files to your data files explicitly with the **Map Role** column property.

Note the following when creating map files:

- Each set of map files that you create must contain a -Name file and a -XY file.
- The first column in both files must be the ascending, numeric Shape ID variable. The -Name file can contain any other columns. The shapes are built by rows. The XY coordinates have to go around the shape rather than just define the convex hull of the shape.
- For the **Map Role** column property, columns that are marked with the **Shape Name Definition** are searched for shape identification and must contain unique values.
- If you import an ESRI® SHP file, it is opened in the correct format. -Name files commonly have a **.dbf** extension. For more information, see "ESRI® Shapefiles" on page 416.
- SAS/GRAPH® software includes a number of map data sets that can be used with JMP. For more information, see "SAS/GRAPH® Map Data Sets" on page 417.

You might want to create choropleth maps of other non-geographic regions (for example, a floor of an office building). Simply, add the two shape files for your non-geographic space. If you do not have XY coordinates, but you do have a graphic image of the space, you can use the Custom Map Creator add-in for JMP. With this add-in, you can trace the outlines of the space and JMP creates the -XY and -Name files for you. You can download this add-in from the JMP File Exchange page.

Map Role

You can specify the attributes and properties of a column in a data table within the Column Info window in Column Properties. The **Map Role** property is set for a column like other column properties in the Column Info window. See *Using JMP* for more information about Column Properties.

Figure 16.10 The Column Info Window

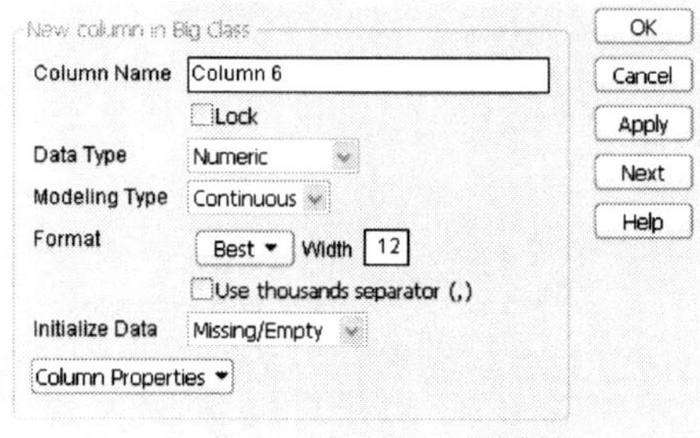

If you have created your own data table that contains boundary data (such as countries, regions, states, provinces, or counties) and you want to see a corresponding map in Graph Builder, use the **Map Role** property within Column Properties. Each pair of map files that you create must contain a -Name file and a -XY file.

Note the following:

- If the custom boundary files reside in the default custom maps directory, then you need to specify only the **Map Role** property in the -Name file.
- If the custom boundary files reside in an alternate location, specify the **Map Role** property in the -Name file and in the data table that you are analyzing.
- The columns that contain the **Map Role** property must contain the same boundary names, but the column names can be different.

To add the Map Role property into the -Name data table:

1. Right-click on the column containing the boundaries and select **Column Properties > Map Role.**
2. Select **Shape Name Definition** below **Map Role.**
3. Click **OK.**
4. Save the data table.

Figure 16.11 Shape Name Definition Example

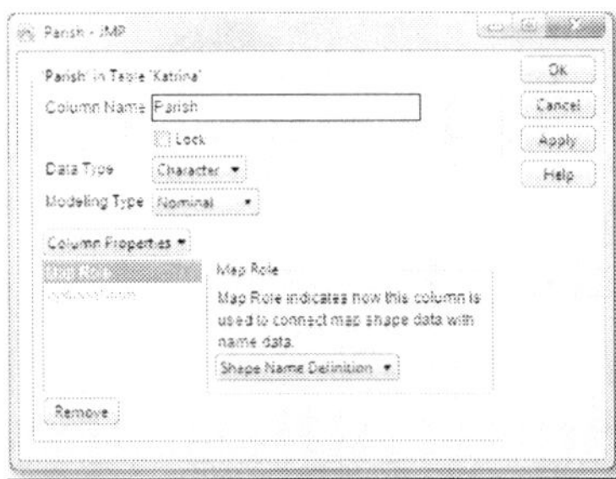

To add the Map Role property into the data table that you are analyzing:

Note: Perform these steps only if your custom boundary files do not reside in the default custom maps directory.

1. Right-click on the column containing the boundaries and select **Column Properties > Map Role.**
2. Select **Shape Name Use** below **Map Role**.
3. Next to **Map name data table**, enter the relative, or absolute path to the **-Name** map data table.

   If the map data table is in the same folder, enter only the filename. Quotes are not required when the path contains spaces.
4. Next to **Shape definition column**, enter the name of the column in the map data table whose values match those in the selected column.
5. Click **OK.**
6. Save the data table.

When you generate a graph in Graph Builder and assign the modified column to the **Map Shape** zone, your boundaries appear on the graph.

Figure 16.12 Shape Name Use Example

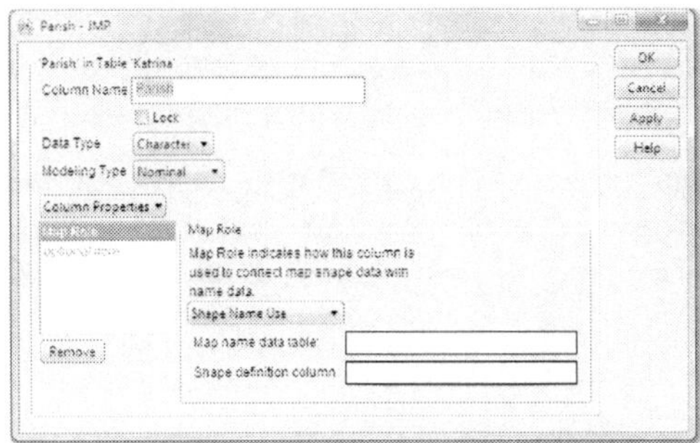

For numeric columns, the Format Menu appears in the Column Info window. Specify the format to tell JMP how to display numbers in the column. Latitude and Longitude for geographic maps are located under Format > Geographic when customizing axes and axes labels.

Geographic Shows latitude and longitude number formatting for geographic maps. Latitude and longitude options include the following:

- DDD (degrees)
- DMM (degrees and minutes)
- DMS (degrees, minutes, and seconds)

In each format, the last field can have a fraction part. You can specify the direction with either a signed degree field or a direction suffix. To show a signed degree field, such as *-59°00'00"*, deselect **Direction Indicator**. To show the direction suffix, such as *59°00'00" S*, select **Direction Indicator**.

To use spaces as field separators, deselect **Field Punctuation**. To use degrees, minutes, and seconds symbols, select **Field Punctuation**.

ESRI® Shapefiles

The ESRI Shapefile (shapefile) is a geospatial vector data format for geographic information systems software. It is developed and regulated by ESRI as a specification among ESRI and other software products. Shapefiles spatially describe geometries (points, polylines, and polygons) that could represent data such as streams, rivers, and lakes. Shapefile is the standard for map data, and each map is actually a set of files with the same name and different extensions. JMP imports two of them:

- a main file (.shp)
- a dBase table (.dbf)

The main file contains sequences of points that make up polygons. When opened with JMP, a .shp file is imported as a JMP table which contains a Shape column to uniquely identify each geographic region, a Part column to indicate discontiguous regions, and the XY coordinates (in latitude and longitude degrees). The Shape column is added during import. You can add formatting and axis settings for the X and Y columns.

The dBase table should have an ID column that maps to the Shape column in the main file, plus any number of columns that provide common names or values to refer to the regions. Add the ID column manually after import.

Opening an .shp or a .dbf file in JMP opens the data as a JMP data table. For an .shp file, each coordinate point is a separate row. JMP only supports two-dimensional .shp files (no elevation information).

Shapefiles store geometrical data types such as points, lines, and polygons. But such data is of limited use without attributes to specify what they represent. Therefore, a table of records stores information for each shape in the shapefile. Shapes, combined with data attributes, create depictions about geographical data that can lead to effective evaluations.

To convert an ESRI shapefile to a JMP map file:

1. Open the .shp file in JMP.

2. Make sure that the Shape column is the first column in the **.shp** file. Add formatting and axis settings for the X and Y columns (optional). Graph Builder uses those settings for the X and Y axes.
3. Save the .shp file as a JMP data table to the Maps folder with a name that ends in -XY.jmp.
4. Open the .dbf file.
5. Add a Shape ID column as the first column in the table. This column should be the row numbers from 1 to n, the number of rows in the data table (Note - You can use Cols >New Column > Initialize Data > Sequence Data).
6. Assign the **Map Role** column property to any column that you use for place names in the Shape role of Graph Builder. To do this, right-click at the top of the column and choose Column Properties > Map Role.
7. Choose **Shape Name Definition** from the drop-down box in the property definition.
8. Save the table as a JMP data table with a name that matches the earlier table and that ends in -Name.jmp.

JMP looks for these files in two locations. One location is shared by all users on a machine. This location is:

- Windows: C:\Program Files\SAS\JMP\<Version Number>\Maps
- Mac: /Library/Application Support/JMP/<Version Number>/Maps

The other location is specific for an individual user:

- On Windows: C:\Users\<user name>\AppData\Local\SAS\JMP\Maps
- On Mac: /Users/<user name>/Library/Application Support/JMP/Maps

SAS/GRAPH® Map Data Sets

SAS/GRAPH software includes a number of map data sets that can be converted for use with JMP. The data sets are in the Maps library. The traditional map data sets contain the XY coordinate data and the feature table contains the common place names. You need to convert both of these files to JMP data tables for use with JMP.

Most of the traditional map data sets have unprojected latitude and longitude variables in radians. The data sets can be used with JMP once they have been converted to degrees and the longitude variable has been adjusted for projection. The following is a DATA step that shows the conversion process for the Belize data set.

```
data WORK.BELIZE;
keep id segment x y;
rename segment=Part;
set maps.belize;
if x NE .;
if y NE .;
y=lat*(180/constant('pi'));
```

```
x=-long*(180/constant('pi'));
run;
```

You can now import the converted file and save it as Belize-XY.jmp.

The next step is to import the matching feature data set (in this case: MAPS.BELIZE2). After importing the feature data set, move the ID column to the first position in the data table. Then assign the Map Role column property to the columns that you use for place names in the Shape role of Graph Builder. To do this, right-click the top of the column and choose **Column Properties** > **Map Role**. Then choose **Shape Name Definition** from the drop-down box in the property definition. For MAPS.BELIZE2, use the IDNAME column. Save the feature data table as Belize-Name.jmp.

To convert SAS maps, download the SAS to JMP Map Converter add-in from the JMP File Exchange. For each map, the add-in reads the data from the two SAS map tables, rearranges and formats the data and then places it into the two JMP map tables.

Background Maps

Adding map images and boundaries to graphs provides visual context to geospatial data. Affixing a background map generates an appealing map, providing your data a geographic context and giving you a whole new way to view your data. For example, you can add a map to a graph that displays an image of the U.S. Another option is displaying the boundaries for each state (when data includes the latitudes and longitudes for the U.S.). There are different types of background maps. Some maps are built into JMP and are delivered as part of the JMP install. Other maps are retrieved from an Internet source, and still other maps are user-defined.

The data should have latitudinal and longitudinal coordinates. Otherwise, the map has no meaning in the context of the data. The x and y axes also have range requirements based on the type of map. These requirements are described in the following sections. Simply plot longitude and latitude on the X and Y axes, and then right-click within the graph and choose **Background Map** from the pop-up menu.

The Background Map window shows two columns of choices: Images and Boundaries. On the left of the window you can choose from two built-in map images, or you can connect to a Web Map Service to retrieve a background image. On the right side of the window, choose to show political boundaries for a number of regions.

Figure 16.13 Example of Background Map Options

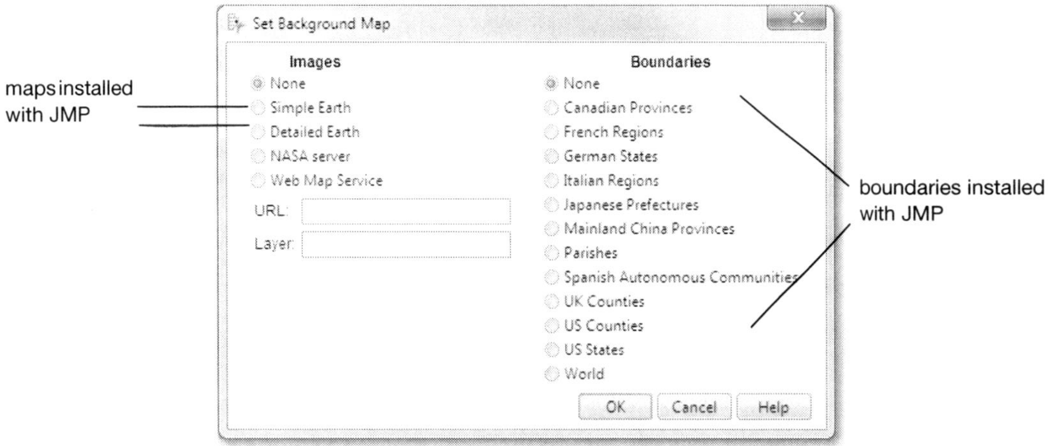

Table 16.1 Descriptions of Background Map Options

Images	
None	Removes the background map that you selected in the Images column.
Simple Earth	Shows a map of basic terrain.
Detailed Earth	Shows a high-resolution map with detailed terrain.
NASA Server	Shows a map from the NASA server. Requires an Internet connection.
Web Map Service	Shows a map from the Uniform Resource Locator (URL) and the layer that you specify. Requires an Internet connection.
Boundaries	
None	Removes the boundaries that you selected in the Boundaries column.
Boundaries for various regions	Shows borders for the map regions, such as Canadian provinces, U.S. counties, U.S. States, and world countries. The list varies based on your location. The maps that you created from ESRI shapefiles are also listed here.

Two tools are especially helpful when you are viewing a map:

- The grabber tool (✋) lets you scroll horizontally and vertically through a map.
- The magnifier tool (🔍) lets you zoom in and out.

Images

Every flat map misrepresents the surface of the Earth in some way. Maps cannot match a globe in truly representing the surface of the entire Earth. A map projection is used to portray all or part of the round Earth on a flat surface. This cannot be done without some distortion. Every projection has its own set of advantages and disadvantages. A map can show one or more, but not all, of the following: true direction, distance, area, or shape. JMP employs a couple of projections (Albers Equal Area Conic and Kavrayskiy VII) for its maps. Within Images, you can choose from two built-in map images, or you can connect to a Web Map Service to retrieve a background image.

Earth Images Installed with JMP

JMP provides two levels of earth imagery; simple and detailed. Both maps show features such as bodies of water and terrain. However, detailed maps show more precise terrain. And with detailed maps, you can zoom in farther, and the map features remain clear. Image maps are raster images. The maps wrap horizontally, so you continue to see map details as you scroll from left to right. The maps do not wrap vertically. Beyond the -90 and 90 y-axis range, a plain background appears instead of the map.

Figure 16.14 Examples of Simple and Detailed Maps

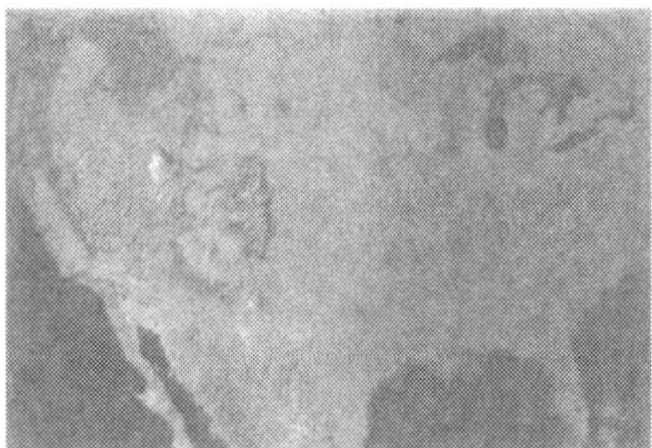

Simple Earth Image

Detailed Earth Image

As its name suggests, Simple Earth is a relatively unadorned image of the earth's geography. It does not show clouds or arctic ice, and it uses a green and brown color scheme for the land and a constant deep blue for water. Detailed Earth has a softer color scheme than Simple Earth, lighter greens and browns for the land, as well as variation in the blue for the water. Detailed Earth also has a slightly higher resolution than Simple Earth. The higher resolution lets you zoom into a graph further with Detailed Earth than with Simple Earth before the quality of the background image begins to blur.

Another feature of Simple Earth and Detailed Earth is the ability to wrap. The Earth is round, and when you cross 180° longitude, the Earth does not end. The longitudinal value continues from -180° and increases. The map wraps continuously in the horizontal direction, much as the Earth does. The background map does not wrap in the vertical direction.

Simple Earth and Detailed Earth both support a geodesic scaling. In Figure 16.14 on page 420, the Earth appears as a rectangle, where the width is twice as wide as the height. If we were to take this rectangle and roll it up, we would have a cylinder. In reality, we know that the Earth does not form a cylinder, but rather a sphere. You can use a geodesic scaling, which transforms the map to a more realistic representation of the Earth. To use the geodesic scaling, change the type of scale on the axes.

To change the axes scale:

1. Right-click over the X or Y axis and then select **Axis Settings...**.
2. Change the Scale to **Geodesic** or **Geodesic US**.

Figure 16.15 Y Axis Setting Window

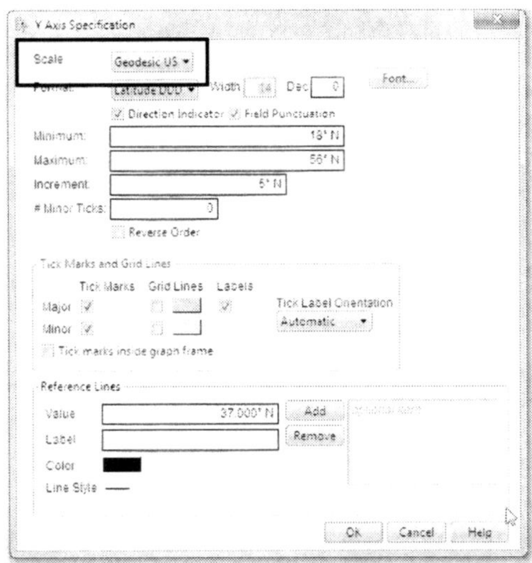

Both choices transform the map to a geodesic scaling. Use Geodesic US if you are viewing a map of the continental US and you want Alaska and Hawaii to be included in the map. It is important to note that you must set the scale to geodesic for both axes to get the transformation. You will not see a change in the map after setting only one of the axes. In the following figure, Simple Earth is used as the background map with the axes set to use a geodesic scale. The axes lines are turned on as well. Notice the longitudinal lines are now curved, instead of straight.

Figure 16.16 Simple Earth Background Map - Axes set to Geodesic Scale

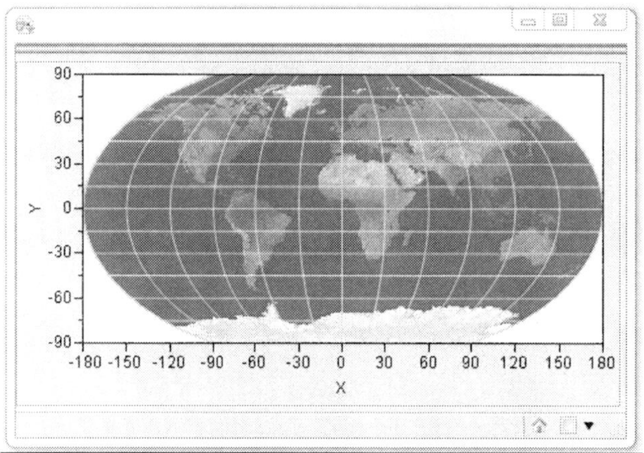

Since Detailed and Simple Earth are built into JMP, these options work any time, without a network connection. However, these images might not be all that you want, or they might not be detailed at the resolution that you need. If this is the case, and if you have an Internet connection, you can connect to a Web Map Service to retrieve a map image that meets your needs.

Maps from the Internet

The National Aeronautics and Space Administration (NASA) and other organizations provide map image data using a protocol called Web Map Service (WMS). These maps have the advantage of showing the most up-to-date geographical information. However, the display of the maps can be slow depending on the response time of the server, and the sites can change or disappear at any time. An Internet connection is required to access the information.

Figure 16.17 Examples of NASA and WMS maps

NASA image

WMS image from the MetaCarta™ Server

The NASA server provides maps for the entire Earth. The following figure displays the Earth using the NASA server as its source for the background map. It also has a boundary map turned on, to show the outlines of the countries.

Figure 16.18 NASA Server Map Example

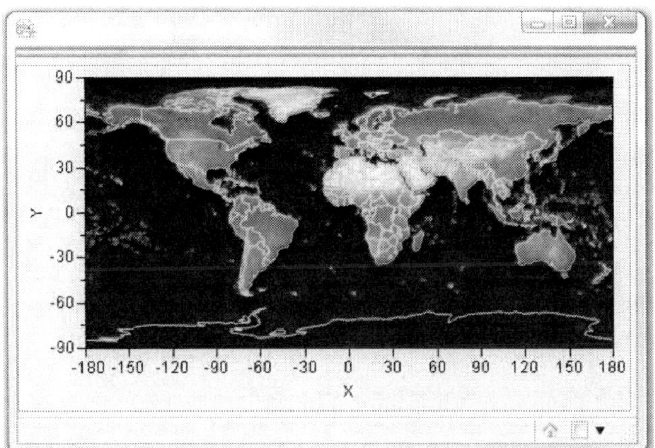

Not only does this server cover the entire Earth, but you can also zoom in on a much smaller area of the Earth and still get a reasonable map. The following figure displays the Colorado River running through the Grand Canyon in Arizona. The Grand Canyon Village is visible in the bottom of the map.

Figure 16.19 NASA Server Map Example - Zoom in on Colorado

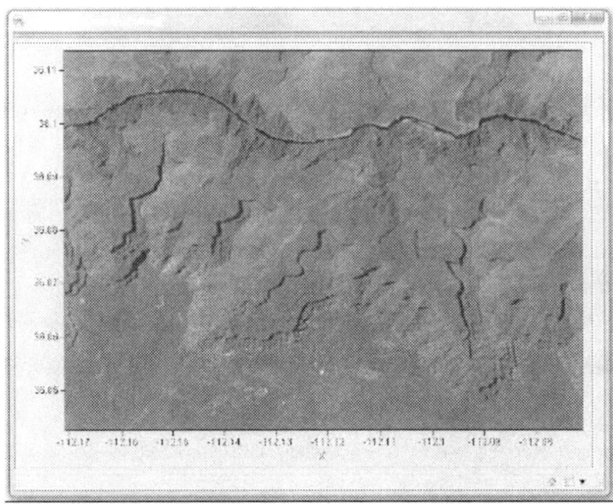

If you look at the axes values, you can see that the area is less than 1/10° by 1/10°. The Simple Earth and Detailed Earth background maps do not display that type of resolution. The NASA server provides a fairly detailed view of any land mass on Earth. Water, however, is simply filled in as black. The NASA server is free to access, but it is also limited in availability. If the server is temporarily unavailable or becomes overloaded with requests, it delivers an error message instead of the requested map.

Another Internet-based option for background maps is a Web Map Service (WMS). The WMS option enables you to specify any server that supports the WMS interface. The NASA server is an example of a WMS server, but we have provided the URL and a layer name for you. With the WMS option, you must know the URL to the WMS server and a layer name supported by the server. Most WMS servers support multiple layers. For example, one layer can show terrain, another layer can show roads, and still another layer can include water, such as rivers and lakes. By specifying the URL for the server and the layer, JMP can make a request to the server and then display the map that is returned.

Unlike with simple and detailed maps, WMS maps do not wrap. You can scroll horizontally and vertically. However, beyond the -180 to 180 (x axis) and -90 to 90 (y axis) ranges, a plain background appears instead of the map. The limits of the axes are used to define the limits of the map that is displayed.

In order to use the WMS option for a background map, you need to decide which WMS server to use. There are many WMS servers freely available from the Internet. Most of them provide maps only for a particular area of the world, and each of them supports their own layers. So you have to search for the appropriate WMS server for your particular situation.

You can search for WMS servers on the Internet using your favorite search engine. Once you find one, you need to discover the layers that it supports. For this, you can use the WMS Explorer add-in. The WMS Explorer add-in generates a list of all the layers available on a server. You can select a layer from the list to see what it looks like. You can download the WMS Explorer add-in from the JMP File Exchange page.

Note: To use the WMS Explorer add-in and the WMS background map capabilities of JMP, your computer must be connected to the Internet.

To locate a server, launch the add-in through the menu items Add-Ins > Map Images > WMS Explorer. The add-in presents a text box for entering the url of a known WMS server. Alternatively, you can make a selection from a drop-down list of pre-discovered WMS servers (the list can be out of date). After specifying a WMS server, select **Get Layers**. Using Get Layers is not necessary if selecting from the drop-down list or if clicking **Enter** after entering a URL. This sends a request to the WMS server for a list of layers that the server supports. The returned list appears in the list box on the left, labeled **Layers.** A map of the world appears as an outline in the graph to the right. Selecting a layer makes a request to the WMS server to return a map, using the specified layer, that represents the entire earth. Selecting a different layer generates a different map.

It is important to note that not all maps are generated to cover the entire earth (for example, some WMS servers might provide mapping data for a particular county, within a state). In that case, it is likely that selecting a layer will not generate any visible map. You might have to zoom in on the appropriate area before any image map is visible. The standard JMP toolbar is available in the add-in window and the zoom tool works just like it does in any JMP window.

The graph is a typical graph in JMP, which means that all the regular JMP controls are available to you. You can adjust the axes or use the zoom tool (found on the hidden menu bar) just as you would in JMP. You can also right-mouse-click to select Size/Scale > Size to Isometric to get your graph back into a proper aspect ratio. You can also select Background Map, where you can adjust the boundary map.

Once a desirable map is determined, note the URL in the text box at the top and the selected layer in the Layers list. This is the information that you need to enter in the background map window when WMS is selected as the type of image background map.

Because requests are being made to a server across the Internet, there are a number of conditions that can generate an error. WMS servers often have limited availability and sometimes are not available at all. Occasionally a WMS server might return a name of a layer that it no longer supports. In these types of cases (and others), a server usually returns an error message in lieu of a map. If that happens, the error message is displayed below the Layers list in an area labeled **Errors**.

Boundaries

JMP can display boundaries (such as U.S. states or French region boundaries). These boundaries draw an outline around a defined area and can be displayed alone on a graph or combined with image data. Several boundaries are installed with JMP. Alternatively, you can create your own boundaries from ESRI shapefiles or from scratch. Because of this, the list of Boundaries that you see in the Set Background Map window can be different.

When you add shape files to the built-in locations in JMP, they are available not only for the Graph Builder platform, but also for the Boundaries option in the Background Map window. In this way, you can add more political boundaries for use with background maps. Boundary-style maps are vector-based shapes.

Figure 16.20 Example of U.S. State Boundaries

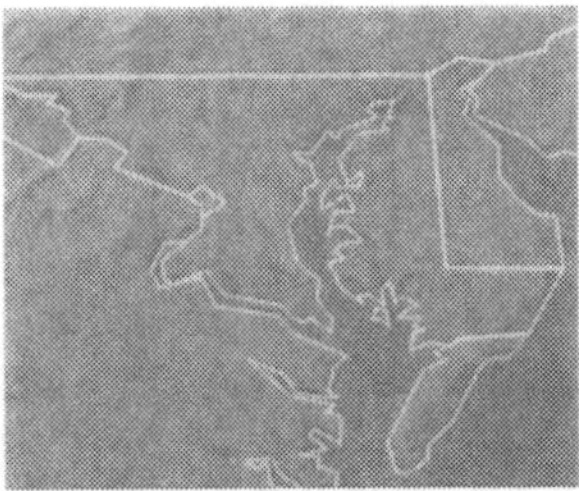

Add a Background Map and Boundaries

To add a background map and boundaries:

1. Right-click a blank area on the graph and select **Background Map** (or select **Graph > Background Map** in Graph Builder).

   The Set Background Map window appears.

Figure 16.21 Set Background Map Window

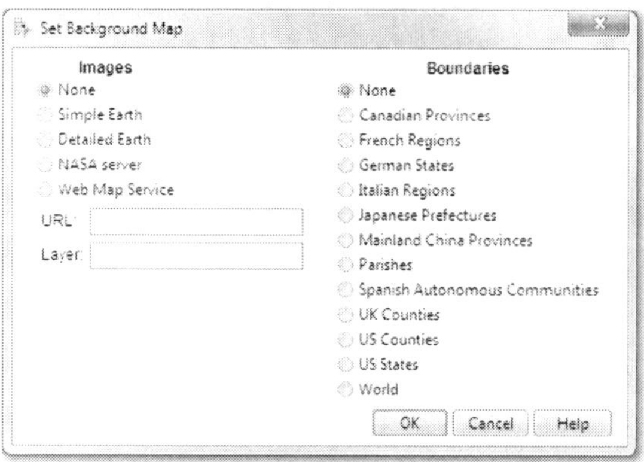

2. To display a background, do one of the following:
   - Select **Simple Earth**, **Detailed Earth** or **NASA server**, in the **Images** column.
   - Select **Web Map Service** and paste a WMS URL next to **URL**. Type the layer identifier next to **Layer**.
3. To display geographic borders on the map, select an option in the **Boundaries** column (If you installed your own boundary shapefiles, they are also listed in this column).
4. Click **OK**.

   If the NASA or WMS map does not appear after you add it, the map server might not be available. View the error log to verify the problem.

Custom Map Images

There are two ways to use your own map images:

- Assign the Map Role column property in a data table and specify the path to the map. For details, see "Map Role" on page 414.
- You can create a map image in a graphics program and drag and drop the image onto a report.

After you add a map image, you can change certain display properties, such as adding state boundaries and setting the transparency level.

JSL Scripting

Background maps can be scripted in JSL. You can write a script that creates a graph, and then turns on the background map in the script. For more information about scripting, see the *Scripting Guide*.

To access the background map functionality through JSL, find an example in the Samples Data directory. For example, run JMP and open the sample data file called **Hurricanes.jmp**. Edit the script called **Bubble Plot with Map**. Look for the command in the script that begins with Background Map.

To figure out the correct parameters for the background map command, simply look at the background map window. The wording is the same as the JSL commands. To create a background map, simply use the command `Background Map()`.

There are two types of background maps that you can specify: `Images()` and `Boundaries()`. Each of these then takes a parameter, which is the name of the map to use. The name is one of the maps listed in the window. For `Images()`, the choices are Simple Earth, Detailed Earth, NASA or Web Map Service. If you elect to use Web Map Service, then there are two additional parameters: the WMS URL and the layer supported by the WMS server. For `Boundaries()`, the choices vary since boundaries can be user-defined. But a typical choice might be World.

To add a background map that uses the Detailed Earth imagery and World as a boundary, the command would look like this (spacing is not important):

```
Background Map ( Images ( "Detailed Earth" ), Boundaries ( "World" ) )
```

To change the script to use a WMS server, the command would look like this:

```
Background Map ( Images ( "Web Map Service", "http://
land.visualiser.nt.gov.au/wms/wms", "NTLIS_Mapserver" ), Boundaries (
"World" ) )
```

You can also create your graph and add a background map through the user interface. Then from the red triangle menu use **Script > Save Script to Script Window** to see the script that is generated.

Figure 16.22 JSL Scripting Example

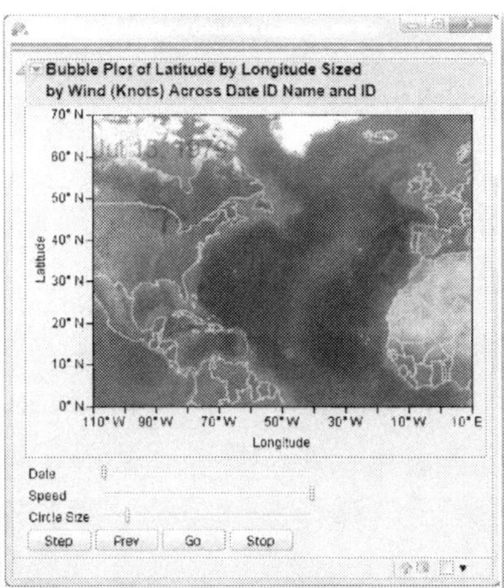

Examples of Creating Maps

The following are examples of creating and using maps to find and display patterns in your geographical data.

Louisiana Parishes Example

In this example you work with custom map files and then create custom maps in two different ways:

- Set up custom map files initially and save them in the predetermined location. JMP finds and uses them in the future with any appropriate data.
- Point to specific predefined map files directly from your data. This step might be required each time you want to specify custom maps.

Set Up Automatic Custom Maps

Suppose that you have downloaded ESRI shapefiles from the Internet and you want to use them as your map files in JMP. The shapefiles are named Parishes.shp and Parishes.dbf. These files contain coordinates and information about the parishes (or counties) of Louisiana.

Save the .shp file with the appropriate name and in the correct directory, as follows:

1. In JMP, open the Parishes.shp file from the following default location:
   - On Windows: C:\Program Files\SAS\JMP\<Version Number>\Samples\Import Data
   - On Mac: /Library/Application Support/JMP/<Version Number>/Samples/Import Data

Note: If you cannot see the file, you might need to change the file type to **All Files**.

   JMP opens the file as Parishes. The .shp file contains the *x* and *y* coordinates.
2. Save the Parishes file with the following name and extension: Parishes-XY.jmp. Save the file here:
   - On Windows: C:\Users\<user name>\AppData\Local\SAS\JMP\Maps
   - On Mac: /Users/<user name>/Library/Application Support/JMP/Maps

Note: If the Maps folder does not already exist, you need to create it.

3. Close the Parishes-XY.jmp file.

Perform the initial setup and save the .dbf file, as follows:

1. Open the Parishes.dbf file from the following default location:
   - On Windows: C:\Program Files\SAS\JMP\<Version Number>\Samples\Import Data
   - On Mac: /Library/Application Support/JMP/<Version Number>/Samples/Import Data

Note: If you cannot see the file, you might need to change the file type to **All Files**.

   JMP opens the file as Parishes. The .dbf file contains identifying information.

2. In the Parishes file, add a new column. Name it Shape ID. Drag and drop it to be the first column.
3. In the first three rows of the Shape ID column, type 1, 2, and 3 (Note - You can also use Cols >New Column > Initialize Data > Sequence Data).
4. Select all three cells, right-click, and select **Fill > Continue sequence to end of table**.

Figure 16.23 Shape ID Column in Parishes File

	Shape ID	PARISH	FIPS	PARISH_FIP	Parish_Num
1	1	Webster	22119.00000000000	119	60
2	2	East Carroll	22035.00000000000	035	18
3	3	West Carroll	22123.00000000000	123	62
4	4	Morehouse	22067.00000000000	067	34
5	5	Union	22111.00000000000	111	56
6	6	Lincoln	22061.00000000000	061	31
7	7	Ouachita	22073.00000000000	073	37
8	8	Richland	22083.00000000000	083	42
9	9	Bienville	22013.00000000000	013	07
10	10	Franklin	22041.00000000000	041	21

5. Right-click on the PARISH column and select **Column Info.**
6. Select **Column Properties > Map Role.**
7. Select **Shape Name Definition.**
8. Click **OK.**
9. Save the Parishes file with the following name and extension: Parishes-Name.jmp. Save the file here:
   - On Windows: C:\Users\<user name>\AppData\Local\SAS\JMP\Maps
   - On Mac: /Users/<user name>/Library/Application Support/JMP/Maps
10. Close the Parishes-Name.jmp file.

Once the map files have been set up, you can use them. The Katrina.jmp data table contains data on Hurricane Katrina's impact by parish. You want to visually see how the population of the parishes changed after Hurricane Katrina. Proceed as follows:

11. Open the Katrina.jmp sample data table.
12. Select **Graph > Graph Builder.**
13. Drag and drop Parish into the **Map Shape** zone.

    The map appears automatically, since you defined the Parish column using the custom map files.
14. Drag and drop Population into the **Color** zone.
15. Drag and drop Date into the **Group X** zone.

Figure 16.24 Population of Parishes Before and After Katrina

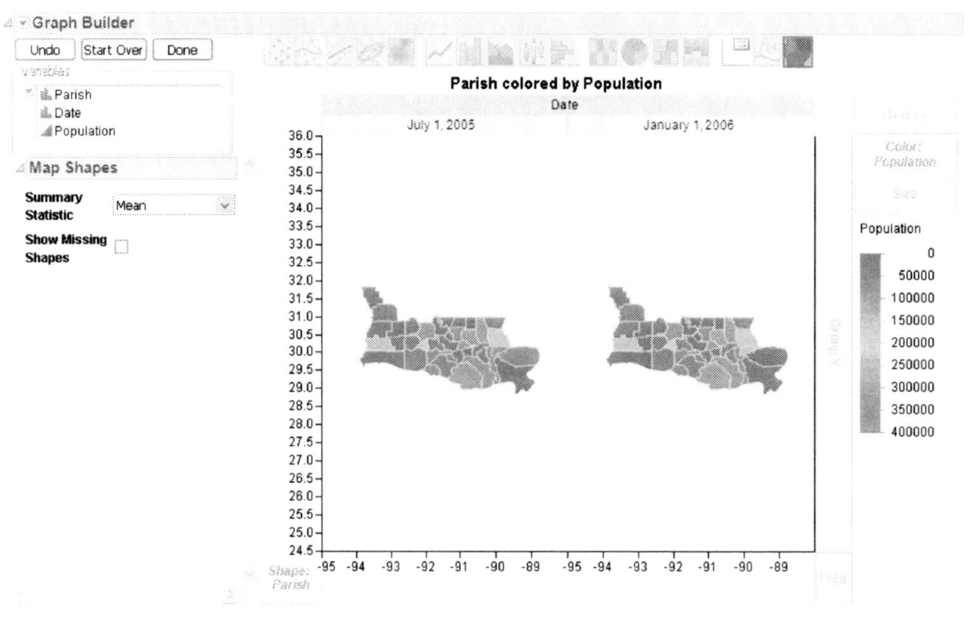

16. Select the **Magnifier** tool to zoom in on the Orleans parish.

Figure 16.25 Orleans Parish

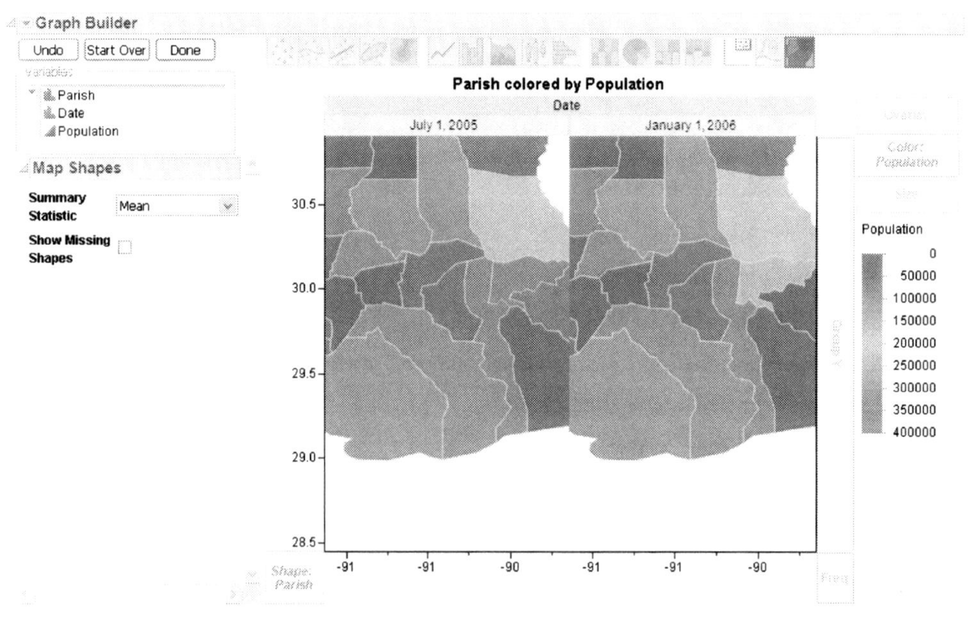

You can clearly see the drop in population as a result of Hurricane Katrina. The population of the Orleans parish went from 437,186 in July 2005 to 158,353 in January 2006.

Point to Specific Map Files Directly from Your Data

As an example of pointing to specific map files directly from your data, say that you already have your custom map files and they are named appropriately. Your map files are US-MSA-Name.jmp and US-MSA-XY.jmp. They are saved in the Sample Data folder.

The PopulationByMSA.jmp data table contains population data from the years 2000 and 2010 for the metropolitan statistical areas (MSAs) of the United States. Set up the **Map Role** column property in the data table, as follows:

1. Open the PopulationByMSA.jmp sample data table.
2. Right-click on the **Metropolitan Statistical Area** column and select **Column Info**.
3. Select **Column Properties > Map Role**.
4. Select **Shape Name Use**.
5. Next to the **Map name data table**, enter the following path:
   - On Windows: C:\Program Files\SAS\JMP\<Version Number>\Samples\Data\US-MSA-Name.jmp
   - On Mac: /Library/Application Support/JMP/<Version Number>/Samples/Data/US-MSA-Name.jmp

   This tells JMP where the data tables containing the map information reside.
6. Next to the **Shape definition column**, type MSA_Name.

   MSA_Name is the specific column within the US-MSA-Name.jmp data table that contains the unique names for each metropolitan statistical area. Notice that the MSA_Name column has the **Shape Name Definition Map Role** property assigned, as part of correctly defining the map files.

Note: Remember, the Shape ID column in the -Name data table maps to the Shape ID column in the -XY data table. This means that indicating where the -Name data table resides links it to the -XY data table, so that JMP has everything that it needs to create the map.

Figure 16.26 Map Role Column Property

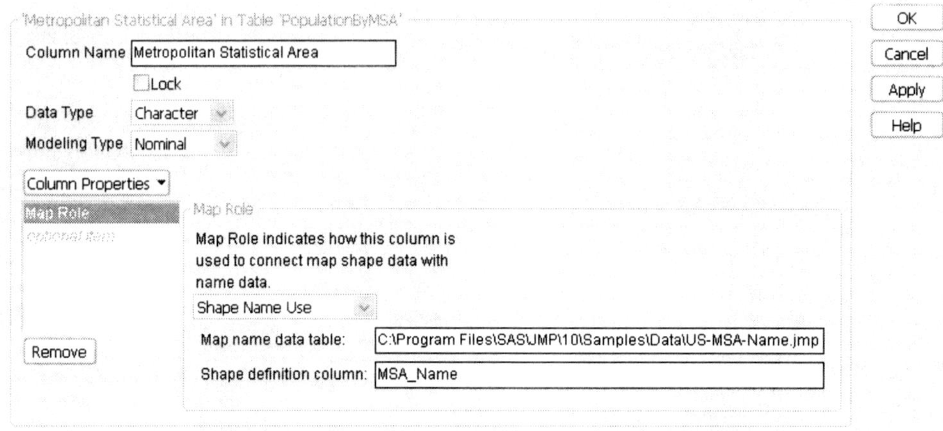

7. Click **OK**.

Once the **Map Role** column property has been set up, you can perform your analysis. You want to visually see how the population has changed in the metropolitan statistical areas of the United States between the years 2000 and 2010.

1. Select **Graph > Graph Builder.**
2. Drag and drop **Metropolitan Statistical Area** into the **Map Shape** zone.

   Since you have defined the **Map Role** column property on this column, the map appears.
3. Drag and drop **Change in Population** to the **Color** zone.

Figure 16.27 Change in Population for Metropolitan Statistical Areas

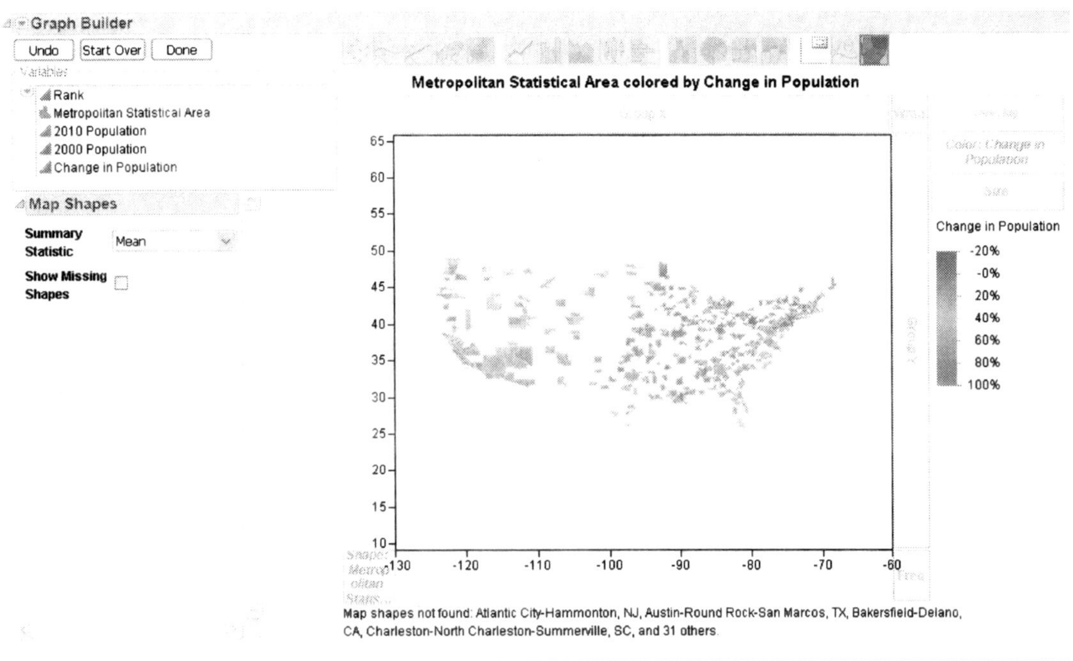

4. Select the **Magnifier** tool to zoom in on the state of Florida.
5. Select the **Arrow** tool and click on the red area.

Figure 16.28 Population Change of Palm Coast, Florida

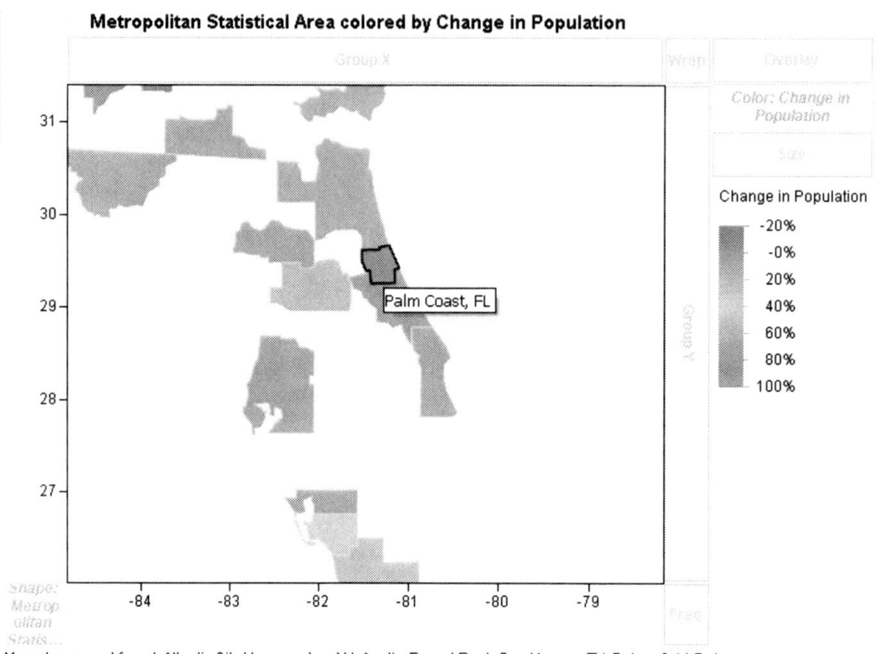

6. Select the **Magnifier** tool and hold down the ALT key while clicking on the map to zoom out.
7. Select the **Magnifier** tool and zoom in on the state of Utah.
8. Select the **Arrow** tool and click on the area that is slightly red.

Figure 16.29 Population Change of St. George, Utah

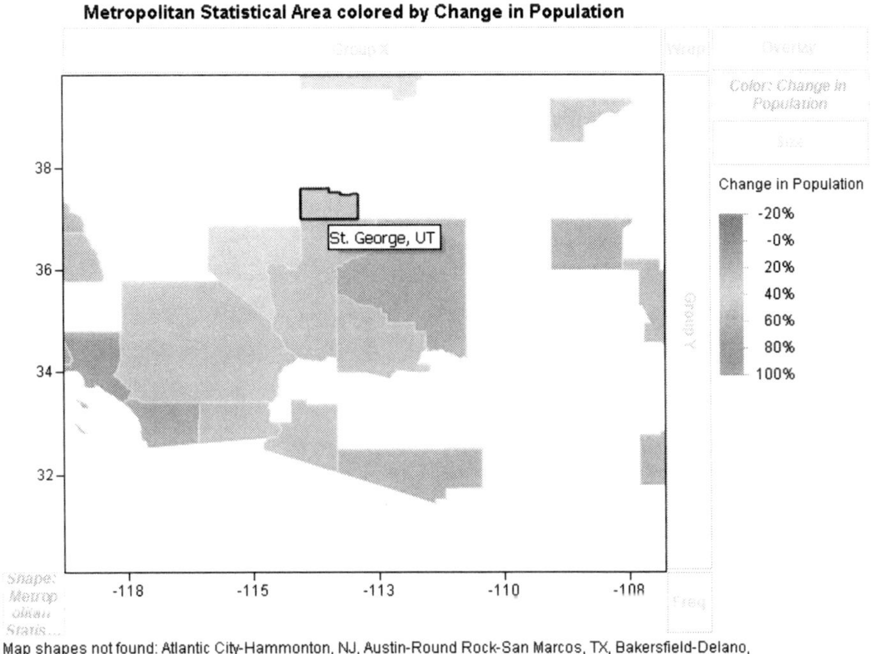

You can see that the areas of Palm Coast, Florida, and St. George, Utah had the most population change between 2000 and 2010. The Palm Coast area saw a population change of 92%, and the St. George area saw a population change of about 53%.

Hurricane Tracking Examples

This example uses the Hurricanes.jmp sample data table, which contains data on hurricanes that have affected the east coast of the United States. Adding a background map helps you see the areas the hurricanes affected. A script has been developed for this example and is part of the data table.

1. Open the Hurricanes.jmp sample data table (C:\Program Files\SAS\JMP\<Version Number>\Samples\Data).
2. Run the Bubble Plot script (**Bubble Plot > Run Script**). The following image appears. The red dot shows the location of Hurricane Claudette on July 15, 1979. Click on the red dot to display the name of the hurricane. The date appears in the upper left corner of the window.

Figure 16.30 Bubble Plot of Hurricanes.jmp

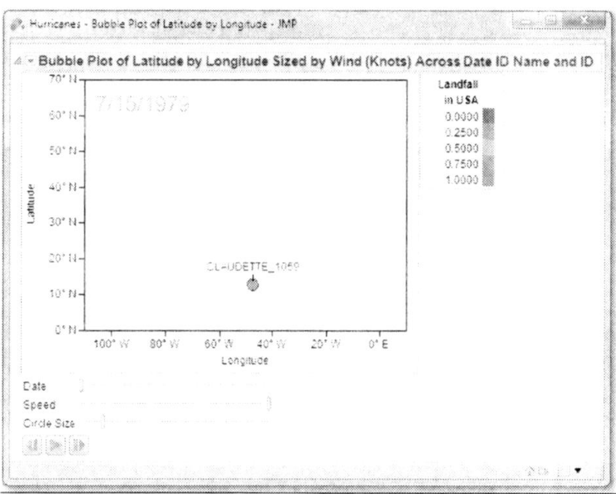

Note that even though the location of the hurricane is plotted, it does not really tell us where it is. The axes information is there (10° North latitude, 50° West longitude) but we need a little more context. It is most likely over the middle of the Atlantic, but is it over a small island? This could make a big difference, especially for the inhabitants of the small island. Obviously, a map in the background of our graph would add a good deal of information.

3. Right-click the map and select **Background Map**. The Set Background Map window appears.

Figure 16.31 Set Background Map Window

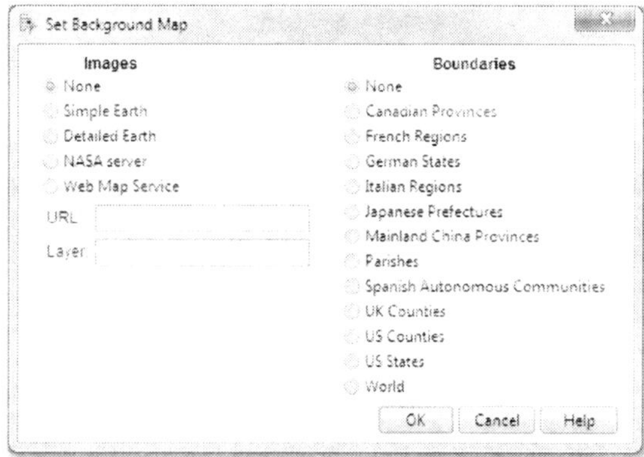

4. Select **Simple Earth** and click **OK**.

Figure 16.32 Bubble Plot of Hurricanes.jmp with Background Map

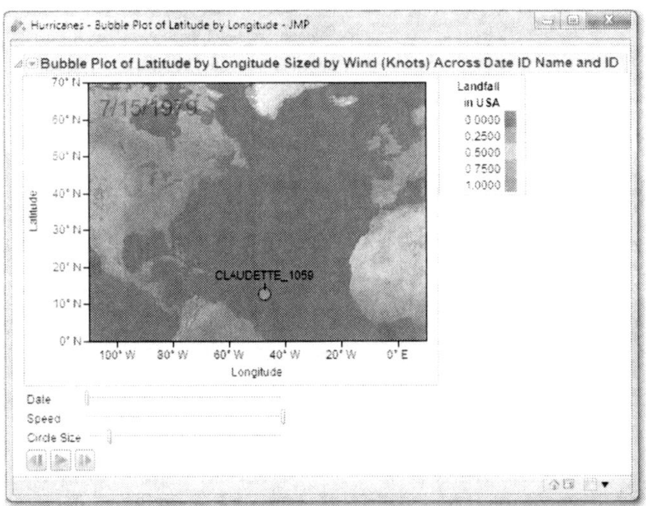

Now the coordinates make geographic sense. Click Run to view the animation of the hurricane data moving over the background map. Experiment with different options and view the displays. Adjust the axes or use the zoom tool to change what part of the world you are viewing. The map adjusts as the view does. You can also right-click the graph and select Size/Scale->Size to Isometric to get the aspect ratio of your graph to be proportional again.

The next example uses the Katrina Data.jmp sample data table, which contains data on hurricane Katrina such as latitude, longitude, date, wind speed, pressure, and status. Adding a background map helps you see the path the hurricane took and impact that it had on land based on size and strength. A script has been developed for this example and is part of the data table.

1. Open the Katrina Data.jmp sample data table (C:\Program Files\SAS\JMP\<Version Number>\Samples\Data).
2. Select **Graph** > **Bubble Plot.**
3. Drag and drop **LAT** onto the **Y** axis.
4. Drag and drop **LON** onto the **X** axis.
5. Drag and drop **Date** onto **Time.**
6. Drag and drop **WIND** onto **Sizes.**
7. Drag and drop **Stat** onto **Coloring.**

Figure 16.33 Bubble Plot Setup of Katrina Data.jmp

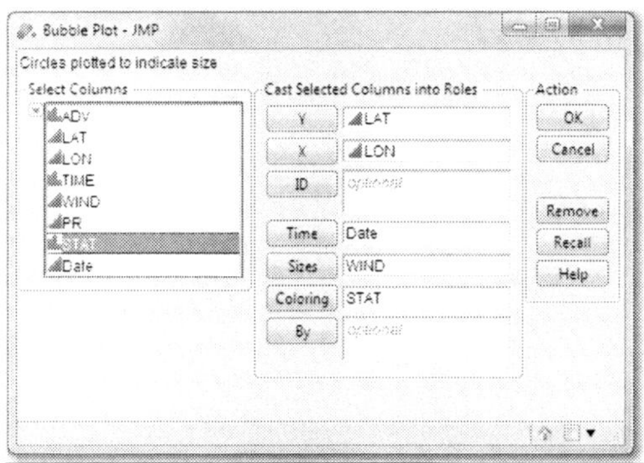

8. Click **OK**.
9. The following image appears. The yellow dot shows the location of Tropical Depression Katrina on August 23, 2005.

Figure 16.34 Bubble Plot of Katrina Data.jmp

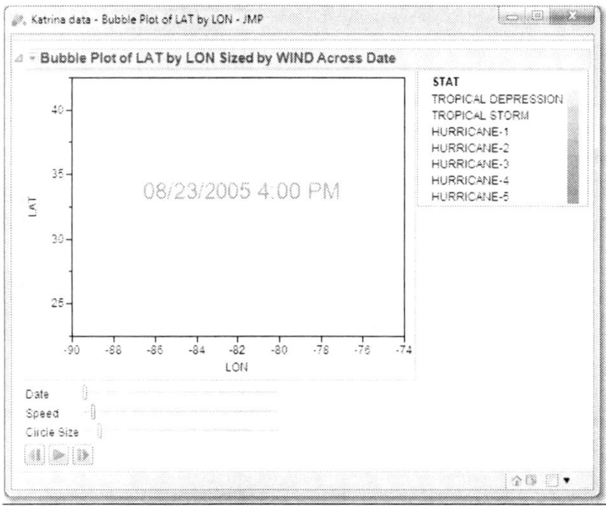

Note that even though the location of the storm is plotted, it does not really tell us where it is. To add more context, add a map in the background.

10. Right-click the graph and select **Background Map**. The Set Background Map window appears.
11. Select **Detailed Earth** and click **OK**.

Figure 16.35 Bubble Plot of Katrina Data.jmp with Background Map

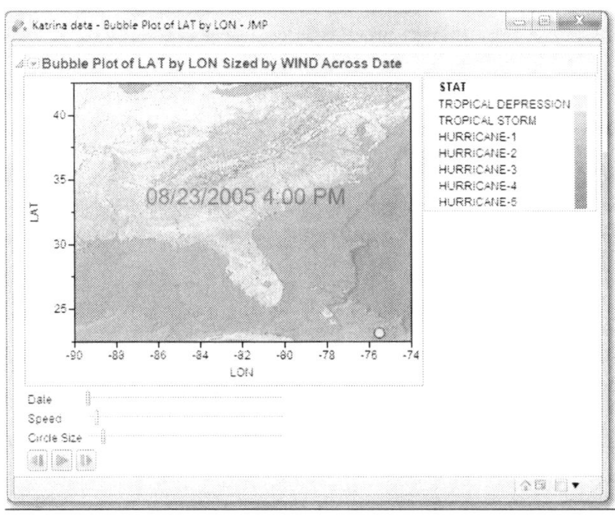

Now the coordinates make geographic sense. You can edit the axes and the size/scale to change the way the graph appears.

12. Right-click the X axis (LON) and select **Axis Settings**. The X Axis Specification window appears.
13. Select **Scale** > **Geodesic US**.
14. Select **Format** > **Geographic** > **Longitude DMM**.
15. Click **OK**.
16. Repeat the same for the Y axis (LAT) except select **Format** > **Geographic** > **Latitude DMM**.

Figure 16.36 X and Y Axes Specification Windows

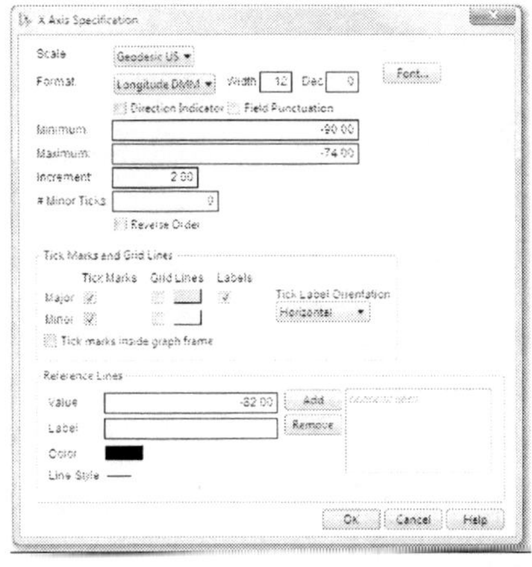

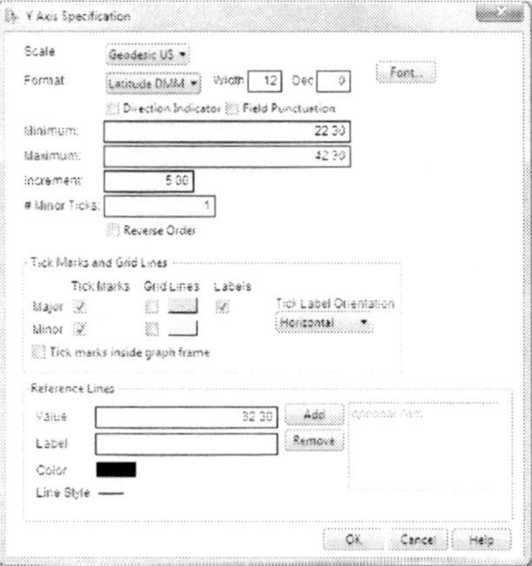

17. Right-click the map and select **Size/Scale** > **Size to Isometric.**

Figure 16.37 Bubble Plot of Katrina Data.jmp with Background Map

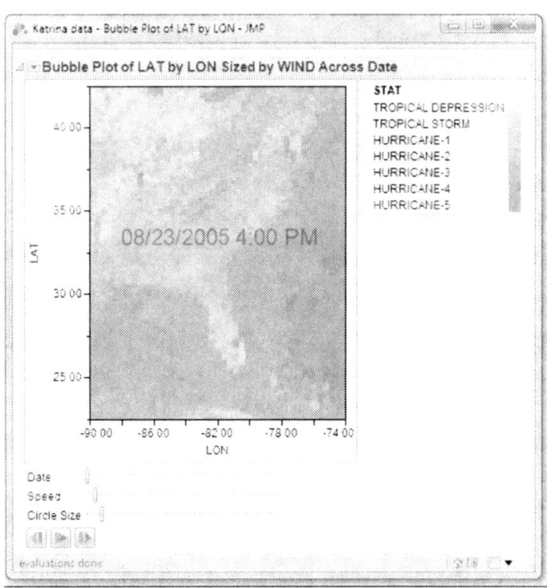

Click Run to view the animation of the hurricane data moving over the background map. You can manipulate the speed and the bubble size. Experiment with different options and view the displays. Adjust the axes or use the zoom tool to change what part of the area you are viewing. Add boundaries to the states. The map adjusts as the view does.

Office Temperature Study

This example demonstrates the creation of a custom background map for an office temperature study and how JMP was used to visualize the results. Data was collected concerning office temperatures for a floor within a building. A map was created for the floor using the Custom Map Creator add-in from the JMP File Exchange page. Using Graph Builder, the office temperature results were then analyzed visually.

The map shown below is the floor, grouped by *time of day*, with color reflecting the *Fahrenheit* value. Exploring data visually in this way can give hints as to what factors are affecting office temperature. Looking at this map, it appears the offices on the east side of the building are warmer in the mornings than they are in the afternoons. On the western side of the building, the opposite appears to be true. From this visualization, we might expect that both of these variables are affecting office temperatures, or perhaps that the interaction between these terms is significant. Such visuals help guide decision-making during the analysis.

Figure 16.38 Room/Office Colored by Fahrenheit and Time of Day

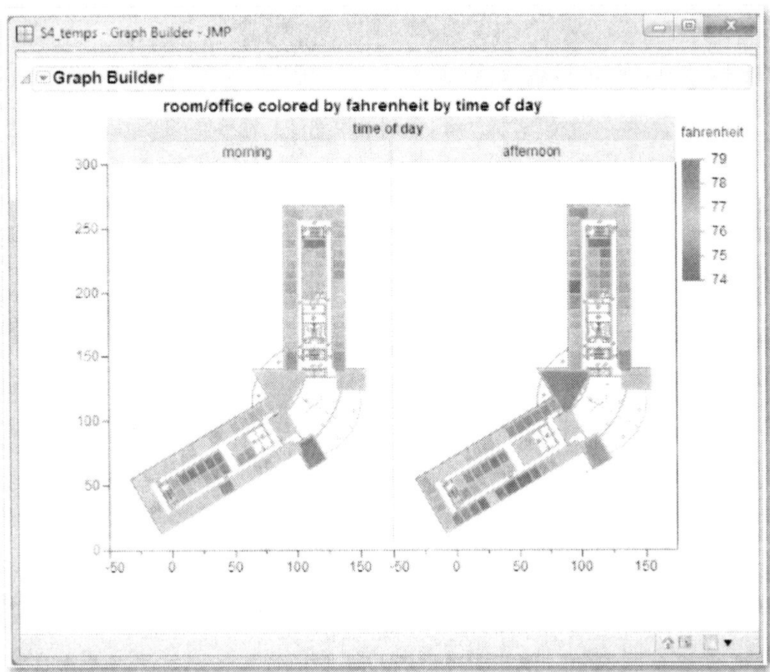

First, data was collected and input into a data table (**S4 Temps.jmp**). Note the **Room/Office** column. It contains the unique names for each office and was assigned the Map Role to correctly define the map files.

Figure 16.39 S4 Temps.jmp Data Table

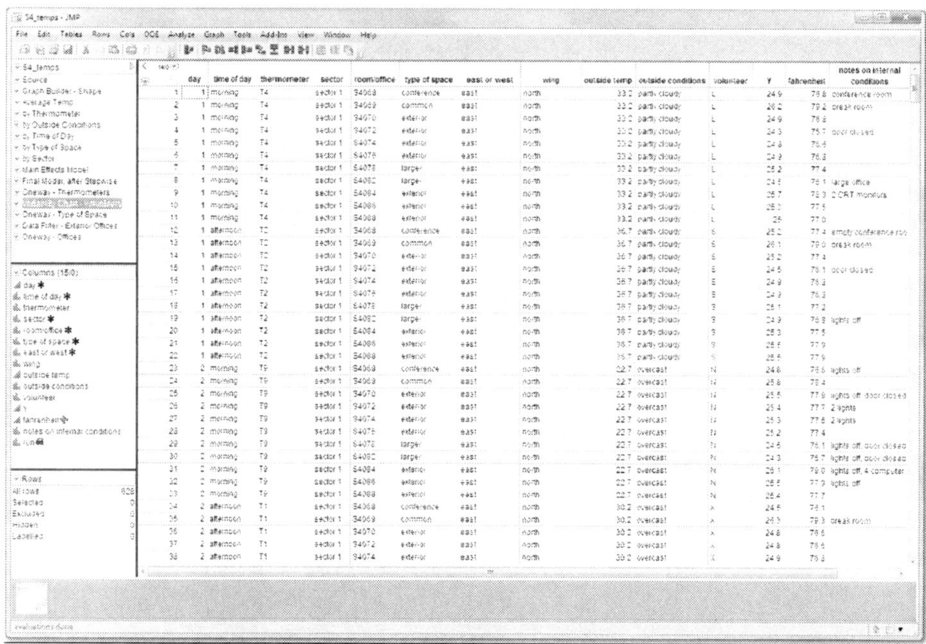

Then, a map of the floor was created using the Custom Map Creator add-in. The add-in creates two tables to define the shapes; an XY table and a Name table. The instructions below describe how it was built.

To create a map of the floor:

1. Launch the add-in through the menu items Add-Ins > Map Shapes > Custom Map Creator. Two tables open in the background followed by the Custom Map Creator Window.
2. Drag a background image into the graph frame. An image of the floor plan was available.
3. Perform any resizing on the background image and graph the frame.
4. Name the table (for example, S4).
5. Click **Next**.
6. Name the shape that you are about to define. For this example, each office was individually named for the map (for example, S4001).
7. Within the graph frame, use your mouse to click all of the boundaries of the shape that you want to define. A line appears that connects all of the boundary points.
8. As soon as you finish defining the boundaries of the shape, click **Next Shape**. Continue adding shapes until you have completed the floor plan. Note that you do not need to connect the final boundary point; the add-in automatically does that for you when you click **Next Shape**.
9. The line size and color can be changed. In addition, checking **Fill Shapes** fills each shape with a random color.

10. Click **Finish**.

The custom map files were created and named appropriately. The map files are S4-Name.jmp and S4-XY.jmp and have been saved in the Samples Data folder.

The S4 Temps.jmp data table contains office data over a three-day period. Set up the **Map Role** column property in the data table, as follows:

1. Open the S4 Temps.jmp sample data table.
2. Right-click on the **Room/Office** column and select **Column Info**.
3. Select **Column Properties > Map Role**.
4. Select **Shape Name Use**.
5. Next to **Map name data table**, enter the following path:
   - On Windows: C:\Program Files\SAS\JMP\<Version Number>\Samples\Data\S4-Name.jmp
   - On Mac: /Library/Application Support/JMP/<Version Number>/Samples/Data/S4-Name.jmp

   This tells JMP where the data tables containing the map information reside.
6. Next to the **Shape definition column**, type room.

   Room is the specific column within the S4-Name.jmp data table that contains the unique names for each office. Notice that the room column has the **Shape Name Definition Map Role** property assigned, as part of correctly defining the map files.

Note: Remember, the Shape ID column in the -Name data table maps to the Shape ID column in the -XY data table. This means that indicating where the -Name data table resides links it to the -XY data table, so that JMP has everything that it needs to create the map.

Figure 16.40 Map Role Column Property

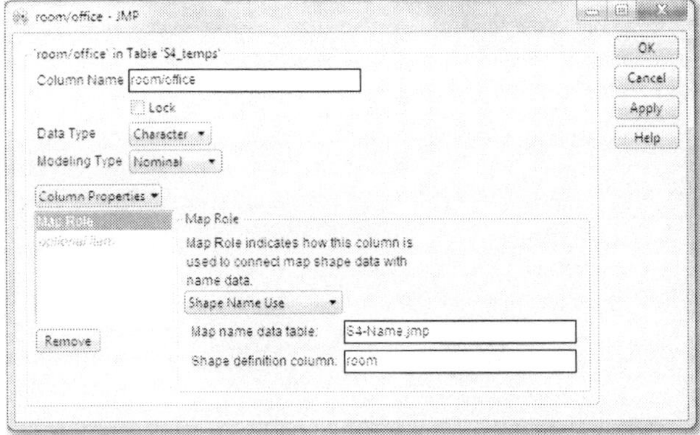

7. Click **OK**.

Once the **Map Role** column property has been set up, you can perform your analysis. You want to visually see the differences in office temperatures throughout the floor.

1. Select **Graph > Graph Builder**.
2. Drag and drop **Room/Office** into the **Map Shape** zone.

   Since you have defined the **Map Role** column property on this column, the map appears.
3. Drag and drop **Fahrenheit** to the **Color** zone.

Figure 16.41 Room/Office Colored by Fahrenheit

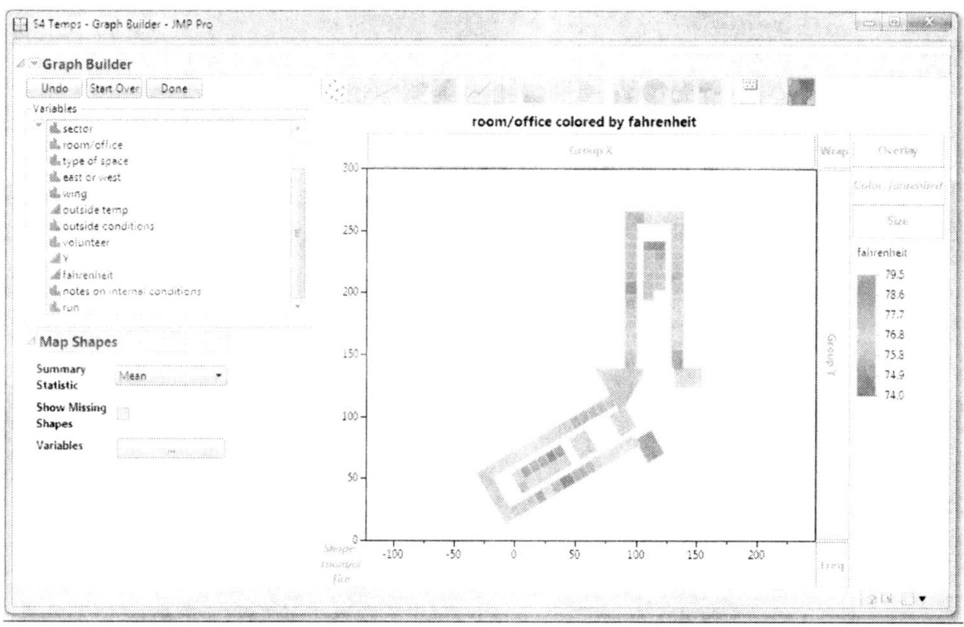

4. Drag and drop **Time of Day** onto the graph.

Figure 16.42 Room/Office Colored by Fahrenheit and Time of Day

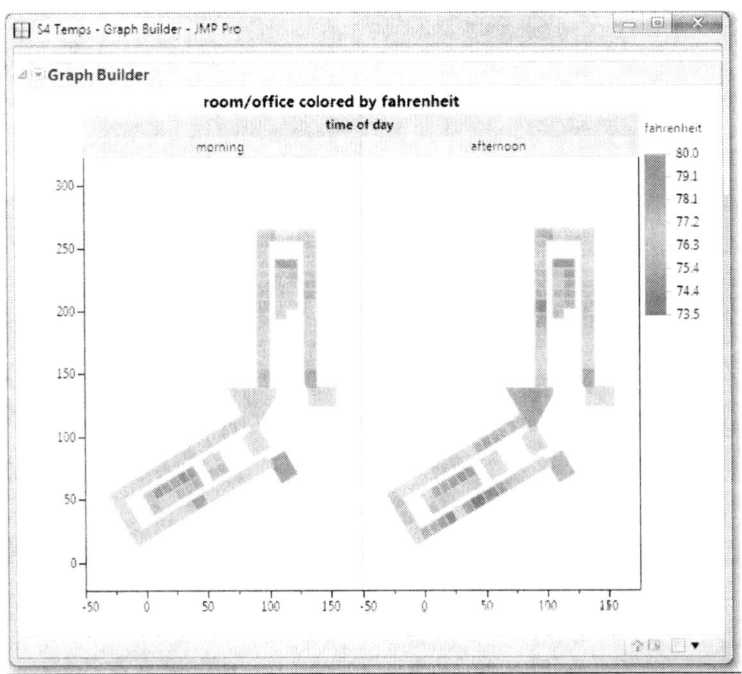

Note that only the offices that were part of the study and were created using the Custom Map Creator add-in are displayed. To add the entire floor plan image, the original floor plan graphic was dragged and dropped onto the Graph Builder window to create Figure 16.43.

To view Figure 16.43, open the S4 Temps.jmp sample data table and run the **by Time of Day** script.

Figure 16.43 Room/Office Colored by Fahrenheit and Time of Day

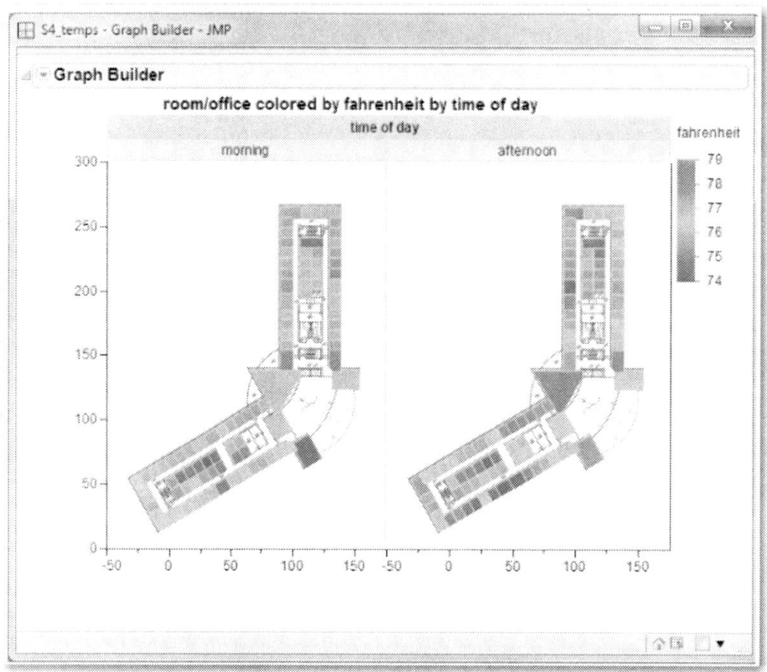

There are several other scripts provided with the data table that you can run to view the various analysis and modeling that can be performed and visually displayed.

Chapter 17

Creating Parallel Plots

Using the Parallel Plot Platform

Using parallel plots, you can visualize each cell in a data table. Parallel plots draw connected line segments that represent each row in a data table. Parallel plots were initially developed by Inselberg (1985) and later popularized by Wegman (1990).

Figure 17.1 Example of a Parallel Plot

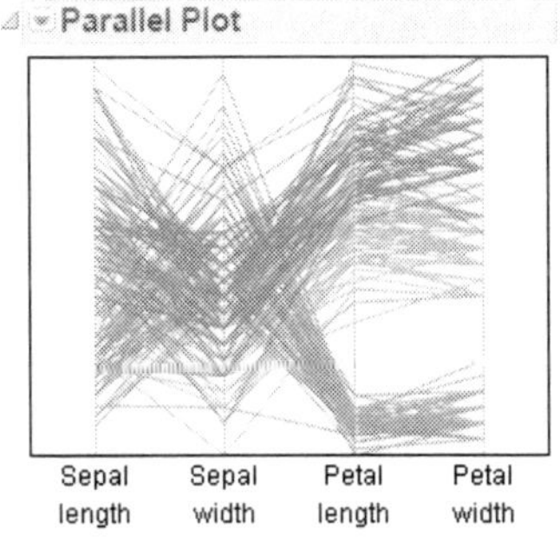

Contents

Example of a Parallel Plot

This example uses the Dogs.jmp sample data table, which contains histamine level measurements for 16 dogs that were given two different drugs. The histamine levels were taken at zero, one, three, and five minutes. Examine the variation in the histamine levels for each drug.

1. Open the Dogs.jmp sample data table.

To see the differences by drug, color the parallel plot lines by drug:

2. Select **Rows > Color or Mark by Column**.
3. Select drug and click **OK**.

   Morphine appears red and trimeth appears blue.

Create the parallel plot:

4. Select **Graph > Parallel Plot**.
5. Select hist0, hist1, hist3, and hist5 and click **Y, Response**.
6. Click **OK**.

   The report window appears.

Figure 17.2 Parallel Plot of Histamine Variables

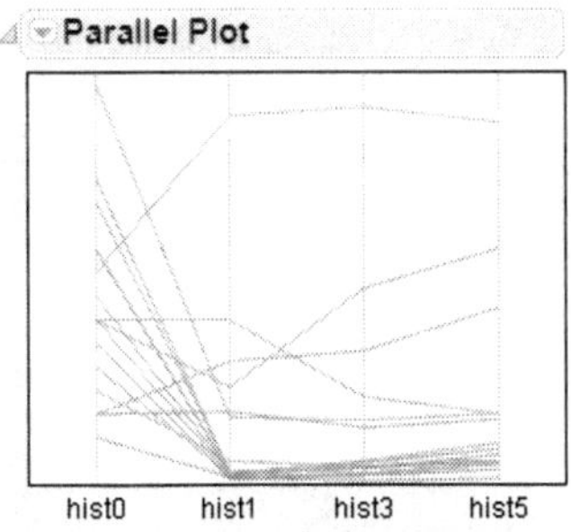

Each connected line segment represents a single observation. Click on a line segment to see which observation (or row) it corresponds to in the data table.

For further exploration, isolate the trimeth values:

7. Select **Rows > Data Filter**.
8. Select drug and click **Add**.
9. Select trimeth.

   Only the trimeth values are highlighted in the parallel plot.

Figure 17.3 Trimeth Values Highlighted

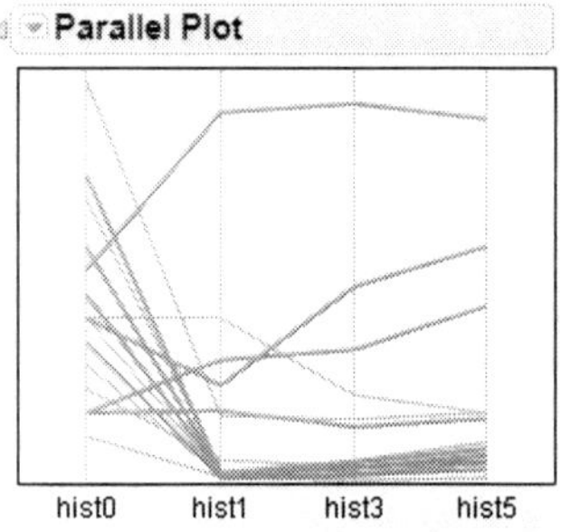

From Figure 17.3, you observe the following about the histamine levels for dogs given trimeth:

- For most of the dogs, the histamine levels had a sharp drop at one minute.
- For four of the dogs, the histamine levels remained high, or rose higher. You might investigate this finding further, to determine why the histamine levels were different for these dogs.

Launch the Parallel Plot Platform

Launch the Parallel Plot platform by selecting **Graph > Parallel Plot**.

Figure 17.4 The Parallel Plot Launch Window

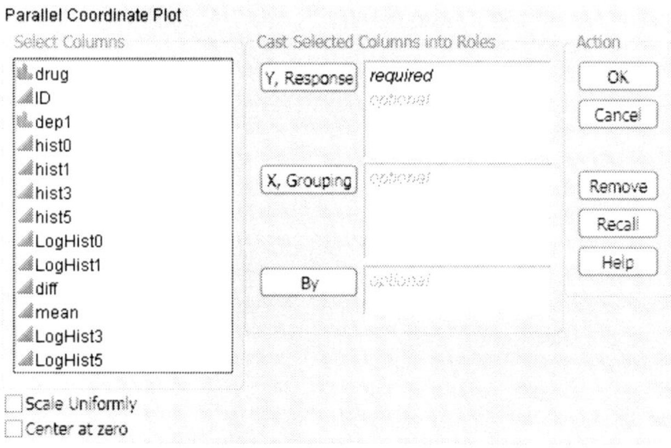

Table 17.1 Description of the Parallel Plot Launch Window

Y, Response	Variables appear on the horizontal axis of the parallel plot. These values are plotted and connected in the parallel plot.

Table 17.1 Description of the Parallel Plot Launch Window *(Continued)*

X, Grouping	Produces a separate parallel plot for each level of the variable.
By	Identifies a column that creates a report consisting of separate analyses for each level of the specified variable.
Scale Uniformly	Represents all variables on the same scale, adding a *y*-axis to the plot. Without this option, each variable is on a different scale. To allow for proper comparisons, select this option if your variables are measured on the same scale.
Center at zero	Centers the parallel plot (not the variables) at zero.

For more information about the launch window, see *Using JMP*.

The Parallel Plot

To produce the plot shown in Figure 17.5, follow the instructions in "Example of a Parallel Plot" on page 449.

Figure 17.5 The Parallel Plot Report

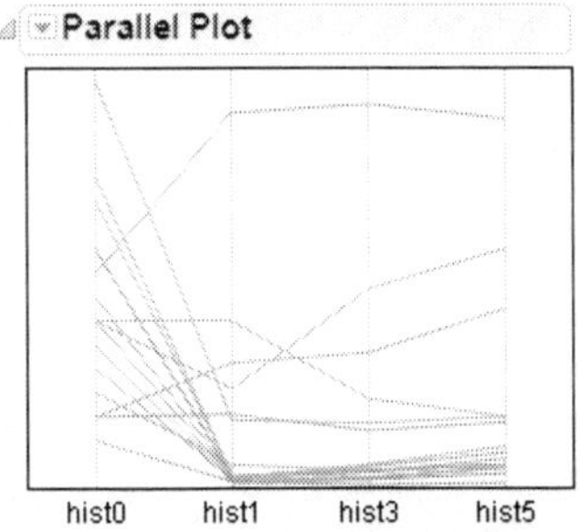

A parallel plot is one of the few types of coordinate plots that show any number of variables in one plot. However, the relationships between variables might be evident only in the following circumstances:

- when the variables are side by side
- if you assign a color to a level of a variable to track groups
- if you select lines to track groups

Note: If the columns in a parallel plot use the **Spec Limits** column property, the specification limits appear as red lines.

Interpreting Parallel Plots

To help you interpret parallel plots, compare the parallel plot with a scatterplot. In each of the following figures, the parallel plot appears on the left, and the scatterplot appears on the right.

Strong Positive Correlation

The following relationship shows a strong positive correlation. Notice the coherence of the lines in the parallel plot.

Figure 17.6 Strong Positive Correlation

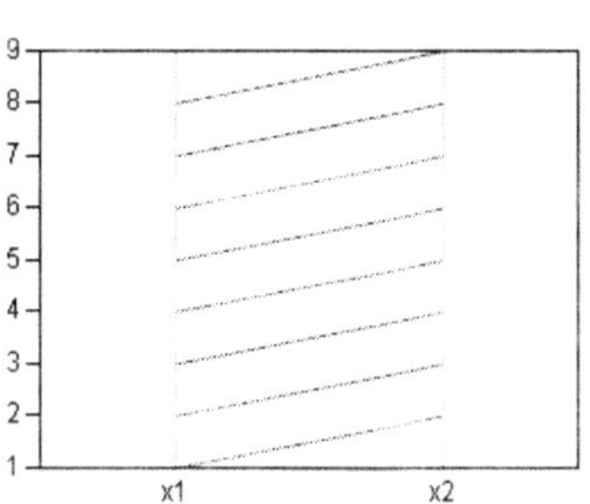

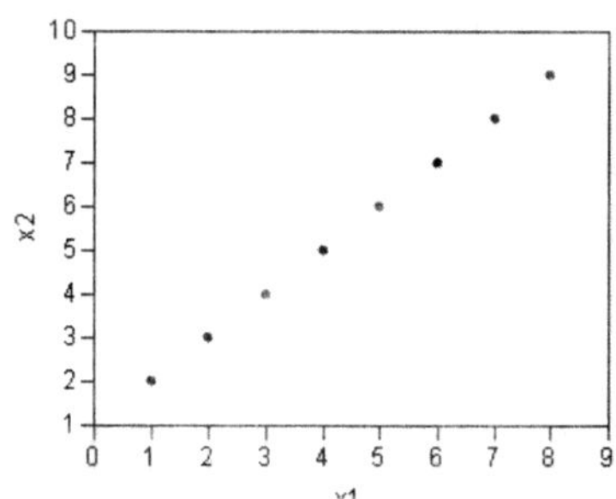

Strong Negative Correlation

A strong negative correlation, by contrast, shows a narrow neck in the parallel plot.

Figure 17.7 Strong Negative Correlation

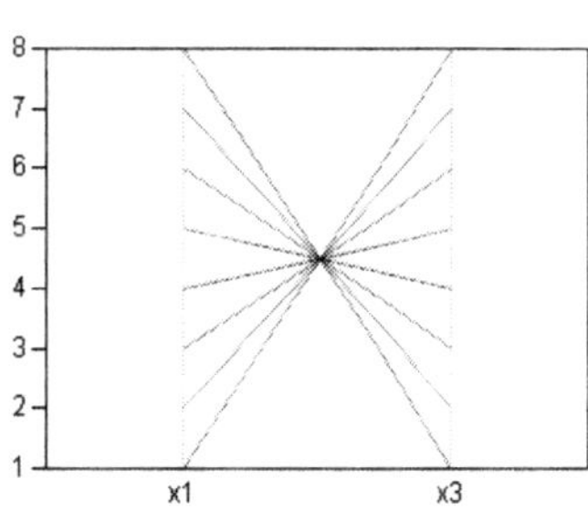

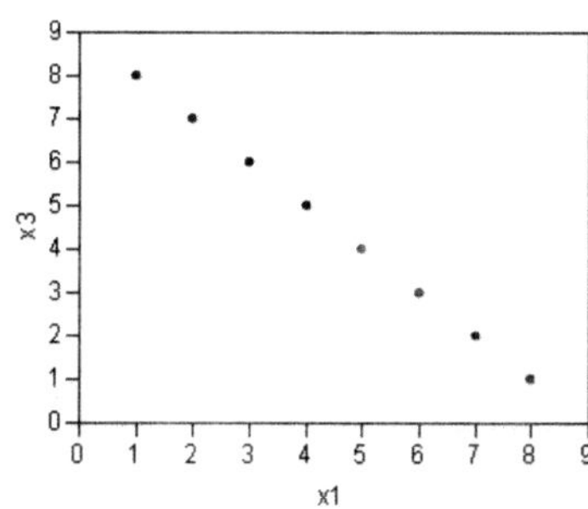

Collinear Groups

Now, consider a case that encompasses both situations: two groups, both strongly collinear. One has a positive slope, the other has a negative slope. In Figure 17.8, the positively sloped group is highlighted.

Figure 17.8 Collinear Groups: Parallel Plot and Scatterplot

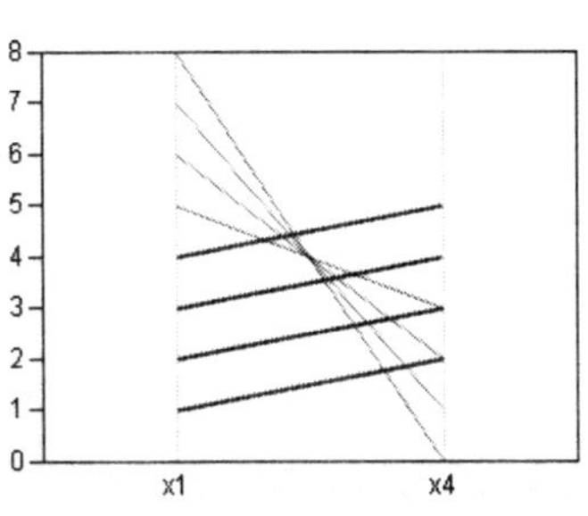

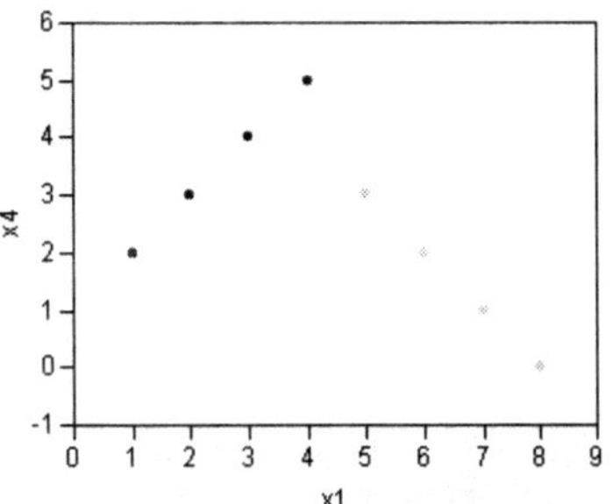

Single Outlier

Finally, consider the case of a single outlier. The parallel plot shows a general coherence among the lines, with a noticeable exception.

Figure 17.9 Single Outlier: Parallel Plot and Scatterplot

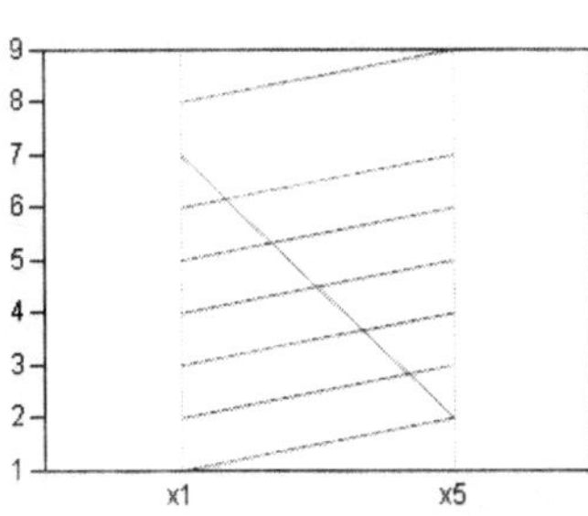

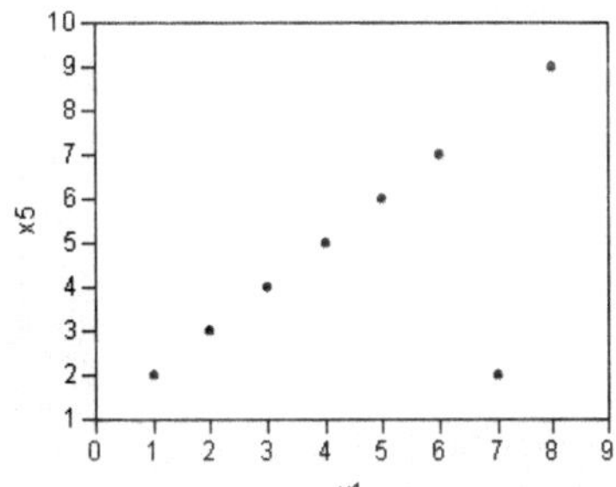

Related Information

- "Additional Examples of the Parallel Plot Platform" on page 454

Parallel Plot Platform Options

The following table describes the options within the red triangle menu for Parallel Plot.

Show Reversing Checkboxes	Reverses the scale for one or more variables.
Script	This menu contains options that are available to all platforms. They enable you to redo the analysis or save the JSL commands for the analysis to a window or a file. For more information, see *Using JMP*.

For details about the context menu options that appear when you right-click on a parallel plot, see *Using JMP*.

Additional Examples of the Parallel Plot Platform

The following examples further illustrate using the Parallel Plots platform.

Examine Iris Measurements

The following example uses the Fisher's Iris data set (Mardia, Kent, and Bibby 1979). The Iris.jmp sample data table contains measurements of the sepal length and width and petal length and width in centimeters for three species of Iris flowers: setosa, versicolor, and virginica. To find characteristics that differentiate the three species, examine these measurements.

Examine Three Species in One Parallel Plot

1. Open the Iris.jmp sample data table.
2. Select **Graph > Parallel Plot**.
3. Select Sepal length, Sepal width, Petal length, and Petal width and click **Y, Response**.
4. Select the **Scale Uniformly** check box.
5. Click **OK**.

   The report window appears.

Figure 17.10 Three Species in One Parallel Plot

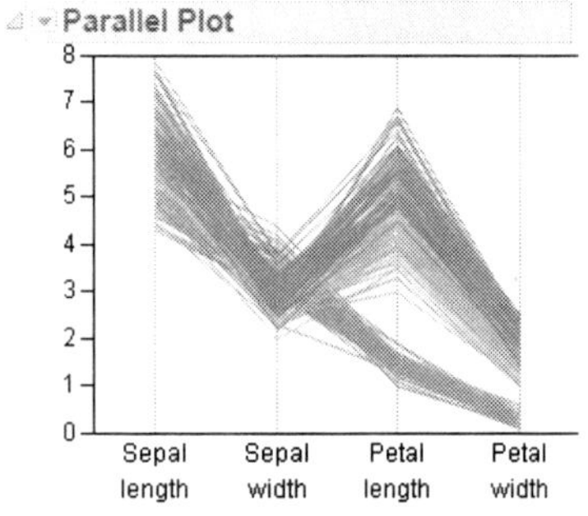

In this parallel plot, the three species are all represented in the same plot. The colors correspond to the three species, as follows:

- Blue corresponds to virginica.
- Green corresponds to versicolor.
- Red corresponds to setosa.

From Figure 17.10, you observe the following:

- For sepal width, the setosa values appear to be higher than the virginica and versicolor values.
- For petal width, the setosa values appear to be lower than the virginica and versicolor values.

Examine Three Species in Different Parallel Plots

1. From the Iris.jmp sample data table, select **Graph > Parallel Plot.**
2. Select Sepal length, Sepal width, Petal length, and Petal width and click **Y, Response**.
3. Select Species and click **X, Grouping**.
4. Click **OK**.

   The report window appears.

Figure 17.11 Three Species in Different Parallel Plots

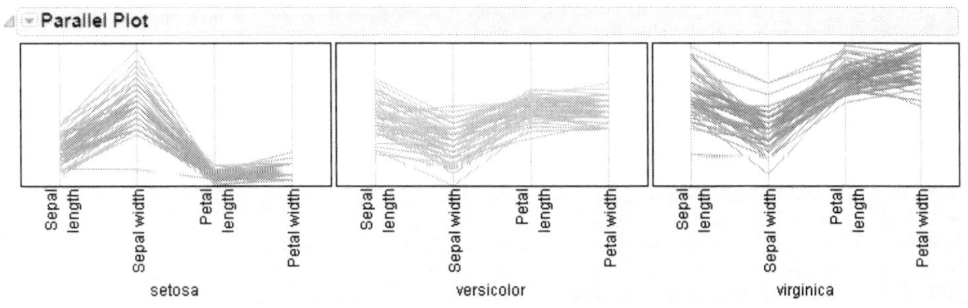

Each species is represented in a separate parallel plot.

Examine Student Measurements

The following example uses the Big Class.jmp sample data table, which contains data on age, sex, height, and weight for 40 students. Examine the relationships between different variables.

1. Open the Big Class.jmp sample data table.
2. Select **Graph > Parallel Plot.**
3. Select height and weight and click **Y, Response**.
4. Select age and click X**, Grouping**.
5. Select sex and click **By**.
6. Select the **Scale Uniformly** check box.
7. Click **OK**.

Figure 17.12 Height and Weight by Sex, Grouped by Age

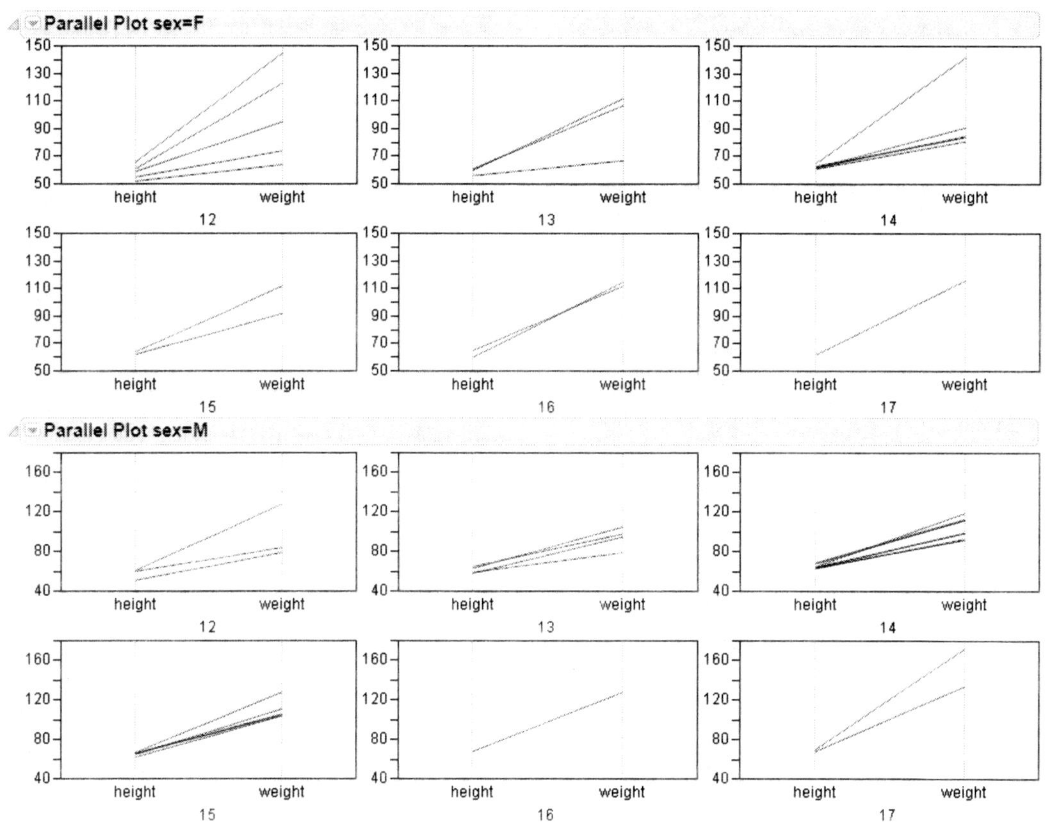

From Figure 17.12, you observe the following:

- Among the 13-year-old females, one female's weight is lower than the other females in her age group. If you click on the line representing the lower weight, the respective individual (Susan) is highlighted in the data table.
- Among the 14-year-old females, one female's weight is higher than the other females in her age group. If you click on the line representing the higher weight, the respective individual (Leslie) is highlighted in the data table.

Chapter 18

Creating Cell Plots

Using the Cell Plot Platform

Using cell plots, you can visualize each cell in a data table. Cell plots are direct representations of a data table, since they draw a rectangular array of cells where each cell corresponds to a data table entry. Cell plots were popularized by genomics applications to browse large numbers of values for gene expression levels.

Figure 18.1 Example of a Cell Plot

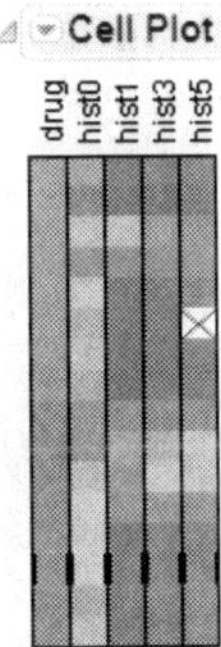

Contents

Example of a Cell Plot

This example uses the **Dogs.jmp** sample data table, which contains histamine level measurements for 16 dogs that were given two different drugs. The histamine levels were taken at zero, one, three, and five minutes. Examine the variation in the histamine levels for each drug.

1. Open the **Dogs.jmp** sample data table.
2. Select the third row from the bottom (row 14).
3. Select **Graph > Cell Plot**.
4. Select **drug**, **hist0**, **hist1**, **hist3**, and **hist5** and click **Y, Response**.
5. Click **OK**.

   The report window appears.

Figure 18.2 Dogs.jmp cell plot

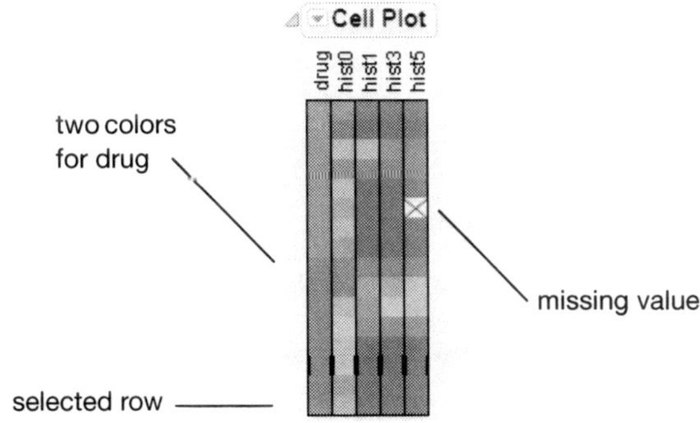

From Figure 18.2, notice the following:

- There are two types of drugs, represented by two distinct colors.
- Histamine levels are assigned colors from a gradient of blue to red.
- Any missing values are delineated by an X.
- The third row from the bottom is selected, and black lines appear next to the cells.

Launch the Cell Plot Platform

Launch the Cell Plot platform by selecting **Graph > Cell Plot**.

Figure 18.3 The Cell Plot Launch Window

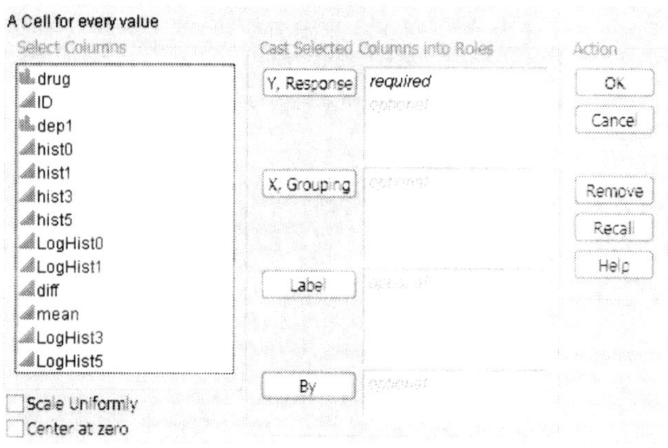

Table 18.1 Description of the Cell Plot Launch Window

Y, Response	Variables appear on the horizontal axis of the cell plot. Each cell represents a value.
X, Grouping	Produces a separate cell plot for each level of the variable.
Label	Labels each row by the specified variable. See "Additional Example of the Cell Plot Platform" on page 463.
By	Identifies a column that creates a report consisting of separate analyses for each level of the variable.
Scale Uniformly	Represents all variables on the same scale. Without this option, each variable is on a different scale.
Center at zero	Centers the cell plot at zero.

For more information about the launch window, see *Using JMP*.

The Cell Plot

To produce the plot shown in Figure 18.4, follow the instructions in "Example of a Cell Plot" on page 459.

Figure 18.4 The Cell Plot Report Window

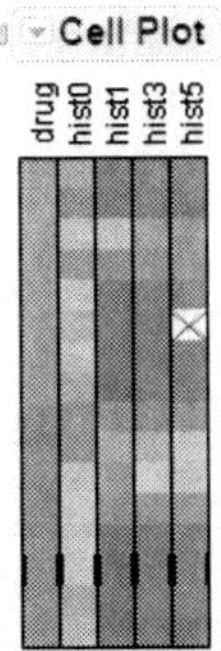

Cell plots are direct representations of a data table, drawn as a rectangular array of cells with each cell corresponding to a data table entry. Colors are assigned to each cell based on the range and type of values found in the column.

- Nominal variables use a distinct color for each level. You can customize nominal and ordinal colors using the **Value Colors** property of data columns, available through the **Column Info** command.
- Continuous variables are assigned a gradient of colors to show the smooth range of values in the variable.
- Ordinal variables are scaled like continuous variables in order.
- When some outliers are present, the scale uses all but the extreme categories for the 90% middle of the distribution, so that the outliers do not overly influence the scale.

The cell plot appears with a one-to-one correspondence of a colored cell representing each data table entry.

Related Information

- "Additional Example of the Cell Plot Platform" on page 463

Cell Plot Platform Options

The following table describes the options within the red triangle menu for Cell Plot.

Legend	Shows or hides a legend.
Arrange Plots	Specifies how many plots to put on the same row before starting the next row of plots. This option is available only if you specify an **X, Grouping** variable.
Script	This menu contains options that are available to all platforms. They enable you to redo the analysis or save the JSL commands for the analysis to a window or a file. For more information, see *Using JMP*.

Context Menu for Cell Plots

The following table describes the options that appear when you right-click on a cell plot.

Note: For details about the context options that appear when you right-click on labels, see the *Using JMP* book.

Graph Type	Determines the appearance of the graph. See "Graph Type" on page 462.
Color Theme	Shows a list of color themes that affect continuous variables in color maps. The default color theme is **Blue to Gray to Red** (corresponding to small values to middle values to large values). Use **White to Black** to create a gray-scale plot. **Note:** To see custom colors, you must first create them. Go to **File > Preferences > Graphs** and click on the **Custom Color Themes** button.
Sort Ascending	Sorts the rows of the plot from lowest to highest by the values of a column. To sort, right-click in the plot under a column and select **Sort Ascending.** The entire plot is rearranged to accommodate the sorting. See "Additional Example of the Cell Plot Platform" on page 463.
Sort Descending	Sorts the rows of the plot from highest to lowest by the values of a column. To sort, right-click in the plot under a column and select **Sort Descending.** The entire plot is rearranged to accommodate the sorting.
No Separator Lines	Draws or removes lines separating the columns.

Graph Type

Use the **Graph Type** option to change the appearance of the graph.

Figure 18.5 Graph Types

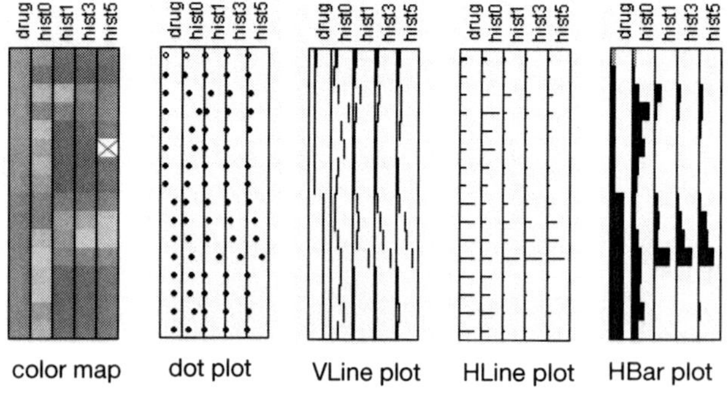

Additional Example of the Cell Plot Platform

This example uses the SAT.jmp sample data table, which contains SAT test scores (divided into verbal and mathematics portions) for all 50 United States.

1. Open the SAT.jmp sample data table.
2. Select **Graph > Cell Plot**.
3. Select all of the Verbal scores for all of the years, and click **Y, Response**.
4. Select all of the Math scores for all of the years, and click **Y, Response**.
5. Select State and click **Label**.
6. Click **OK**.

   The report window appears.
7. Right-click on the plot under 2004 Verbal (the top left cell) and select **Sort Ascending**.

   This sorts the cell plot by the verbal scores for 2004.

Figure 18.6 Cell Plot for SAT Scores

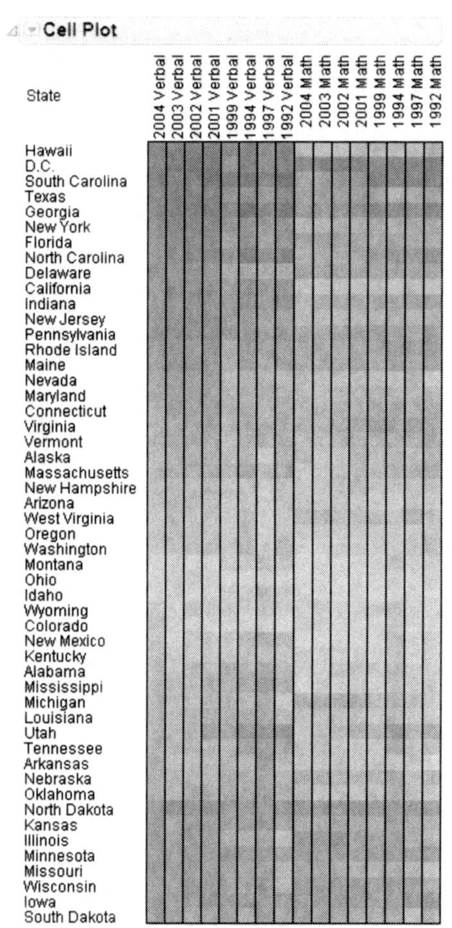

From Figure 18.6, you notice the following:

- Hawaii has the lowest verbal scores for 2004, and South Dakota has the highest verbal scores for 2004.
- There is a contrast between Hawaii's math and verbal scores. Hawaii has average math scores (represented by gray color values) but low verbal scores (represented by blue color values). Hawaii appears to be an outlier, since it has a strikingly different pattern for its math scores and its verbal scores.
- There is very little contrast between North Dakota's math and verbal scores. North Dakota's math and verbal scores are generally high (represented by red color values).

For a description of color themes, see "Context Menu for Cell Plots" on page 462.

Chapter 19

Creating Tree Maps

Using the Tree Map Platform

Tree maps are useful for observing patterns among groups that have many levels. Tree maps are like bar charts that have been folded over in two dimensions so that there is no unused space. Rather than drawing a single bar for each measurement, a tree map can show the magnitude of a measurement by varying the size or color of a rectangular area. Use tree maps when your data contains many categories, to visualize many groups.

Tree maps were named and popularized by Ben Schneiderman, who has an extensive Web site about the idea (http://www.cs.umd.edu/hcil/treemap-history/index.shtml)

Figure 19.1 Example of a Tree Map

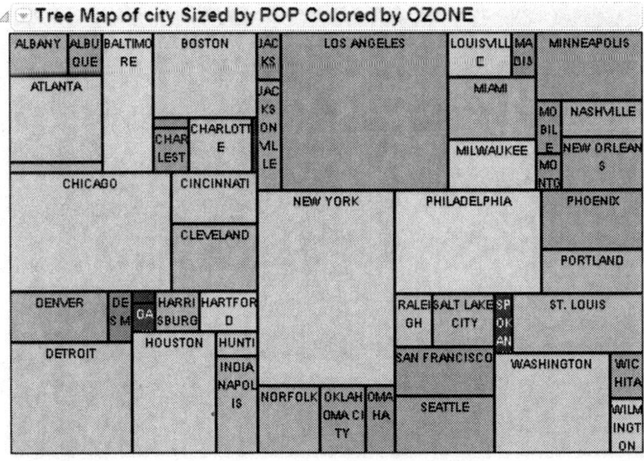

Contents

Example of Tree Maps

Tree maps can be useful in cases where histograms or bar charts are ineffective. This example uses the Cities.jmp sample data table, which contains meteorological and demographic statistics for 52 cities. Compare the bar chart to the tree map.

Create a bar chart representing ozone levels in each of the 52 cities:

1. Open the Cities.jmp sample data table.
2. Select **Graph > Chart**.

   The launch window appears.
3. Select OZONE and click **Statistics.**
4. Select **Mean**.
5. Select city and click **Categories, X, Levels**.
6. Click **OK**.

   The report window appears.

Figure 19.2 Ozone Levels in a Bar Chart

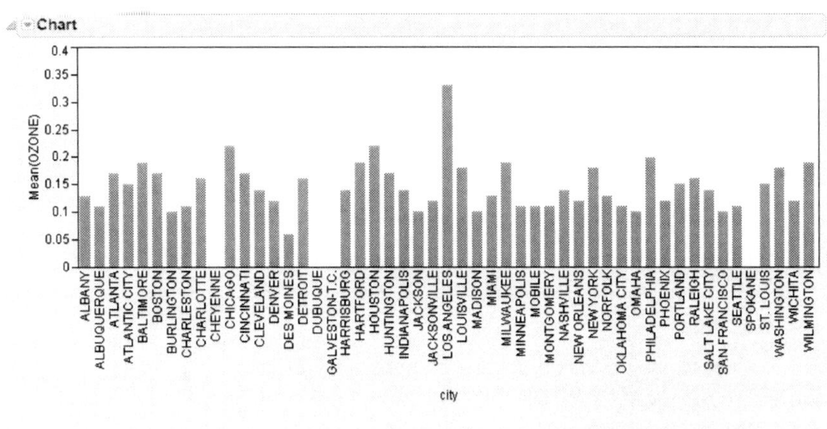

Although it is easy to see that there is a single large measurement, each bar looks similar. Subtle distinctions are difficult to see.

Create a tree map representing ozone levels in each of the 52 cities:

1. Return to the Cities.jmp sample data table.
2. Select **Graph > Tree Map**.

   The launch window appears.
3. Select POP (population) and click **Sizes**.
4. Select city and click **Categories**.

5. Select OZONE and click **Coloring**.
6. Click **OK**.

   The report window appears.

Figure 19.3 Ozone Levels in a Tree Map

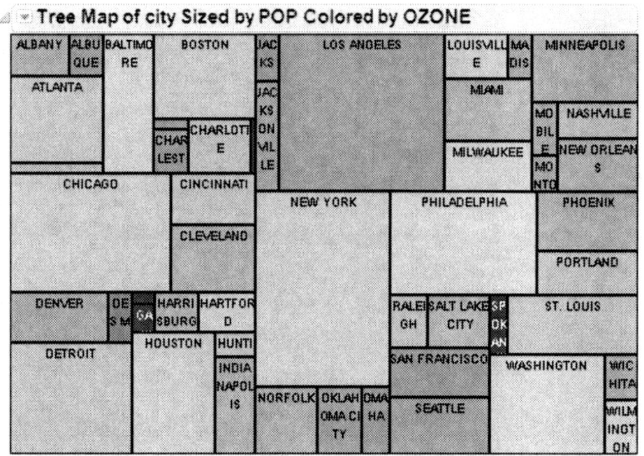

Compare the bar chart to the tree map. Because the tree map folds the data over two dimensions (size and color), each city's data looks more distinctive than it did in the bar chart.

Note the following about this tree map:

- The magnitude of the ozone level for each city is represented by color.
- Each rectangle is colored based on a continuous color spectrum, with bright blue on the lowest end and bright red on the highest end. In this tree map, Des Moines has the lowest ozone levels, and Los Angeles has the highest ozone levels.
- Ozone levels somewhere in the middle decrease the intensity of the color, so pale blue, and pink indicate levels that are closer to the mean ozone level.
- Cities colored black have missing ozone values.

Launch the Tree Map Platform

Launch the Tree Map platform by selecting **Graph** > **Tree Map**.

Figure 19.4 The Tree Map Launch Window

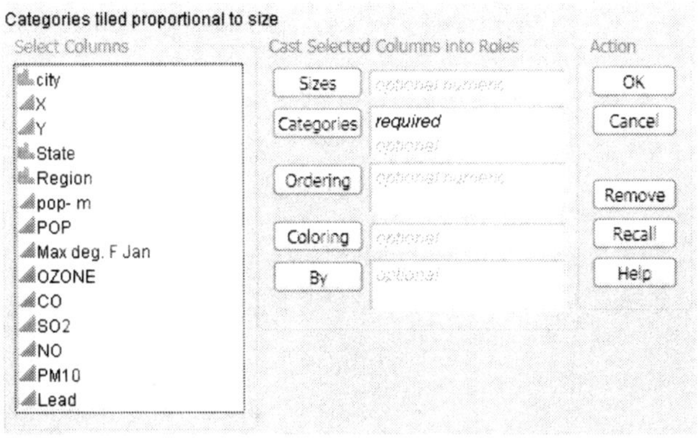

Table 19.1 Description of the Tree Map Launch Window

Sizes	Determines the size of the rectangles based on the values of the specified variable. See “Sizes” on page 469.
Categories	(Required) Specifies the category that comprises the tree map. See “Categories” on page 470.
Ordering	Changes the ordering from alphabetical (where values progress from the top left to the lower right) to order by the specified variable. You can specify more than one ordering variable. See “Ordering” on page 470.
Coloring	Colors the rectangles corresponding to the levels of the specified variable. • If the variable is continuous, the colors are based on a continuous color spectrum. • If the variable is categorical, the default colors are selected in order from JMPs color theme. See “Coloring” on page 470.
By	Identifies a column that creates a report consisting of separate tree maps for each level of the variable.

For more information about the launch window, see *Using JMP*.

Sizes

If you want the size of the rectangles to correspond to the levels of a variable, specify a **Sizes** variable. The rectangle size is proportional to the sum of the **Sizes** variable across all of the rows corresponding to a category. If you do not specify a **Sizes** variable, the rectangle size is proportional to the number of rows for each category.

Categories

The only required variable role for the Tree Map platform is **Categories**. If you specify only a **Categories** variable and no other variables, the rectangles in the tree map have these attributes:

- They are colored from a rotating color theme.
- They are arranged alphabetically.
- They are sized by the number of occurrences in each group.

If you specify two **Categories** variables, the tree map is grouped by the first variable, and sorts within groups by the second variable. For example, using the Cities.jmp sample data table, specify Region and city (in that order) as the **Categories** variables.

Ordering

By default, the rectangles in a tree map appear in alphabetical order, with values progressing from the top left to the lower right. To change this ordering, specify an **Ordering** variable. When an **Ordering** variable is specified, the rectangles appear with the values progressing from the bottom left to the upper right.

If you specify a single **Ordering** variable, the rectangles are clustered, with the high levels or large values together, and the low levels or small values together.

If you specify two **Ordering** variables, the tree map arranges the rectangles horizontally by the first ordering variable, and vertically by the second ordering variable. This approach can be useful for geographic data.

Coloring

If you specify a **Coloring** variable, the colors of the rectangles correspond to the levels of the variable.

- If the variable is continuous, the colors are based on a continuous color theme setting. The default color theme is **Blue to Gray to Red**. The color of each value is based on the average value of all of the rows. Blue represents the lowest values, and red represents the highest values. The color is most intense at the extremes of the variable, and paler colors correspond to levels that are close to the mean. For example, see "The Tree Map Window" on page 471.
- If the variable is categorical, the colors are based on a categorical color theme. The default color theme is **JMP Default**. For example, see "Example of a Categorical Coloring Variable" on page 476.

Note: If you have used the **Value Colors** column property to color a column, that property determines the colors of the categories.

- If you do not specify a **Coloring** variable, colors are chosen from a rotating color palette.

Related Information

- "Example Using a Sizes Variable" on page 473
- "Example Using an Ordering Variable" on page 474

- "Example Using Two Ordering Variables" on page 474
- "Example of a Continuous Coloring Variable" on page 475
- "Example of a Categorical Coloring Variable" on page 476
- "Tree Map Platform Options" on page 472

The Tree Map Window

To produce the plot shown in Figure 19.5, follow the instructions in "Example of Tree Maps" on page 467.

Figure 19.5 The Tree Map Window

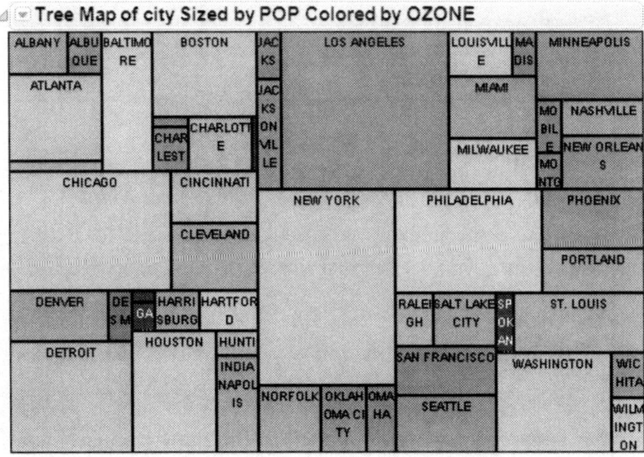

Tip: To zoom in on the tree map, use the **Magnifier** tool or press the Z key.

Tree map rectangles can have the following attributes:

- Categories add labels to the rectangles. You can specify one or two categories. You can show or hide labels using the **Show Labels** option. See "Context Menu" on page 472.
- Rectangle size is determined by one of the following:
  - The **Sizes** variable, if you specify one.
  - If you do not specify a **Sizes** variable, size is determined by the frequency of the category.
- Rectangle color is determined by one of the following:
  - If the variable is continuous, the colors are based on a continuous color theme setting. The default color theme is **Blue to Gray to Red**. The color of each value is based on the average value of all of the rows. Blue represents the lowest values, and red represents the highest values. The color is most intense at the extremes of the variable, and paler colors correspond to levels that are close to the mean. For example, see "Example of a Continuous Coloring Variable" on page 475.

- If the variable is categorical, the colors are based on a categorical color theme. The default color theme is **JMP Default**. For example, see "Example of a Categorical Coloring Variable" on page 476.
- If you do not specify a **Coloring** variable, colors are chosen from a rotating color theme.
- If you have used the **Value Colors** column property to color a column, that property determines the colors of the categories.

• The order of the rectangles is determined by one of the following:
  - The **Ordering** variable, if you specify one.
  - If you do not specify an **Ordering** variable, the order is alphabetical by default, with values progressing from the top left to the lower right.

Related Information

- "Additional Examples of the Tree Map Platform" on page 473

Tree Map Platform Options

The following table describes the options within the red triangle menu next to Tree Map.

Change Color Column	Change the column that is currently used to color the rectangles.
Color Theme	Change the colors representing the high, middle, and low values of the color column. This option is available only if you have specified a **Coloring** variable.
Color Range	Specify the range that you want applied to the color gradient. The default low value is the column minimum, and the default high value is the column maximum. This option is available only if you have specified a continuous column as the **Coloring** variable.
Legend	Shows or hides a legend that defines the coloring used on the tree map. This option is available only if you have specified a **Coloring** variable.
Script	This menu contains options that are available to all platforms. They enable you to redo the analysis or save the JSL commands for the analysis to a window or a file. For more information, see *Using JMP*.

Context Menu

The following table describes the options available by right-clicking on the tree map.

Suppress Box Frames	Suppresses the black lines outlining each box.
Ignore Group Hierarchy	Flattens the hierarchy and sorts by the **Ordering** columns without using grouping, except to define cells.

Show Labels	Shows or hides the **Categories** labels. If you have specified two **Categories**, the secondary labels are hidden or shown. If you hide the **Categories** labels, hover over a rectangle to show the primary or secondary category label.
Show Group Labels	Shows or hides the group labels. This option is available only if more than one variable is assigned to the **Categories** role.
Group Label Background	Adjust the transparency of the group label. This option is available only if more than one variable is assigned to the **Categories** role.

Additional Examples of the Tree Map Platform

The following examples further illustrate using the Tree Map platform.

Example Using a Sizes Variable

Using the Cities.jmp sample data table, specify a **Sizes** variable to see a tree map of city sizes based on population.

1. Open the Cities.jmp sample data table.
2. Select **Graph > Tree Map**.
3. Select POP (population) and click **Sizes**.
4. Select city and click **Categories**.
5. Click **OK**.

   The report window appears.

Figure 19.6 POP as Sizes Variable

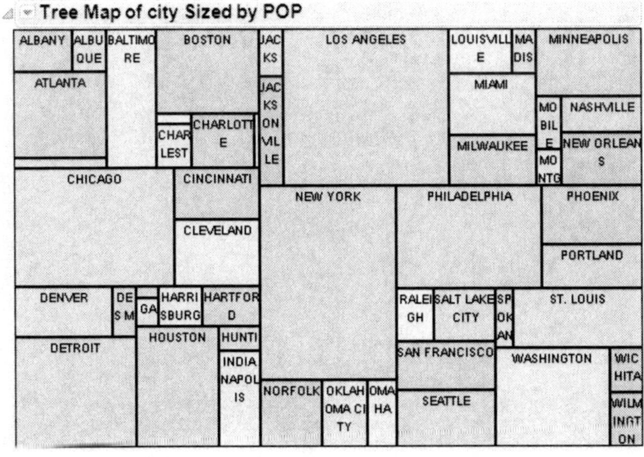

- Cities with large populations are represented by large rectangles, such as New York and Los Angeles.
- Cities with smaller populations are represented by smaller rectangles, such as Cheyenne and Dubuque.

Some rectangles are too small to show their labels (for example, Cheyenne and Dubuque). Click on a small rectangle to select the corresponding row in the data table, where you can see the label.

Example Using an Ordering Variable

For example, using the Cities.jmp sample data table, specify an **Ordering** variable as follows:

1. Open the Cities.jmp sample data table.
2. Select **Graph > Tree Map**.
3. Select city and click **Categories**.
4. Select POP (population) and click **Ordering**.
5. Click **OK**.

   The report window appears. Since you specified an **Ordering** variable, the cities in the tree map are ordered from bottom left (small cities) to upper right (big cities).

Figure 19.7 POP as Ordering Variable

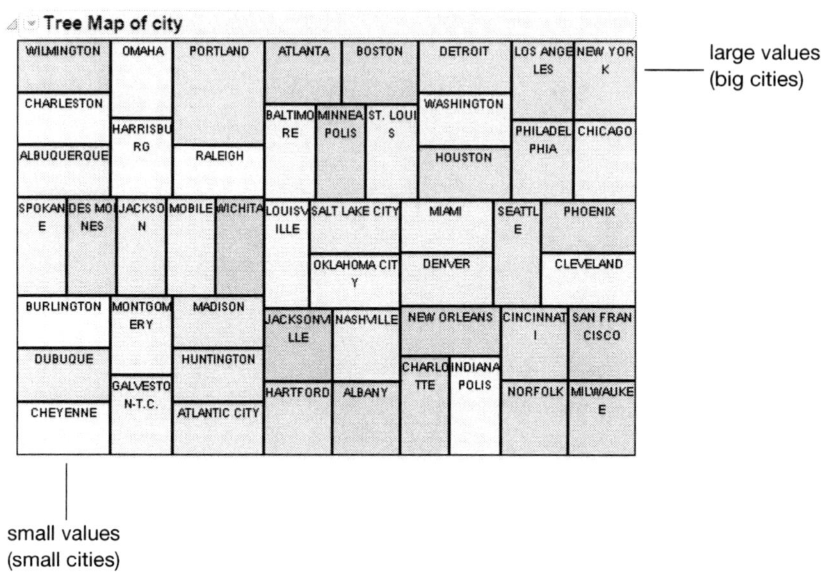

Example Using Two Ordering Variables

For example, in the Cities.jmp sample data table, the X and Y columns correspond to the geographic location of the cities. Specify the X and Y columns as **Ordering** variables:

1. Open the Cities.jmp sample data table.

2. Select **Graph > Tree Map**.
3. Select city and click **Categories**.
4. Select X and click **Ordering**.

   The X variable corresponds to the western and eastern US states.
5. Select Y and click **Ordering**.

   The Y variable corresponds to the northern and southern US states.
6. Click **OK**.

   The report window appears.

Figure 19.8 Tree Map with Two Ordering Variables

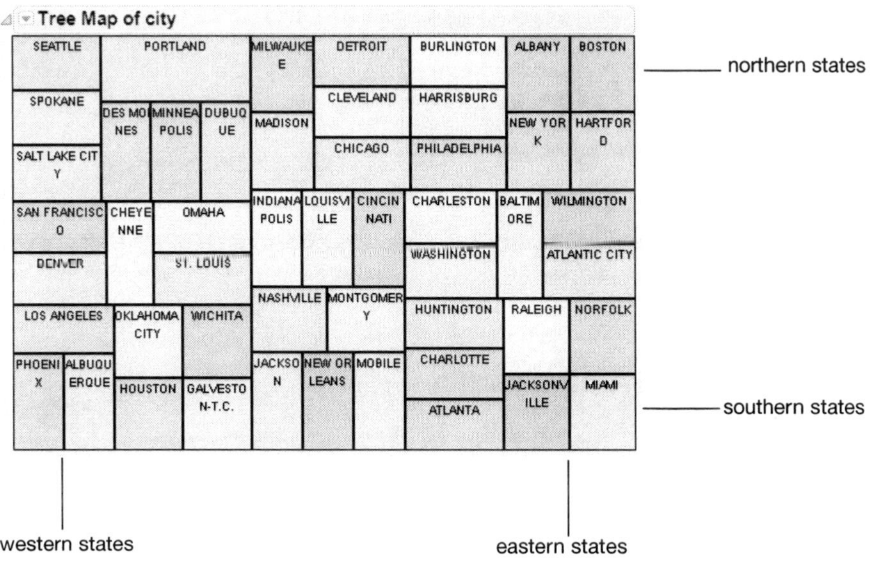

The western and eastern US states are arranged horizontally, and the northern and southern US states are arranged vertically.

Example of a Continuous Coloring Variable

For example, using the Cities.jmp sample data table, specify a continuous **Coloring** variable as follows:

1. Open the Cities.jmp sample data table.
2. Select **Graph > Tree Map**.
3. Select city and click **Categories**.
4. Select OZONE and click **Coloring**.
5. Click **OK**.

The report window appears.

Figure 19.9 City Colored by OZONE

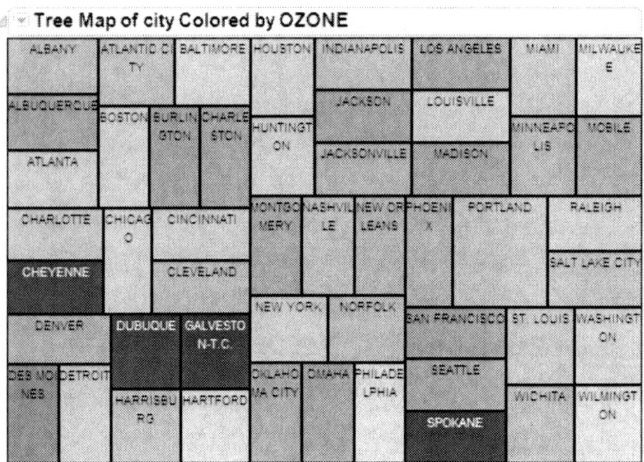

Note that the size of the rectangles is still based on the number of occurrences of the **Categories** variable, but the colors are mapped to ozone values. The high ozone value for Los Angeles clearly stands out. Missing values appear as black rectangles.

Example of a Categorical Coloring Variable

For example, using the Cities.jmp sample data table, specify a categorical **Coloring** variable as follows:

1. Open the Cities.jmp sample data table.
2. Select **Graph > Tree Map**.
3. Select city and click **Categories**.
4. Select Region and click **Coloring**.
5. Click **OK**.

   The report window appears.

Figure 19.10 City Colored by Region

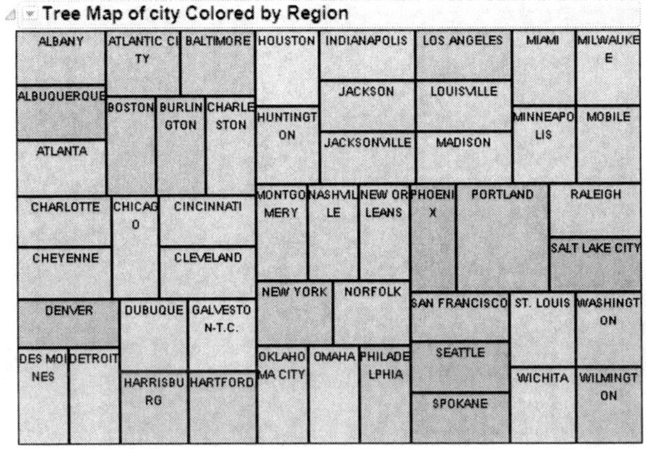

All of the cities belonging to the same region are colored the same color. The colors are chosen from JMPs color theme.

Examine Pollution Levels

Using the Cities.jmp sample data table, examine the distribution of different pollution measurements (ozone and lead) across selected cities in the United States.

First, examine ozone levels in the selected cities.

1. Open the Cities.jmp sample data table.
2. Select **Graph > Tree Map**.
3. Select POP (population) and click **Sizes.**
4. Select city and click **Categories.**
5. Select X and Y and click **Ordering.**
6. Select OZONE and click **Coloring.**
7. Click **OK.**

   The report window appears.

Figure 19.11 OZONE Levels for Selected Cities

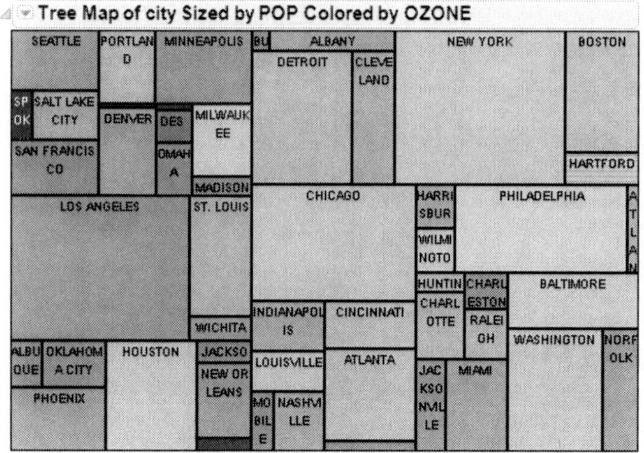

From Figure 19.11, you observe the following:

- Los Angeles has a high level of ozone. It is also a western state, and that is reflected by its positioning in the tree map.
- Chicago and Houston have slightly elevated ozone levels.
- New York and Washington have slightly lower-than-average ozone levels.

Next, examine lead levels in the selected cities. Perform the same steps as before, but substitute Lead with OZONE for the **Coloring** variable.

Figure 19.12 Lead Levels for Selected Cities

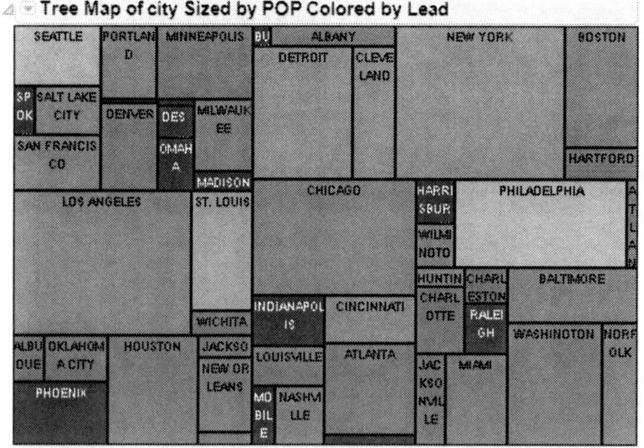

From Figure 19.12, you observe the following:

- Cleveland and Seattle have high levels of lead. Cleveland is near the middle of the country, and Seattle is in the northwest. These locations are reflected by their positioning in the tree map.
- Interestingly, most other cities have rather low lead levels.
- Raleigh and Phoenix have missing values for lead measurements.

Examine Causes of Failure

This example uses the Failure3.jmp sample data table, which contains the common causes of failure during the fabrication of integrated circuits. Examine the causes of failure and when it occurs.

1. Open the Failure3.jmp sample data table, located in the Quality Control folder.
2. Select **Graph > Tree Map**.
3. Select N and click **Sizes**.
4. Select failure and clean and click **Categories**.
5. Select clean and click **Coloring**.
6. Click **OK**.

   The report window appears.

Figure 19.13 Failure Modes

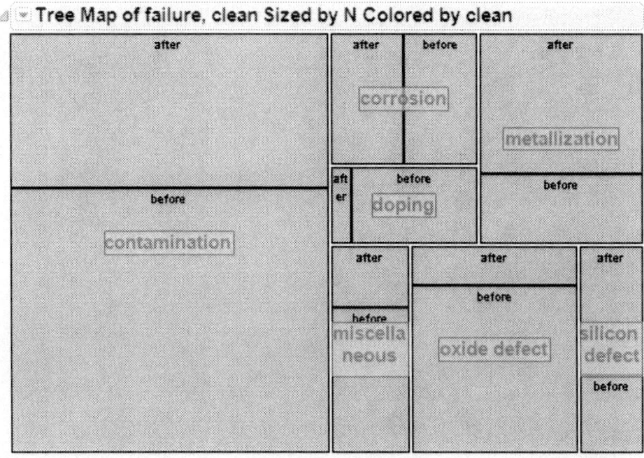

From Figure 19.13, you observe the following:

- Contamination is the biggest cause of failure.
- Contamination occurs more often before the circuits were cleaned, rather than after they were cleaned.

Examine Patterns in Car Safety

This example uses the Cars.jmp sample data table, which contains impact measurements of crash-test dummies in automobile safety tests. Compare these measurements for different automobile makes and models during the years 1990 and 1991.

1. Open the Cars.jmp sample data table.

Filter the data to show only the years 1990 and 1991 and create a subset of the Cars.jmp data table.

2. Select **Rows > Data Filter**.
3. Select Year.
4. Click **Add**.
5. Select 90 and 91.

   The rows corresponding to 1990 and 1991 are highlighted in the data table.
6. Select **Tables > Subset**.
7. Ensure that **Selected Rows** is selected and click **OK**.

   A new data table (Subset of Cars) appears that contains only the data corresponding to the years 1990 and 1991.

Now, using the Subset of Cars data table, create the tree map.

1. Select **Graph > Tree Map**.
2. Select Wt (weight) and click **Sizes**.
3. Select Make and Model and click **Categories**.
4. Select L Leg and click **Coloring**.

   L Leg represents a measurement of injuries resulting from the deceleration speed of the left leg, where more deceleration causes more injury.
5. Click **OK**.

   The report window appears.

Figure 19.14 Left Leg Deceleration Injuries

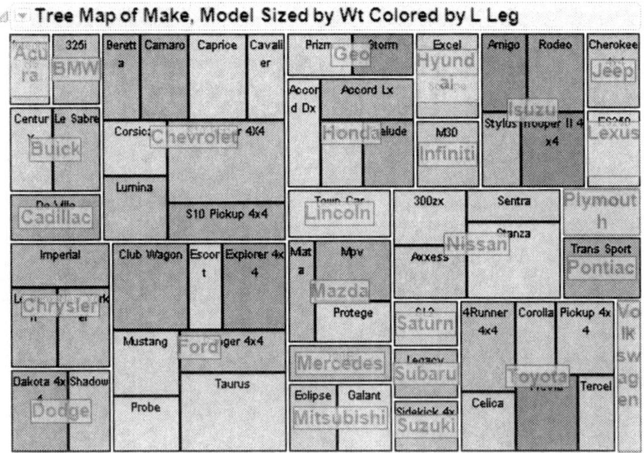

From Figure 19.14, you can see that the Club Wagon and S10 Pickup 4x4 have the largest number of left leg deceleration injuries.

You can examine other safety measurements without re-launching the Tree Map platform, as follows:

1. From the red triangle menu, select **Change Color Column**.
2. Select Head IC.
3. Click **OK**.

   The tree map updates to reflect head injuries instead of left leg injuries.

Figure 19.15 Head Injuries

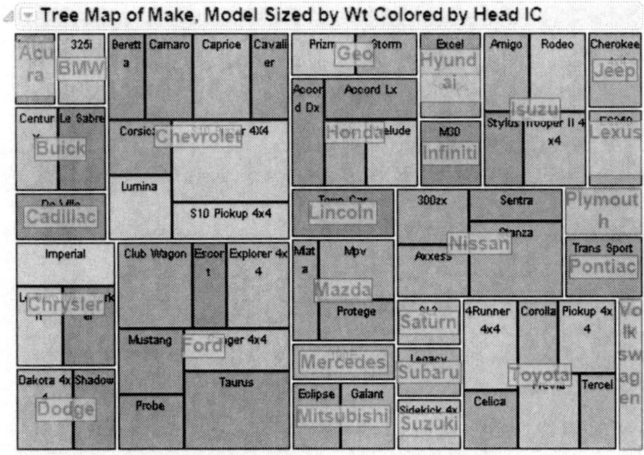

From Figure 19.15, you notice the following:

- Although the S10 Pickup 4x4 had a high number of left leg deceleration injuries, it has a lower number of head injuries.
- The Club Wagon still has a high number of head injuries, in addition to the high number of left leg deceleration injuries.
- The Trooper II 4x4 had a low number of left leg deceleration injuries, but it has a high number of head injuries.

Chapter 20

Creating Scatterplot Matrices

Using the Scatterplot Matrix Platform

Using the Scatterplot Matrix platform, you can assess the relationships between multiple variables simultaneously. A scatterplot matrix is an ordered collection of bivariate graphs. For further analysis, you can customize the scatterplots with density ellipses for all of your data, or for only groups of your data.

Figure 20.1 Example of a Scatterplot Matrix

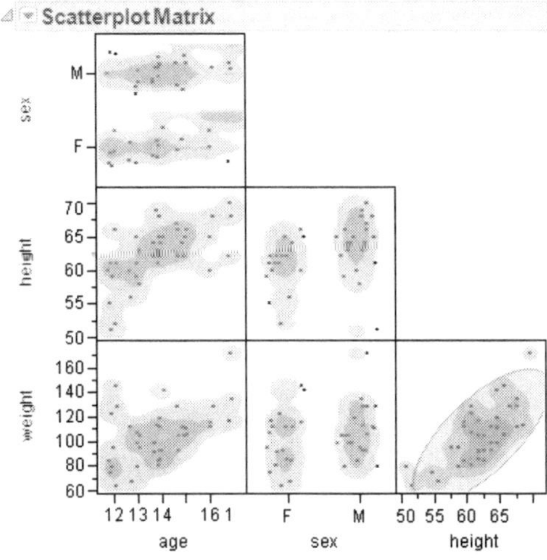

Contents

Example of a Scatterplot Matrix

This example shows you how to create a scatterplot matrix.

1. Open the Students.jmp sample data table.
2. Select **Graph > Scatterplot Matrix**.
3. Select age, sex, height, and weight and click **Y, Columns**.
4. Click **OK**.

Figure 20.2 Example of a Scatterplot Matrix

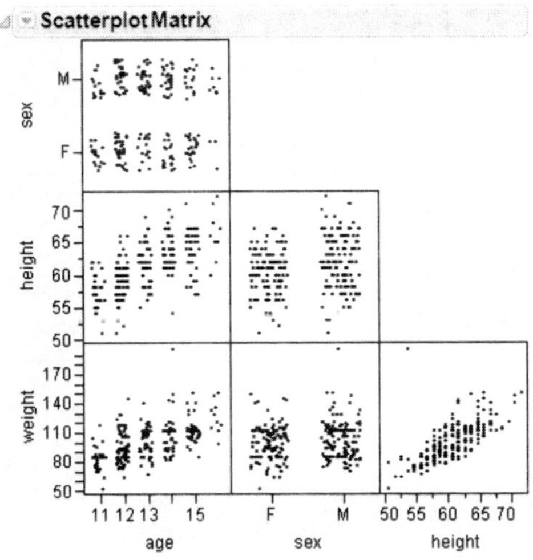

In this example, you can see that the graph for weight versus height is different from the graph for sex versus age. If you turn off jitter by clicking on the red triangle menu and selecting **Points Jittered**, the difference becomes even more pronounced.

Figure 20.3 Example of a Scatterplot Matrix with No Jitter

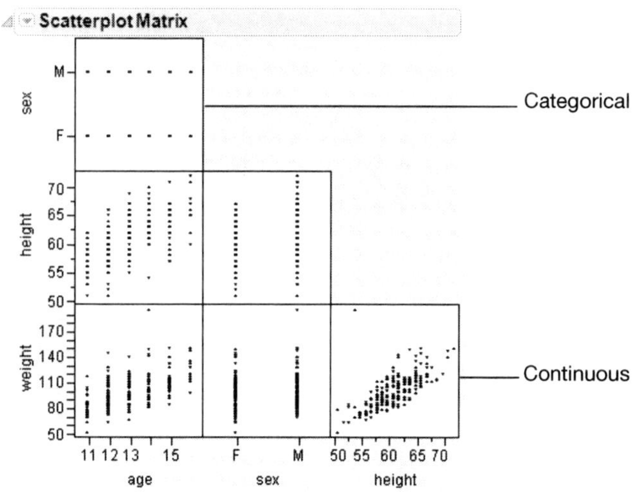

The weight versus height graph shows continuous data, and the sex versus age graph shows categorical data.

Launch the Scatterplot Matrix Platform

Launch the Scatterplot Matrix platform by selecting **Graph > Scatterplot Matrix**.

Figure 20.4 The Scatterplot Matrix Launch Window

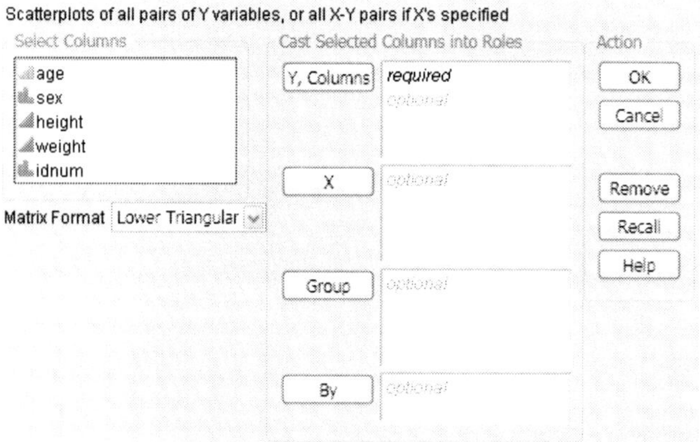

Table 20.1 Description of the Scatterplot Matrix Launch Window

Y, Columns, X	• If you assign variables to the **Y, Columns** role only, they appear on both the horizontal and vertical axes. • If you assign variables to both the **Y, Columns** and **X** role, then the **Y, Columns** variables appear on the vertical axis. The **X** variables appear on the horizontal axis. This approach enables you to produce rectangular matrices, or matrices that have different, yet overlapping, sets of variables forming the axes of the matrix.
Group	If you assign a variable to the **Group** role, you can add shaded density ellipses for each level of the **Group** variable. See "Example Using a Grouping Variable" on page 490.
By	This option produces a separate scatterplot matrix for each level of the **By** variable. If two **By** variables are assigned, a separate graph for each possible combination of the levels of both **By** variables is produced.
Matrix Format	The **Matrix Format** can be one of three arrangements: **Upper Triangular**, **Lower Triangular**, or **Square**. See "Change the Matrix Format" on page 487.

For more information about the launch window, see *Using JMP*.

Change the Matrix Format

The **Matrix Format** can be one of three arrangements: **Upper Triangular**, **Lower Triangular**, or **Square**.

Figure 20.5 Examples of Matrix Formats

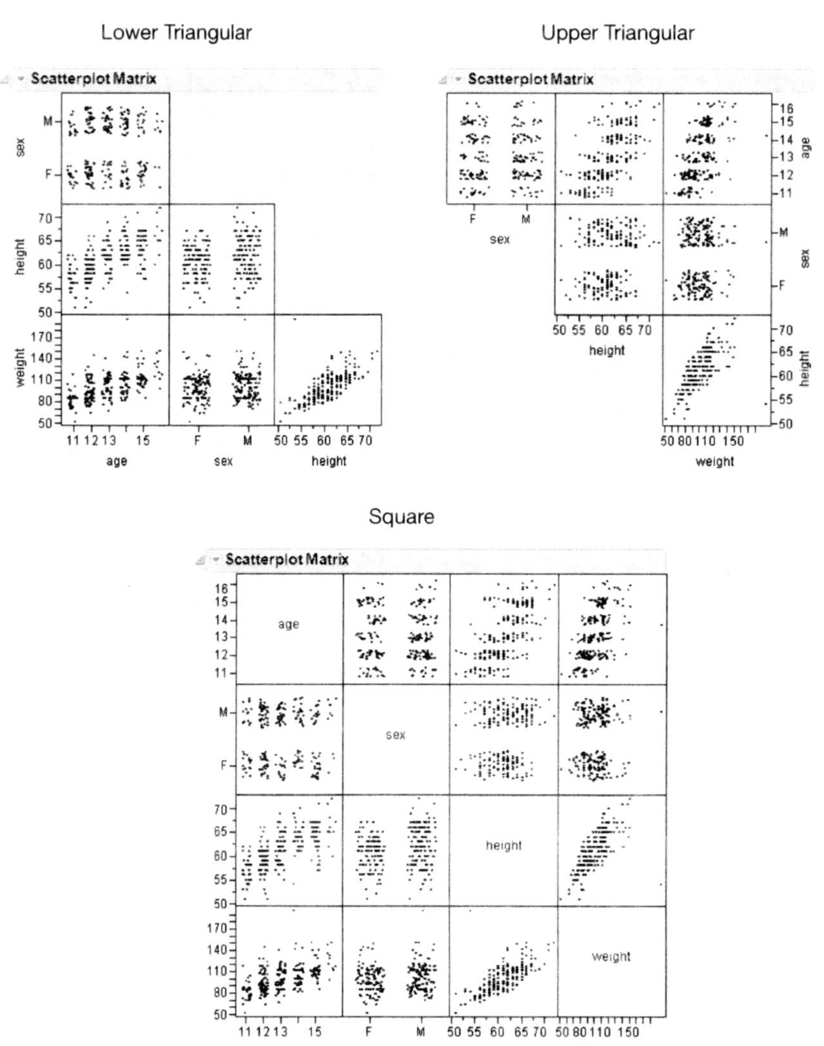

The Scatterplot Matrix Window

The Scatterplot Matrix window shows an ordered grouping of bivariate graphs. In each graph, you can examine the relationships between each pair of variables.

Follow the instructions in "Example of a Scatterplot Matrix" on page 485 to produce the plot shown in Figure 20.6.

Note: For information about additional options, see "Scatterplot Matrix Platform Options" on page 489.

Figure 20.6 Example of a Scatterplot Matrix Window

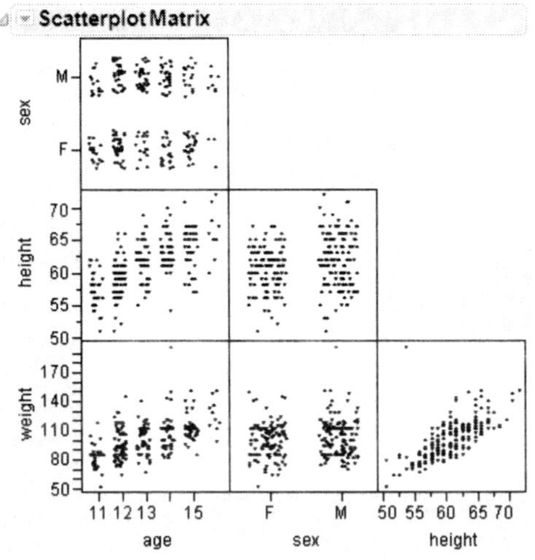

Replace variables in the plot by dragging and dropping a variable, in one of two ways: swap existing variables by dragging and dropping a variable from one axis to the other axis; or, click on a variable in the Columns panel of the associated data table and drag it onto an axis. This feature is not available for matrices in the Square format.

Scatterplot Matrix Platform Options

The following table describes the options within the red triangle menu next to Scatterplot Matrix.

Show Points	Shows or hides the points in the scatterplots.
Points Jittered	Turns the jittering of the points in the scatterplot on or off.
Fit Line	Fits a simple regression line and its mean confidence interval to the scatterplots.
Density Ellipses	Shows or hides the outline of the density ellipses. See "Example Using a Grouping Variable" on page 490.
Shaded Ellipses	Colors the area within each ellipse. See "Example Using a Grouping Variable" on page 490.
Ellipses Coverage	Enables you to choose an α-level for the ellipses to cover.
Ellipses Transparency	Enables you to choose the transparency of the shaded ellipses, where 0 is completely transparent and 1 is completely opaque.

Nonpar Density	Shows or hides the nonparametric density, which represents the areas where the data points are the most dense. The nonparametric density estimation is helpful when you have a lot of points and the density of the points is difficult to see. There are two quantile density contours. One contour includes 50% of the points, and the other contour includes 100% of the points.
Group By	In the Group By window, you can perform the following actions: • If you did not select a **Group** variable in the launch window, you can add one now. • If you did select a **Group** variable in the launch window: – you can remove the existing **Group** variable. – you can replace the **Group** variable. See "Example Using a Grouping Variable" on page 490.
Script	This menu contains options that are available to all platforms. They enable you to redo the analysis or save the JSL commands for the analysis to a window or a file. For more information, see *Using JMP*.

Example Using a Grouping Variable

This example shows you how to create a scatterplot matrix using a grouping variable.

1. Open the Iris.jmp sample data table.
2. Select **Graph > Scatterplot Matrix**.
3. Select Sepal length, Sepal width, Petal length, and Petal width and click **Y, Columns**.
4. Select Species and click **Group**.
5. Click **OK**.

Figure 20.7 Initial Example Using a Grouping Variable

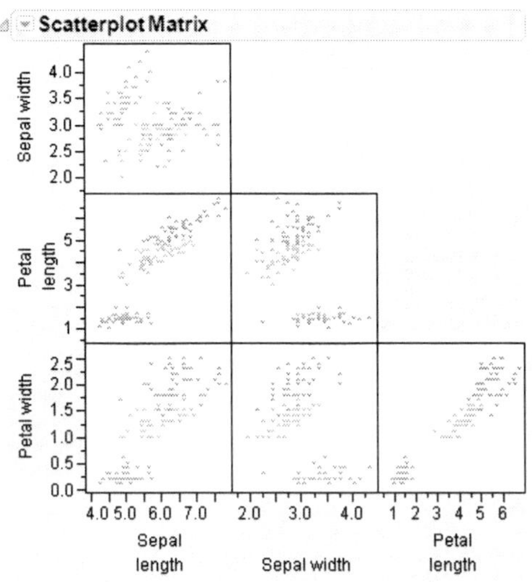

To make the groupings stand out, proceed as follows:

6. From the red triangle menu, select **Density Ellipses**.
7. From the red triangle menu, select **Shaded Ellipses**.

Figure 20.8 Example of a Scatterplot Matrix with Ellipses

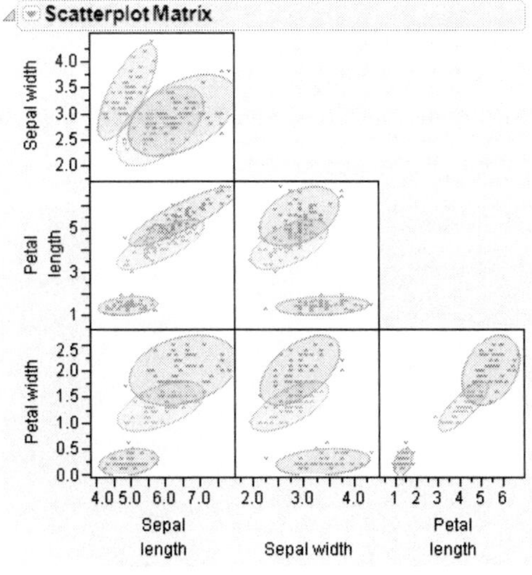

Create a Grouping Variable

If your data does not already have a grouping variable, you can create one using the Cluster platform. Using the Iris.jmp data, assume that the Species column does not exist. You know that the data comes from three species of Iris flowers, so you want to create three clusters within a group.

Proceed as follows:

1. Using the Iris.jmp sample data table, select **Analyze > Multivariate Methods > Cluster**.
2. Select Sepal length, Sepal width, Petal length, and Petal width and click **Y, Columns**.
3. Click **OK**.
4. From the red triangle menu, select **Number of Clusters**.
5. Type 3 to represent the three different Iris species.
6. Click **OK**.
7. From the red triangle menu, select **Save Clusters**.
8. Close the Hierarchical Cluster report window, and go back to the Iris.jmp data table.

   You can see that a Cluster column has been added to the Iris.jmp data table.
9. Perform the Scatterplot Matrix analysis. Follow the instructions in the section "Example Using a Grouping Variable" on page 490, but use Cluster as the grouping variable.

Figure 20.9 Example of a Scatterplot Matrix Using a Cluster Variable

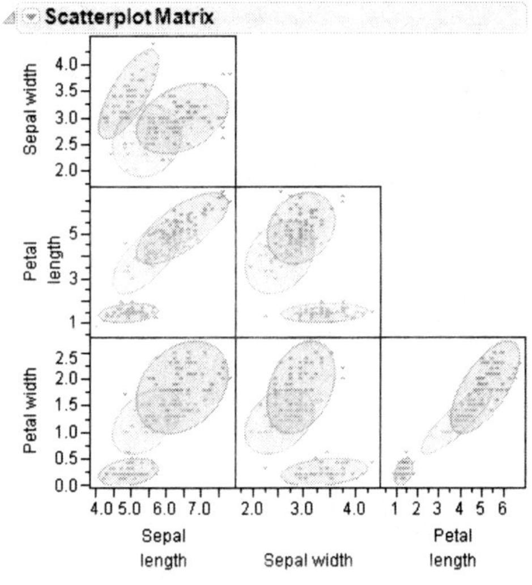

Chapter **21**

Creating Ternary Plots

Using the Ternary Plot Platform

The **Ternary Plot** command in the **Graph** menu produces a three-axis plot.

Ternary plots are a way of displaying the distribution and variability of three-part compositional data. (For example, the proportion of sand, silt, and clay in soil or the proportion of three chemical agents in a trial drug.) You can use data expressed in proportions or use absolute measures.

The ternary display is a triangle with sides scaled from 0 to 1. Each side represents one of the three components. A point is plotted so that a line drawn perpendicular from the point to each leg of the triangle intersect at the component values of the point.

Figure 21.1 Examples of Ternary Plots

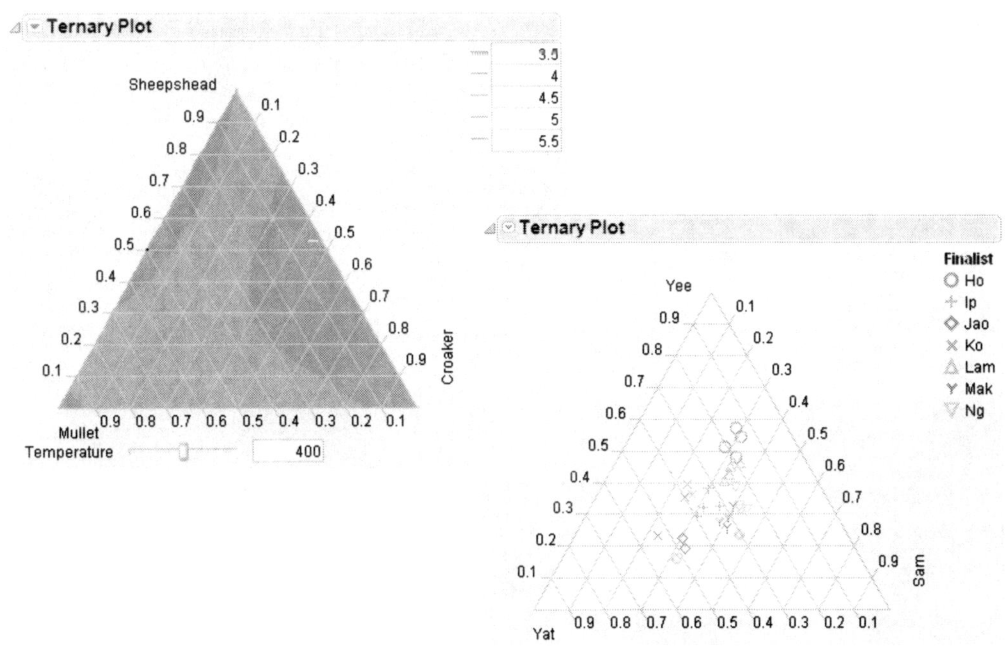

Contents

Example of a Ternary Plot

This example uses the Pogo Jumps.jmp sample data table. The data, adapted from Aitchison (1986), show measurements for pogo jumps of seven finalists in the 1985 Hong Kong Pogo-Jump Championship. A single pogo jump is the total jump distance in three consecutive bounces, referred to as *yat*, *yee*, and *sam*.

1. Open the Pogo Jumps.jmp sample data table.
2. From the **Graph** menu, select **Ternary Plot**.
3. Select Yat, Yee, and Sam and click **X, Plotting**.
4. Click **OK**.

Figure 21.2 Example of a Ternary Plot

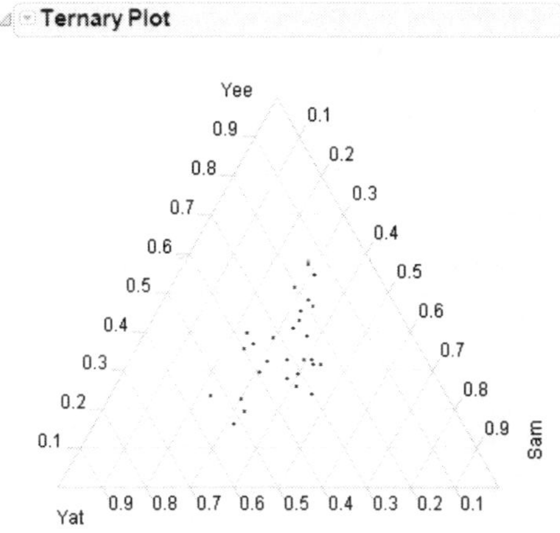

Use the crosshairs tool to determine exact coordinates of points within the plot.

Figure 21.3 Using the Crosshairs Tool

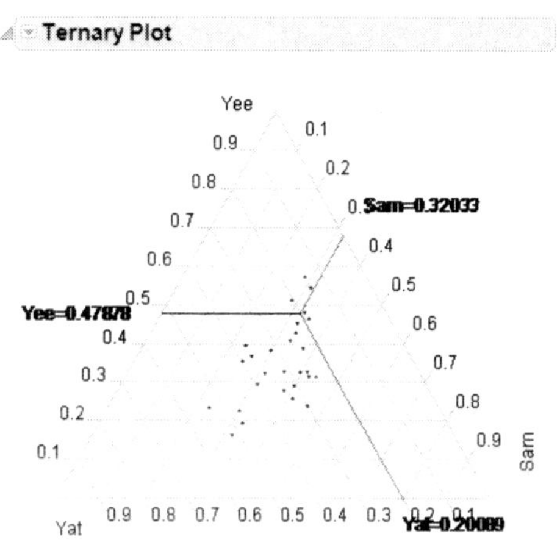

To get a better idea of how the three bounces contribute to total distance, assign each contestant's points a different color and marker.

1. Right-click on the plot and select **Row Legend**.
2. Select **Finalist** in the column list box.

   Colors should be automatically set to **JMP Default.**
3. Select **Standard** from the **Markers** menu.
4. Click **OK**.
5. To make the markers easier to see, right-click on the plot and select **Marker Size** > **3, Large**.

Figure 21.4 Pogo Data Colored by Finalist

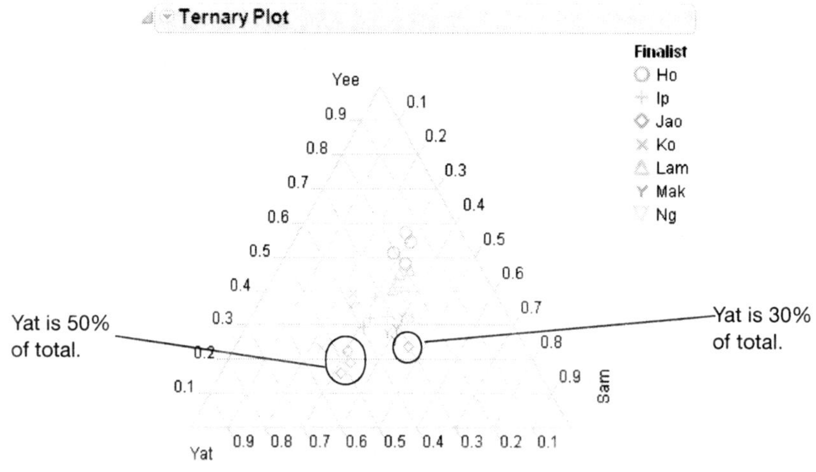

Note that most of the finalists are consistent in the composition of total distance. However, two finalists, Jao and Ko, both have one jump that is not consistent with their other jumps. For example, for three of Jao's jumps, the Yat composed about 50% of the total distance, but for the other jump, the Yat composed only 30% of the total distance. That jump is not consistent with the others. A similar observation can be made about Ko's jumps.

Launch the Ternary Plot Platform

Launch Ternary Plot by selecting **Graph > Ternary Plot**.

Figure 21.5 The Ternary Plot Launch Window

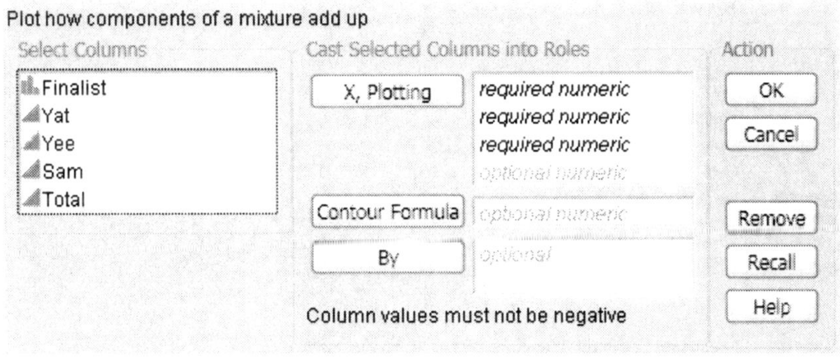

Table 21.1 Description of the Ternary Plot Launch Window

X, Plotting	Assign three columns to generate a ternary plot. If you assign more than 3 variables to the **X, Plotting** role, a matrix of ternary plots is created. A separate variable is assigned to the first two axes of a plot, with the third axis being the sum of the other variables. If necessary, the variables are scaled so they sum to 1.
Contour Formula	To plot contours of a response surface, assign a column containing a formula to the **Contour Formula** role. If you have variables in a Contour formula that are not listed as **X, Plotting** variables, JMP appends sliders below the plot so that the values can be interactively adjusted. See "Example Using a Contour Function" on page 500.
By	This option produces a separate graph for each level of the By variable.

For more information about the launch window, see *Using JMP*.

The Ternary Plot

Follow the instructions in "Example of a Ternary Plot" on page 495 to produce the plot shown in Figure 21.6.

Each of the three sides of a ternary plot represents a proportion of 0%, with the point of the triangle opposite that base representing a proportion of 100%. As a proportion increases in any one sample, the point representing that sample moves from the base to the opposite point of the triangle.

Figure 21.6 The Ternary Plot

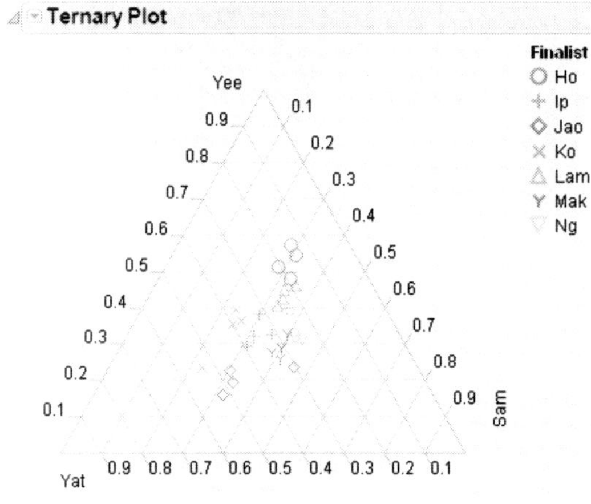

Mixtures and Constraints

Ternary Plot uses the Mixture column property to shade the portion of the graph that is out of bounds. The only constraints that the Ternary plot recognizes are the mixture sum and the mixture bounds. The Ternary plot does not recognize a general linear constraint like the Mixture Profiler does. For information about setting the Mixture column property in the Column Info window, see the *Using JMP* book.

Related Information

- "Example Using Mixture Constraints" on page 499

Ternary Plot Platform Options

The red triangle menu next to Ternary Plot contains options to modify the plot and add a 3D graph.

Table 21.2 Descriptions of Ternary Plot Platform Options

Ref Labels	Shows or hides the tick marks and tick labels on the three axes. The default plot shows the labels.

Table 21.2 Descriptions of Ternary Plot Platform Options *(Continued)*

Ref Lines	Shows or hides the triangular grid lines. The default plot shows the grid lines.
Contour Fill	Allows filling of contours if a contour formula is specified in the plot. You can select **Lines Only**, **Fill Above**, or **Fill Below**. The default platform shows lines only.
3D Graph	Available only when four or more X variables are assigned. To see this option on the menu, press the SHIFT key and click on the red triangle menu. Adds a Ternary 3D graph to the report window showing your data points in a three-dimensional scatterplot.
Show Points	Shows or hides the plotted points. The default plot shows the points.
Script	This menu contains commands that are available to all platforms. They enable you to redo the analysis or save the JSL commands for the analysis to a window or a file. For more information, see *Using JMP*.

Additional Examples of the Ternary Plot Platform

This section contains additional examples illustrating ternary plots.

Example Using Mixture Constraints

1. Open the Plasticizer.jmp sample data table.

The p1, p2, and p3 columns all have Mixture Column Properties defined.

2. From the **Graph** menu, select **Ternary Plot.**
3. Select p1, p2, and p3 and click **X, Plotting**.
4. Click **OK**.

Figure 21.7 Mixture Constraints in a Ternary Plot

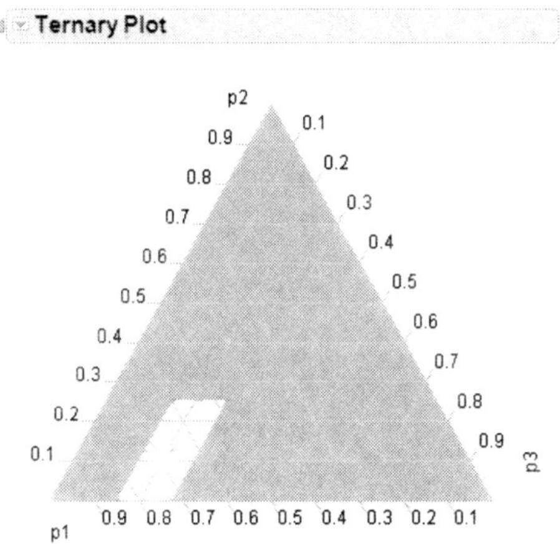

For more information about mixtures, see the section on the Mixture Profiler in the Profilers chapter in the *Modeling and Multivariate Methods* book.

Example Using a Contour Function

The data in Fish Patty.jmp is adapted from Cornell (1990) and comes from an experiment to optimize the texture of fish patties. The columns Mullet, Sheepshead, and Croaker represent what proportion of the patty came from those fish types. The column Temperature represents the oven temperature used to bake the patties. The column Rating is the response and is a measure of texture acceptability, where higher is better. A response surface model was fit to the data and the prediction formula was stored in the column Predicted Rating. (For more information, see the section on Mixture Profilers in the *Modeling and Multivariate Methods* book.)

1. Open the Fish Patty.jmp sample data table.
2. From the **Graph** menu, select **Ternary Plot.**
3. Select Mullet, Sheepshead, and Croaker and click **X, Plotting.**
4. Select Predicted Rating and click **Contour Formula.**
5. Click **OK.**
6. From the red triangle menu, select **Contour Fill > Fill Above.**

Figure 21.8 Ternary Plot with a Contour Formula

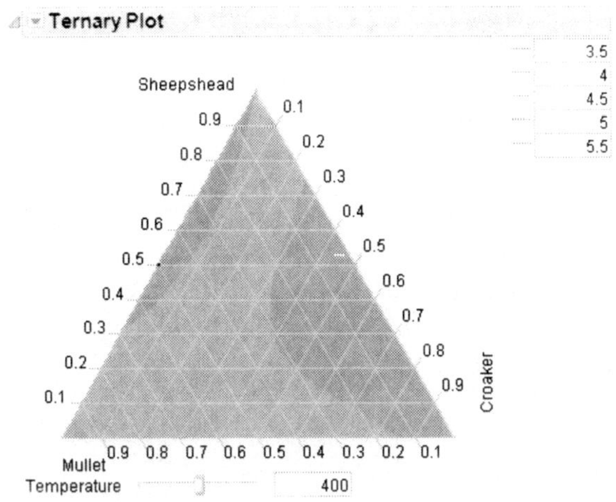

The manufacturer wants the rating to be at least 5. You can drag the slider for Temperature and see the contours for the Predicted Rating change. Each point represents a mixture of the three fish. Any given mixture of fish types receives different ratings according to the temperature at which the patties are baked.

In this example, the red shaded area shows the mixture of fish that results in a rating of 5 to 5.5. Any purple areas show the mixture of fish that results in a rating of 5.5 and above. At 400 degrees, a mixture of mostly sheepshead and mullet with very little croaker results in a rating of 5 and above.

References

Abernethy, Robert B. (1996) *The New Weibull Handbook*. Published by the author: 536 Oyster Road North Palm Beach, Florida 33408.

Agresti, A. (1984), *Analysis of Ordinal Categorical Data*, New York: John Wiley and Sons, Inc.

Agresti, A. (1990), *Categorical Data Analysis*, New York: John Wiley and Sons, Inc.

Agresti, A., and Coull, B. (1998), "Approximate is Better Than 'Exact' for Interval Estimation of Binomial Proportions," *The American Statistician*, 52, 119–126

Aitken, M. (1987) "Modelling Variance Heterogeneity in Normal Regression Using GLIM," *Applied Statistics* 36:3, 332–339.

Akaike, H. (1974), "Factor Analysis and AIC," *Pschychometrika*, 52, 317–332.

Akaike, H. (1987), "A new Look at the Statistical Identification Model," *IEEE Transactions on Automatic Control*, 19, 716–723

American Society for Quality Statistics Division (2004), *Glossary and Tables for Statistical Quality Control*, Fourth Edition, Milwaukee: Quality Press.

Anderson, T.W. (1971), *The Statistical Analysis of Time Series*, New York: John Wiley and Sons.

Anderson, T. W. (1958) *An Introduction to Multivariate Statistical Analysis*. New York: John Wiley & Sons.

Andrews, D.F. and A. M. Herzberg (1985), *Data: A Collection of Problems from Many Fields for the Student and Research Worker*. New York: Springer-Verlag.

Ashton, W.D. (1972), "The Logit Transformation," *Griffin's Statistical Monographs*, New York: Hafner Publishing.

Atkinson, A.C. (1985), *Plots, Transformations and Regression*, Oxford: Clarendon Press.

Barlow, R.E., Bartholomew, D.J., Bremner, J.M., and Brunk, H.D. (1972), *Statistical Inference under Order Restrictions*, New York: John Wiley and Sons, Inc.

Barrentine (1991), *Concepts for R&R Studies*, Milwaukee, WI: ASQC Quality Press.

Bartlett, M.S. and D.G. Kendall (1946), "The Statistical Analysis of Variances–Heterogeneity and the Logarithmic Transformation," *JRSS* Suppl 8, 128–138.

Bartlett, M.S. (1966), *An Introduction to Stochastic Processes*, Second Edition, Cambridge: Cambridge University Press.

Bates, D.M. and Watts, D.G. (1988), *Nonlinear Regression Analysis & its Applications*. New York, John Wiley and Sons.

Beaton, A.E. and Tukey, J.W. (1974), "The Fitting of Power Series, Meaning Polynomials, Illustrated on Band–Spectroscopic Data," *Technometrics* 16, 147–180.

Becker, R.A. and Cleveland, W.S. (1987), "Brushing Scatterplots," *Technometrics*, 29, 2.

Berger, R.L., and Hsu, J.C. (1996), "Bioequivalence Trails, Intersection-Union Tests and Equivalence Confidence Sets," *Statistical Science*, 11, 283–319.

Belsley, D.A., Kuh, E., and Welsch, R.E. (1980), *Regression Diagnostics*, New York: John Wiley and Sons.

Ben-Akiva, M. and Lerman, S.R. (1985). *Discrete Choice Analysis: Theory and Application to Travel Demand*, Cambridge: MIT Press.

Benzecri, J.P. (1973), "L'Analyse des Donnees," *l'analyse des Correspondances*, Paris: Dunod.

Bissell, A. F. (1990), "How Reliable is Your Capability Index?", *Applied Statistics*, 30, 331-340.

Bowman, A. and Foster, P. (1992) "Density Based Exploration of Bivariate Data," Dept. of Statistics, Univ. of Glasgow, Tech Report No 92–1.

Bowman, A. and Schmee, J. (2004)"Estimating Sensitivity of Process Capability Modeled by a Transfer Function" Journal of Quality Technology, v36, no.2 (April)

Box, G. E. P. (1954). "Some Theorems on Quadratic Forms Applied in the Study of Analysis of Variance Problems, II: Effects of Inequality of Variance and Correlation Between Errors in the Two-Way Classification". *Annals of Mathematical Statistics*, 1, 69-82.

Box, G.E.P. (1988), "Signal–to–Noise Ratio, Performance Criteria, and Transformations," *Technometrics* 30, 1–40.

Box, G.E.P. and Cox, D.R. (1964), "An Analysis of Transformations," *JRSS* B26, 211–243.

Box, G.E.P. and Draper, N.R. (1969), *Evolutionary Operation: A Statistical Method for Process Improvement*, New York: John Wiley and Sons.

Box, G.E.P. and Draper, N.R. (1987), *Empirical Model–Building and Response Surfaces*, New York: John Wiley and Sons.

Box, G.E.P. and Meyer, R.D. (1986), "An analysis of Unreplicated Fractional Factorials," *Technometrics* 28, 11–18.

Box, G.E.P. and Meyer, R. D. (1993), "Finding the Active Factors in Fractionated Screening Experiments.", *Journal of Quality Technology* Vol.25 #2: 94–105.

Box, G.E.P., Hunter,W.G., and Hunter, J.S. (1978), *Statistics for Experimenters*, New York: John Wiley and Sons, Inc.

Brown, M.B. and Benedetti, J.K. (1977), "Sampling Behavior of Tests for Correlation in Two-Way Contingency Tables," *Journal of the American Statistical Association* 72, 305-315.

Brown, M.B. and Forsythe, A.B. (1974a), "The Small Sample Behavior of Some Statistics Which Test the Equality of Several Means," *Technometrics* 16:1, 129–132.

Brown, M.B. and Forsythe, A.B. (1974), "Robust tests for the equality of variances" *Journal of the American Statistical Association*, 69, 364–367.

Byrne, D.M. and Taguchi, G. (1986), *ASQC 40th Anniversary Quality Control Congress Transactions*, Milwaukee, WI: American Society of Quality Control, 168–177.

Carroll, R.J. and Ruppert, D. (1988), *Transformation and Weighting in Regression*, New York: Chapman and Hall.

Carroll, R.J., Ruppert, D. and Stefanski, L.A. (1995), *Measurement Error in Nonlinear Models*, New York: Chapman and Hall.

Chou, Y.M., Owen, D.B., and Borrego, S.A. (1990), "Lower confidence limits on process capability indices," *Journal of Quality Technology*, 22(3): 223-229.

Cleveland, W.S. (1979). "Robust Locally Weighted Regression and Smoothing Scatterplots." *Journal of the American Statistical Association* 74 (368): 829–836.

Cobb, G.W. (1998), *Introduction to Design and Analysis of Experiments*, Springer-Verlag: New York.

Cohen, J. (1960), "A coefficient of agreement for nominal scales," *Education Psychological Measurement*, 20: 37–46.

Cole, J.W.L. and Grizzle, J.E. (1966), "Applications of Multivariate Analysis of Variance to Repeated Measures Experiments," *Biometrics*, 22, 810–828.

Cochran, W.G. and Cox, G.M. (1957), *Experimental Designs*, Second Edition, New York: John Wiley and Sons.

Conover, W. J. (1972). "A Kolmogorov Goodness-of-fit Test for Discontinuous Distributions". *Journal of the American Statistical Association* 67: 591–596.

Conover, W.J. (1980), *Practical Nonparametric Statistics*, New York: John Wiley and Sons, Inc.

Cook, R.D. and Weisberg, S. (1982), *Residuals and Influence in Regression*, New York: Chapman and Hall.

Cook, R.D. and Weisberg, S. (1983), "Diagnostics for heteroscedasticity in regression" *Biometrika* 70, 1–10.

Cornell, J.A. (1990), *Experiments with Mixtures*, Second Edition, New York: John Wiley and Sons.

Cox, D.R. (1970), *The Analysis of Binary Data*, London: Metheun.

Cox, D.R. (1972), "Regression Models And Life-tables", *Journal Of The Royal Statistical Society Series B-statistical Methodology*. 34 (2): 187–220, 1972.

Cronbach, L.J. (1951), "Coefficient Alpha and the Internal Structure of Tests," *Psychometrika*, 16, 297–334.

Daniel C. and Wood, F. (1980), *Fitting Equations to Data*, Revised Edition, New York: John Wiley and Sons, Inc.

Daniel, C. (1959), "Use of Half–normal Plots in Interpreting Factorial Two–level Experiments," *Technometrics*, 1, 311–314.

Davis, H.T. (1941), *The Analysis of Economic Time Series*, Bloomington, IN: Principia Press.

DeLong, E., Delong, D, and Clarke-Pearson, D.L. (1988), "Comparing the Areas Under Two or more Correlated Receiver Operating Characteristic Curves: A Nonparametric Approach," *Biometrics* 44, 837–845.

Derringer, D. and Suich, R. (1980), "Simultaneous Optimization of Several Response Variables," *Journal of Quality Technology*, 12:4, 214–219.

Devore, J. L. (1995), *Probability and Statistics for Engineering and the Sciences*, Duxbury Press, CA.

Do, K-A, and McLachlan G.J. (1984). Estimation of mixing proportions: a case study. *Journal of the Royal Statistical Society, Series C*, 33: 134-140.

Draper, N. and Smith, H. (1981), *Applied Regression Analysis*, Second Edition, New York: John Wiley and Sons, Inc.

Dunn, O.J. (1964), "Multiple Comparisons Using Rank Sums," *Technometrics* 6, 241-252.

Dunnett, C.W. (1955), "A multiple comparison procedure for comparing several treatments with a control" *Journal of the American Statistical Association*, 50, 1096–1121.

Dwass, M. (1955), "A Note on Simultaneous Confidence Intervals," *Annals of Mathematical Statistics* 26: 146–147.

Eppright, E.S., Fox, H.M., Fryer, B.A., Lamkin, G.H., Vivian, V.M., and Fuller, E.S. (1972), "Nutrition of Infants and Preschool Children in the North Central Region of the United States of America," World *Review of Nutrition and Dietetics*, 14.

Eubank, R.L. (1999), *Nonparametric Regression and Spline Smoothing*, Second Edition, Boca Raton, Florida: CRC.

Farebrother, R.W. (1981), "Mechanical Representations of the L1 and L2 Estimation Problems," *Statistical Data Analysis*, 2nd Edition, Amsterdam, North Holland: edited by Y. Dodge.

Firth, D. (1993), "Bias Reduction of MaximumLikelihood Estimates," Biometrika 80:1, 27–38.

Fisher, L. and Van Ness, J.W. (1971), "Admissible Clustering Procedures," *Biometrika*, 58, 91–104.

Fisherkeller, M.A., Friedman, J.H., and Tukey, J.W. (1974), "PRIM–9: An Interactive Multidimensional Data Display and Analysis System," SLAC–PUB–1408, Stanford, California: Stanford Linear Accelerator Center.

Fleiss, J. L. (1981). Statistical Methods for Rates and Proportions. New York: John Wiley and Sons.

Fleiss, J.L., Cohen J., and Everitt, B.S. (1969), "Large-Sample Standard Errors of Kappa and Weighted Kappa," *Psychological Bulletin*, 72: 323–327.

Florek, K., Lukaszewicz, J., Perkal, J., and Zubrzycki, S. (1951a), "Sur La Liaison et la Division des Points d'un Ensemble Fini," *Colloquium Mathematicae*, 2, 282–285.

Foster, D.P., Stine, R.A., and Waterman, R.P. (1997), *Business Analysis Using Regression*, New York, Springer-Verlag.

Foster, D.P., Stine, R.A., and Waterman, R.P. (1997), *Basic Business Statistics*, New York, Springer-Verlag.

Friendly, M. (1991), "Mosaic Displays for Multiway Contingency Tables," *New York University Department of Psychology Reports*: 195.

Fuller, W.A. (1976), *Introduction to Statistical Time Series*, New York, John Wiley and Sons.

Fuller, W.A. (1987), *Measurement Error Models*, New York, John Wiley and Sons.

Gabriel, K.R. (1982), "Biplot," *Encyclopedia of Statistical Sciences*, Volume 1, eds. N.L.Johnson and S. Kotz, New York: John Wiley and Sons, Inc., 263–271.

Gallant, A.R. (1987), *Nonlinear Statistical Models*, New York, John Wiley and Sons.

Giesbrecht, F.G. and Gumpertz, M.L. (2004). *Planning, Construction, and Statistical Analysis of Comparative Experiments*. New York: John Wiley & Sons.

Goodman, L.A. and Kruskal, W.H. (1979), *Measures of Association for Cross Classification*, New York: Springer-Verlag (reprint of JASA articles).

Goodnight, J.H. (1978), "Tests of Hypotheses in Fixed Effects Linear Models," *SAS Technical Report R–101*, Cary: SAS Institute Inc, also in Communications in Statistics (1980), A9 167–180.

Goodnight, J.H. and W.R. Harvey (1978), "Least Square Means in the Fixed Effect General Linear Model," *SAS Technical Report R–103*, Cary NC: SAS Institute Inc.

Greenacre, M.J. (1984), *Theory and Applications of Correspondence Analysis*, London: Academic Press.

Greenhouse, S. W. and Geiser, S. (1959). "On Methods in the Analysis of Profile Data." *Psychometrika*, 32, 95–112.

Gupta, S.S. (1965), On Some Multiple Decision (selection and ranking), Rules., *Technometrics* 7, 225–245.

Haaland, P.D. (1989), *Experimental Design in Biotechnology*, New York: Marcel Dekker, Inc.

Hahn, G.J. (1976), "Process Improvement Through Simplex EVOP," *Chemical Technology* 6, 243–345.

Hahn, G. J. and Meeker, W. Q. (1991) *Statistical Intervals: A Guide for Practitioners*. New York: Wiley.

Hajek, J. (1969), *A Course in Nonparametric Statistics*, San Francisco: Holden–Day.

Harrell, F. (1986), "The Logist Procedure," *SUGI Supplemental Library User's Guide*, Version 5 Edition, Cary, NC: SAS Institute Inc.

Harris, R.J. (1975), *A Primer of Multivariate Statistics*, New York: Academic Press.

Hartigan, J.A. (1975), *Clustering Algorithms*, New York, John Wiley and Sons.

Hartigan, J.A. (1981), "Consistence of Single Linkage for High–Density Clusters," *Journal of the American Statistical Association*, 76, 388–394.

Hartigan, J.A. and Kleiner, B. (1981), "Mosaics for Contingency Tables," *Proceedings of the 13th Symposium on the Interface between Computer Science and Statistics*, Ed. Eddy, W. F., New York: Springer–Verlag, 268–273.

Harvey, A.C. (1976), "Estimating Regression Models with Multiplicative Heteroscedasticity," *Econometrica* 44–3 461–465.

Hauck, W.W. and Donner, A. (1977), "Wald's Test as Applied to Hypotheses in Logit Analysis," *Journal of the American Statistical Association*, 72, 851–863.

Hawkins, D.M. (1974), "The Detection of Errors in Multivariate Data Using Principal Components," *Journal of the American Statistical Association*, 69.

Hayashi, C. (1950), "On the Quantification of Qualitative Data from the Mathematico–Statistical Point of View," *Annals of the Institute of Statistical Mathematics*, 2:1, 35–47.

Hayter, A.J. (1984), "A proof of the conjecture that the Tukey–Kramer multiple comparisons procedure is conservative," *Annals of Mathematical Statistics*, 12 61–75.

Heinze, G. and Schemper, M. (2002), "A Solution to the Problem of Separation in Logistic Regression," *Statistics in Medicine* 21:16, 2409–2419.

Henderson, C.R. (1984), *Applications of Linear Models in Animal Breeding*, Univ. of Guelph.

Hocking, R.R. (1985), *The Analysis of Linear Models*, Monterey: Brooks–Cole.

Hoeffding, W (1948), "A Non-Parametric Test of Independence", Annals of Mathematical Statistics, 19, 546–557.

Hoffman, Heike (2001). "Generalized Odds Ratios for Visual Modeling," Journal of Computational and Graphical Statistics, 10:4 pp. 628–640.

Holland, P.W. and Welsch, R.E. (1977), "Robust Regression Using Iteratively Reweighted Least Squares," *Communications Statistics: Theory and Methods*, 6, 813–827.

Hooper, J. H. and Amster, S. J. (1990), "Analysis and Presentation of Reliability Data," *Handbook of Statistical Methods for Engineers and Scientists*, Harrison M. Wadsworth, editor. New York: McGraw Hill.

Hosmer, D.W. and Lemeshow, S. (1989), *Applied Logistic Regression*, New York: John Wiley and Sons.

"Hot Dogs," (1986), *Consumer Reports* (June), 364–367.

Hsu, J. (1981), "Simultaneous confidence intervals for all distances from the 'best'," *Annals of Statistics*, 9, 1026–1034.

Hsu, J. (1984), "Constrained two–sided simultaneous confidence intervals for multiple comparisons with the 'best'," *Annals of Statistics*, 12, 1136–1144.

Hsu, J. (1989), "Multiple Comparison Procedures" ASA Short Course notes, Columbus OH: Ohio State University.

Hsu, J. (1989), *Tutorial Notes on Multiple Comparisons*, American Statistical Association, Washington, DC.

Hsu, J.C. (1996), *Multiple Comparisons: Theory and Methods*, Chapman and Hall.

Hunter, J.S. (1985), "Statistical Design Applied to Product Design," *Journal of Quality Technology*, 17, 210–221.

Huynh, H. and Feldt, L. S. (1970). "Conditions under which Mean Square Ratios in Repeated Measurements Designs have Exact *F*-Distributions." *Journal of the American Statistical Association*, 65, 1582–1589.

Huynh, H. and Feldt, L. S. (1976). "Estimation of the Box Correction for Degrees of Freedom from Sample Data in the Randomized Block Split Plot Designs." Journal of Educational Statistics, 1, 69–82.

Iman, R.L. (1974), "Use of a t-statistic as an Approximation to the Exact Distribution of Wilcoxon Signed Ranks Test Statistic," *Communications in Statistics—Simulation and Computation*, 795–806.

Inselberg, A. (1985) "The Plane with Parallel Coordinates." *Visual Computing* 1. pp 69–91.

Jardine, N. and Sibson, R. (1971), *Mathematical Taxonomy*, New York: John Wiley and Sons.

John, P.W.M. (1971), *Statistical Design and Analysis of Experiments*, New York: Macmillan Publishing Company, Inc.

Johnson, M.E. and Nachtsheim, C.J. (1983), "Some Guidelines for Constructing Exact D–Optimal Designs on Convex Design Spaces," *Technometrics* 25, 271–277.

Johnson, N.L. (1949). *Biometrika*, 36, 149-176.

Judge, G.G., Griffiths,W.E., Hill,R.C., and Lee, Tsoung–Chao (1980), *The Theory and Practice of Econometrics*, New York: John Wiley and Sons.

Kalbfleisch, J.D. and Prentice, R.L. (1980), *The Statistical Analysis of Failure Time Data*, New York: John Wiley and Sons.

Kalbfleisch, J.D. and Prentice, R.L. (2002), 2nd Edition, *The Statistical Analysis of Failure Time Data*, New York: John Wiley and Sons, pp 71-73.

Kackar, R.N. and Harville, D.A. (1984), Approximations for standard errors of estimators of fixed and random effects in mixed linear models, *Journal of the American Statistical Association*, 79, 853-862.

Kaiser, H.F. (1958), "The varimax criterion for analytic rotation in factor analysis" *Psychometrika*, 23, 187–200.

Kendall, M. and Stuart, A. (1979), *The Advanced Theory of Statistics, Volume 2*, New York: Macmillan Publishing Company, Inc.

Kenward, M.G. and Roger, J.H. (1997). Small sample inference for fixed effects from restricted maximum likelihood. *Biometrics*, 53, 983-997.

Khuri, A.I. and Cornell J.A. (1987), *Response Surfaces: Design and Analysis*, New York: Marcel Dekker.

Koehler, M.G., Grigoras, S., and Dunn, J.D. (1988), "The Relationship Between Chemical Structure and the Logarithm of the Partition Coefficient," *Quantitative Structure Activity Relationships*, 7.

Kohonen, Teuvo. (1989) *Self-Organization and Associative Memory.* Berlin: Springer.

Kramer, C.Y. (1956), "Extension of multiple range tests to group means with unequal numbers of replications," *Biometrics*, 12, 309–310.

Lawless, J.F. (1982), *Statistical Models and Methods for Lifetime Data*, New York: John Wiley and Sons.

Lawless, J.F. (2003), *Statistical Models and Methods for Lifetime Data, 2nd Edition*, pp. 33–34. New York: John Wiley and Sons.

Lebart, L., Morineau, A., and Tabaard, N. (1977), *Techniques de la Description Statistique*, Paris: Dunod.

Lee, E.T. (1980), *Statistical Methods for Survival Data Analysis*, Belmont CA, Lifetime Learning Publications, a Division of Wadsworth, Inc.

Lenth, R.V. (1989), "Quick and Easy Analysis of Unreplicated Fractional Factorials," *Technometrics*, 31, 469–473.

Leven, J.R., Serlin, R.C., and Webne-Behrman, L. (1989), "Analysis of Variance Through Simple Correlation," *American Statistician*, 43, (1), 32–34.

Levene, H. (1960), "Robust tests for the equality of variances" In I. Olkin (ed), *Contributions to probability and statistics*, Stanford Univ. Press.

Linnerud: see Rawlings (1988).

Louviere, J.J., Hensher, D.A. and Swait, J.D. (2000), *Stated Choice Methods: Analysis and Application*, Cambridge: Cambridge University Press.

Lucas, J.M. (1976), "The Design and Use of V–Mask Control Schemes," *Journal of Quality Technology*, 8, 1–12.

MacQueen, J.B. (1967) (1967) "Some Methods for Classification and Analysis of Multivariate Observations," *Proceedings of the fifth Berkeley Symposium on Mathematical Statistics and Probability*, 1, 281–297.

Mallows, C.L. (1967), "Choosing a Subset Regression," unpublished report, Bell Telephone Laboratories.

Mallows, C.L. (1973), "Some Comments on Cp," *Technometrics*, 15, 661–675.

Mardia, K.V., Kent, J.T., and Bibby J.M. (1979). *Multivariate Analysis*, New York: Academic Press.

Marsaglia, G. (1996) DIEHARD: A Battery of Tests of Randomness". http://stat.fsu.edu/~geo.

Mason, R.L. and Young, J.C. (2002), *Multivariate Statistical Process Control with Industrial Applications*, Philadelphia: ASA-SIAM.

Matsumoto, M. and Nishimura, T. (1998)"Mersenne Twister: A 623-Dimensionally Equidistributed Uniform Pseudo-Random Number Generator", *ACM Transactions on Modeling and Computer Simulation*, Vol. 8, No. 1, January 1998, 3–ƒ30.

McFadden, D. (1974), "Conditional Logit Analysis of Qualitative Choice Behavior," in P. Zarembka, ed., *Frontiers in Econometrics*, pp. 105–142.

McLachlan, G.J. and Krishnan, T. (1997), *The EM Algorithm and Extensions*, New York: John Wiley and Sons.

McCullagh, P. and Nelder, J.A. (1983), *Generalized Linear Models*, London: Chapman and Hall Ltd.

McCulloch, C.E., Searle, S.R., and Neuhaus, J.M. (2008), *Generalized, Linear, and Mixed Models,* New York: John Wiley and Sons.

McQuitty, L.L. (1957), "Elementary Linkage Analysis for Isolating Orthogonal and Oblique Types and Typal Relevancies," *Educational and Psychological Measurement*, 17, 207–229.

Meeker, W.Q. and Escobar, L.A. (1998), *Statistical Methods for Reliability Data*, pp. 60–62, New York: John Wiley and Sons.

Myers, R.H. (1976), *Response Surface Methodology*, Boston: Allyn and Bacon.

Myers, R.H. (1988), *Response Surface Methodology*, Virginia Polytechnic and State University.

Myers, R.H. (1989), *Classical and Modern Regression with Applications*, Boston: PWS-KENT.

Meyer, R.D., Steinberg, D.M., and Box, G. (1996), "Follow-up Designs to Resolve Confounding in Multifactor Experiments," *Technometrics*, Vol. 38, #4, p307

Miller, A.J. (1990), *Subset Selection in Regression*, New York: Chapman and Hall.

Milligan, G.W. (1980), "An Examination of the Effect of Six Types of Error Perturbation on Fifteen Clustering Algorithms," *Psychometrika*, 45, 325–342.

Milliken, G.A. and Johnson, E.J. (1984), *Analysis of Messy Data Volume I: Design of Experiments*, New York: Van Nostrand Reinhold Company.

Montgomery, D.C. and Peck, E.A. (1982), *Introduction to Linear Regression Analysis*, New York: John Wiley and Sons.

Montgomery, D. C. (1991), "Using Fractional Factorial Designs for Robust Process Development," *Quality Engineering*, 3, 193–205.

Montgomery, D. C. (1996) *Introduction to Statistical Quality Control*, 3rd edition. New York: John Wiley.

Montgomery, D.C. (2001), *Introduction to Statistical Quality Control*, 4th Edition New York: John Wiley and Sons.

Moore, D.S. and McCabe, G. P. (1989), *Introduction to the Practice of Statistics*, New York and London: W. H. Freeman and Company.

Mosteller, F. and Tukey, J.W. (1977), *Data Analysis and Regression*, Reading, MA: Addison–Wesley.

Muller, K.E. and Barton, C.N. (1989), "Approximate Power for Repeated–measures ANOVA Lacking Sphericity," *Journal of the American Statistical Association*, 84, 549–555.

Myers, R. H. and Montgomery, D. C. (1995), *Response Surface Methodology*, New York: John Wiley and Sons.

Nelder, J.A. and Mead, R. (1965), "A Simplex Method for Function Minimization," *The Computer Journal*, 7, 308–313.

Nelder, J.A. and Wedderburn, R.W.M. (1983), "Generalized Linear Models," *Journal of the Royal Statistical Society*, Series A, 135, 370–384.

Nelson, F. (1976), "On a General Computer Algorithm for the Analysis of Model with Limited Dependent Variables," *Annals of Economic and Social Measurement*, 5/4.

Nelson, L. (1984), "The Shewhart Control Chart—Tests for Special Causes," *Journal of Quality Technology*, 15, 237–239.

Nelson, L. (1985), "Interpreting Shewhart X Control Charts," *Journal of Quality Technology*, 17, 114–116.

Nelson, P.R., Wludyka, P.S., and Copeland, K.A.F. (2005), *The Analysis of Means: A Graphical Method for Comparing Means, Rates, and Proportions*, Philadelphia: Society for Industrial and Applied Mathematics.

Nelson, W.B. (1982), *Applied Life Data Analysis*, New York: John Wiley and Sons.

Nelson, W.B. (1990), *Accelerated Testing: Statistical Models, Test Plans, and Data analysis*, New York: John Wiley and Sons.

Nelson, W.B. (2003), *Recurrent Events Data Analysis for Product Repairs, Disease Recurrences, and Other Applications*, Philadelphia: Society for Industrial Mathematics.

Neter, J., Wasserman, W. and Kutner, M.H. (1990), *Applied Linear Statistical Models*, Third Edition, Boston: Irwin, Inc.

Nunnaly, J. C. (1978) Psychometric Theory, 2nd Ed., New York: McGraw-Hill.

O'Brien, R.G. (1979), "A general ANOVA method for robust tests of additive models for variances," *Journal of the American Statistical Association*, 74, 877–880.

O'Brien, R., and Lohr, V. (1984), "Power Analysis For Linear Models: The Time Has Come," *Proceedings of the Ninth Annual SAS User's Group International Conference*, 840–846.

Odeh, R. E. and Owen, D. B. (1980) *Tables for Normal Tolerance Limits, Sampling Plans, and Screening*. New York: Marcel Dekker, Inc.

Olejnik, S.F. and Algina, J. (1987), "Type I Error Rates and Power Estimates of Selected Parametric and Nonparametric Tests of Scale," *Journal of Educational Statistics* 12, 45–61.

Olson, C.L. (1976), "On Choosing a Test Statistic in MANOVA," *Psychological Bulletin* 83, 579–586.

Patterson, H. D. and Thompson, R. (1974). Maximum likelihood estimation of components of variance. *Proc. Eighth International Biochem. Conf.*, 197–209.

Piepel, G.F. (1988), "Programs for Generating Extreme Vertices and Centroids of Linearly Constrained Experimental Regions," *Journal of Quality Technology* 20:2, 125–139.

Plackett, R.L. and Burman, J.P. (1947), "The Design of Optimum Multifactorial Experiments," *Biometrika*, 33, 305–325.

Poduri, S.R.S. Rao (1997), *Variance Components: Mixed Models, Methodologies and Applications (Monographs on Statistics and Applied Probability)*, New York, Chapman & Hall.

Portnoy, Stephen (1971), "Formal Bayes Estimation with Application to a Random Effects Model", *The Annals of Mathematical Statistics*, Vol. 42, No. 4, pp. 1379–1402.

Prentice, R.L. (1973), "Exponential survivals with censoring and explanatory variables," *Biometrika*, 60:2, 279–288.

Ratkowsky, D.A. (1990), *Handbook of Nonlinear Regression Models*, New York, Marcel Dekker, Inc.

Rawlings, J.O. (1988), *Applied Regression Analysis: A Research Tool*, Pacific Grove CA: Wadsworth and Books/Cole.

Reinsch, C.H. (1967), *Smoothing by Spline Functions*, Numerische Mathematik, 10, 177–183.

Robertson, T., Wright, F.T., and R.L. Dykstra, R.L (1988), *Order Restricted Statistical Inference*, New York: John Wiley and Sons, Inc.

Rodriguez, R.N. (1990), "Selected SAS/QC Software Examples, Release 6.06," SAS Institute Inc., Cary, NC.

Rodriguez, R.N. (1991), "Applications of Computer Graphics to Two Basic Statistical Quality Improvement Methods," National Computer Graphics Association Proceedings, 17–26.

Rosenbaum, P.R. (1989), "Exploratory Plots for Paired Data," *American Statistician*, 108–109.

Rousseuw, P.J. and Leroy, A.M. (1987), *Robust Regression and Outlier Detection*, New York: John Wiley and Sons.

Royston, J.P. (1982), "An Extension of Shapiro and Wilk's W Test for Normality to Large Samples," *Applied Statistics* 31, 115–124.

Sahai, Hardeo (1974), "Some Formal Bayes Estimators of Variance Components in the Balanced Three-Stage Nested Random Effects Model", *Communication in Statistics – Simulation and Computation*, 3:3, 233–242.

Sall, J.P. (1990), "Leverage Plots for General Linear Hypotheses," *American Statistician*, 44, (4), 308–315.

Santer, T., Williams, B., and Notz, W. (2003), *The Design and Analysis of Computer Experiments*, New York, Springer-Verlag New York, Inc.

SAS Institute Inc. (1995), *SAS/QC Software: Usage and References, Version* 6, 1st Ed., Vol. 1, SAS Institute Inc., Cary, NC.

SAS Institute Inc. (1986), *SAS/QC User's Guide*, Version 5 Edition, SAS Institute Inc., Cary, NC.

SAS Institute Inc. (1987), *SAS/STAT Guide for Personal Computers*, Version 6 Edition, Cary NC: SAS Institute Inc.

SAS Institute Inc. (1999), *SAS/ETS User's Guide, Version 8*, Cary NC: SAS Institute Inc.

SAS Institute Inc. (1989), "SAS/ Technical Report P–188: SAS/QC Software Examples, Version 6 Edition," SAS Institute Inc., Cary, NC.

SAS Institute Inc. (1999), *SAS/STAT User's Guide*, Version 8, Cary, NC: SAS Institute Inc.

SAS Institute Inc. (1996), "SAS/STAT Software, Changes and Enhancements through Version 6.11, The Mixed Procedure, Cary, NC: SAS Institute Inc.

SAS Institute Inc. (2004), *SAS/STAT User's Guide*, Version 9.1, Cary, NC: SAS Institute Inc.

Satterthwaite, F.E., (1946), "An approximate distribution of Estimates of Variance Components," *Biometrics Bulletin*, 2, 110–114.

Scheffé, H. (1958) "Experiments with Mixtures", *Journal of the Royal Statistical Society B* v20, 344–360.

Searle, S. R, Casella, G. and McCulloch, C. E. (1992) *Variance Components*. New York: John Wiley and Sons.

Seber, G.A.F. (1984), *Multivariate Observations*, New York: John Wiley and Sons, 413–416.

Seder, L.A. (1950) "Diagnosis with Diagrams, Part I and Part II, Industrial Quality Control," 6, 11–19, 7–11 reprinted in Quality Engineering 2, 505–530 (1990).

Shapiro, S.S. and Wilk, M.B. (1965), "An Analysis of Variance Test for Normality (complete samples)," *Biometrika* 52, 591–611.

Slifker, J. F. and Shapiro, S. S. (1980). *Technometrics*, 22, 239-246.

Sneath, P.H.A. (1957) "The Application of Computers to Taxonomy," *Journal of General Microbiology*,17, 201–226.

Snedecor, G.W. and Cochran, W.G. (1967), *Statistical Methods*, Ames, Iowa: Iowa State University Press.

Snedecor, G.W. and Cochran, W.G. (1980), *Statistical Methods*, 7th edition, Ames, Iowa: Iowa State University Press.

Snee, R.D. (1979), "Experimental Designs for Mixture Systems with Multicomponent Constraints," *Commun. Statistics*, A8(4), 303–326.

Snee, R.D. and Marquardt, D.W. (1974), "Extreme Vertices Designs for Linear Mixture Models," *Technometrics* 16, 391–408.

Snee, R.D. and Marquardt D.W. (1975), "Extreme vertices designs for linear mixture models," *Technometrics* 16 399–408.

Sokal, R.R. and Michener, C.D. (1958), "A Statistical Method for Evaluating Systematic Relationships," *University of Kansas Science Bulletin*, 38, 1409–1438.

Somers, R.H. (1962), "A New Asymmetric Measure of Association for Ordinal Variables," *American Sociological Review*, 27, 799-811.

Spendley, W., Hext, G.R., and Minsworth, F.R. (1962), "Sequential Application of Simplex Designs in Optimization and Evolutionary Operation," *Technometrics* 4, 441–461.

Stevens, J.P. (1979), "Comment on Olson: Choosing a Test Statistic in Multivariate Analysis of Variance," *Psychological Bulletin*, 86, 355–360.

Stevens, J.P. (1986), *Applied Multivariate Statistics for the Social Sciences*, New Jersey: Laurence Erlbaum Associates.

Stone, C. and Koo, C.Y. (1986). "Additive Splines in Statistics," *In Proceedings of the Statistical Computing Section,* 45-48, Amer. Statist. Assoc., Washington, DC.

Sullivan, J.H. and Woodall, W.H. (2000), "Change-point detection of mean vector or covariance matrix shifts using multivariate individual observations," *IIE Transactions*, 32, 537-549.

Tan, Charles Y., and Iglewicz, Boris (1999), "Measurement-methods Comparisons and Linear Statistical Relationship," *Technometrics*, 41:3, 192–201.

Taguchi, G. (1976), "An Introduction to Quality Control," Nagoya, Japan: Central Japan Quality Control Association.

Tobin, J. (1958), "Estimation of Relationships for Limited Dependent Variables," *Econometrica*, 26 24–36.

Tamhane, A. C. and Dunlop, D. D. (2000) Statistics and Data Analysis. Prentice Hall.

Tobias, P.A. and Trindade, D.C. (1995), *Applied Reliability, 2nd Edition.* New York: Van Nostrand Reinhold Company.

Train, K.E. (2003), *Discrete Choice Methods wih Simulation*, Cambridge: Cambridge University Press.

Tukey, J. (1953), "A problem of multiple comparisons," Dittoed manuscript of 396 pages, Princeton University.

Tukey, J. (1991), "The Philosophy of Multiple Comparisons," *Statistical Science*, 6, 100–116.

Umetrics (1995), *Multivariate Analysis (3-day course)*, Winchester, MA.

Wadsworth, H. M., Stephens, K., Godfrey, A.B. (1986) *Modern Methods for Quality Control and Improvement*. John Wiley & Sons.

Walker, S.H. and Duncan, D.B. (1967), "Estimation of the Probability of an Event as a Function of Several Independent Variables," *Biometrika* 54.

Wegman, E. J. (1990) "Hyperdimensional Data Analysis using Parallel Coordinates." *Journal of the American Statistical Association* 85, pp. 664–675.

Welch, B.L. (1951), "On the comparison of several mean values: an alternative approach," *Biometrika* 38, 330–336.

Western Electric Company (1956), *Statistical Quality Control Handbook*, currently referred to as the *AT&T Statistical Quality Control Handbook.*

Westgard, J.O. (2002), *Basic QC Practices, 2nd Edition*. Madison, Wisconsin: Westgard QC Inc.

Winer, B.J. (1971), *Statistical Principals in Experimental Design*, Second Edition, New York: McGraw–Hill, Inc.

Wludyka, P.S. and Nelson, P.R. (1997), "An Analysis-of-Means- Type Test for Variances From Normal Populations", *Technometrics*, 39:3, 274-285.

Wold, S. (1994), "PLS for Multivariate Linear Modeling", *QSAR: Chemometric Methods in Molecular Design. Methods and Principles in Medicinal Chemistry.*

Wolfinger, R., Tobias, R., and Sall, J. (1994). Computing Gaussian likelihoods and their derivatives for general linear mixed models. *SIAM J. Sci. Comput.* 15, 6 (Nov. 1994), 1294-1310.

Wright, S.P. and R.G. O'Brien (1988), "Power Analysis in an Enhanced GLM Procedure: What it Might Look Like," *SUGI 1988, Proceedings of the Thirteenth Annual Conference*, 1097–1102, Cary NC: SAS Institute Inc.

Ye, K. and Hamada, M. (2000) "Critical Values of the Lenth Method for Unreplicated Factorial Designs," *Journal of Quality Technology*, 32, 57-66.

Ye, K. Q., and Hamada, M.(2001) "A Step-Down Lenth Method for Analyzing Unreplicated Factorial Designs." *Journal of Quality Technology*, 33, 140-152.

Index

JMP Basic Analysis and Graphing

Numerics

A

B

C

D

N

O

P

U

V

W-Z

CPSIA information can be obtained at www.ICGtesting.com
Printed in the USA
LVOW111309230413

330514LV00005B/238/P